Barcode in Back

Textiles for cold weather apparel

The Textile Institute and Woodhead Publishing

The Textile Institute is a unique organisation in textiles, clothing and footwear. Incorporated in England by a Royal Charter granted in 1925, the Institute has individual and corporate members in over 90 countries. The aim of the Institute is to facilitate learning, recognise achievement, reward excellence and disseminate information within the global textiles, clothing and footwear industries.

Historically, The Textile Institute has published books of interest to its members and the textile industry. To maintain this policy, the Institute has entered into partnership with Woodhead Publishing Limited to ensure that Institute members and the textile industry continue to have access to high calibre titles on textile science and technology.

Most Woodhead titles on textiles are now published in collaboration with The Textile Institute. Through this arrangement, the Institute provides an Editorial Board which advises Woodhead on appropriate titles for future publication and suggests possible editors and authors for these books. Each book published under this arrangement carries the Institute's logo.

Woodhead books published in collaboration with The Textile Institute are offered to Textile Institute members at a substantial discount. These books, together with those published by The Textile Institute that are still in print, are offered on the Woodhead web site at: www.woodheadpublishing.com. Textile Institute books still in print are also available directly from the Institute's web site at: www.textileinstitutebooks.com.

A list of Woodhead books on textiles science and technology, most of which have been published in collaboration with the Textile Institute, can be found on pages xv–xix.

Woodhead Publishing in Textiles: Number 93

Textiles for cold weather apparel

Edited by
J. Williams

The Textile Institute

CRC Press
Boca Raton Boston New York Washington, DC

WOODHEAD PUBLISHING LIMITED
Oxford   Cambridge   New Delhi

Published by Woodhead Publishing Limited in association with The Textile Institute
Woodhead Publishing Limited, Abington Hall, Granta Park, Geat Abington
Cambridge CB21 6AH, UK
www.woodheadpublishing.com

Woodhead Publishing India Private Limited, G-2, Vardaan House, 7/28 Ansari Road,
Daryaganj, New Delhi – 110002, India
www.woodheadpublishingindia.com

Published in North America by CRC Press LLC, 6000 Broken Sound Parkway, NW,
Suite 300, Boca Raton, FL 33487, USA

First published 2009, Woodhead Publishing Limited and CRC Press LLC
© Woodhead Publishing Limited, 2009
The authors have asserted their moral rights.

This book contains information obtained from authentic and highly regarded sources. Reprinted material is quoted with permission, and sources are indicated. Reasonable efforts have been made to publish reliable data and information, but the authors and the publishers cannot assume responsibility for the validity of all materials. Neither the authors nor the publishers, nor anyone else associated with this publication, shall be liable for any loss, damage or liability directly or indirectly caused or alleged to be caused by this book.

Neither this book nor any part may be reproduced or transmitted in any form or by any means, electronic or mechanical, including photocopying, microfilming and recording, or by any information storage or retrieval system, without permission in writing from Woodhead Publishing Limited.

The consent of Woodhead Publishing Limited does not extend to copying for general distribution, for promotion, for creating new works, or for resale. Specific permission must be obtained in writing from Woodhead Publishing Limited for such copying.

Trademark notice: product or corporate names may be trademarks or registered trademarks, and are used only for identification and explanation, without intent to infringe.

British Library Cataloguing in Publication Data
A catalogue record for this book is available from the British Library

Library of Congress Cataloging in Publication Data
A catalog record for this book is available from the Library of Congress

Woodhead Publishing Limited ISBN 978-1-84569-411-1 (book)
Woodhead Publishing Limited ISBN 978-1-84569-717-4 (e-book)
CRC Press ISBN 978-1-4398-0123-9
CRC Press order number: N10018

The publishers' policy is to use permanent paper from mills that operate a sustainable forestry policy, and which has been manufactured from pulp which is processed using acid-free and elemental chlorine-free practices. Furthermore, the publishers ensure that the text paper and cover board used have met acceptable environmental accreditation standards.

Typeset by Godiva Publishing Services Ltd, Coventry, West Midlands, UK
Printed by TJ International Limited, Padstow, Cornwall, UK

Contents

Contributor contact details		xi
Woodhead Publishing in Textiles		xv
Introduction		xxi

Part I Material and design issues in cold weather clothing

1 Comfort and thermoregulatory requirements in cold weather clothing — 3
R. ROSSI, Empa Materials Science and Technology, Switzerland

1.1	Introduction	3
1.2	Human thermoregulation in the cold	4
1.3	Clothing and comfort	6
1.4	Thermal and tactile comfort in the cold	7
1.5	New trends in thermoregulatory textiles for cold protection	14
1.6	References	15

2 Thermal insulation properties of textiles and clothing — 19
G. SONG, University of Alberta, Canada

2.1	Introduction	19
2.2	Thermal comfort	20
2.3	Heat transfer in fabrics	21
2.4	Moisture transport in fabrics	23
2.5	Fibre properties and thermal insulation	24
2.6	Yarn/fabric structure and thermal insulation	25
2.7	Predicting heat and moisture transfer in fabrics	27
2.8	Conclusions	30
2.9	References	30

3	Assessing fabrics for cold weather apparel: the case of wool	33
	R. M. LAING, University of Otago, New Zealand	
3.1	Introduction	33
3.2	Developments and demonstration of efficacy of wool apparel	35
3.3	Summary and future trends	48
3.4	Sources of further information and advice	51
3.5	References	52
4	Coating and laminating fabrics for cold weather apparel	56
	R. LOMAX, Baxenden, a Chemtura Company, UK	
4.1	Introduction	56
4.2	Historical aspects and evolution of the modern industry	57
4.3	Breathable membranes	61
4.4	Manufacture and properties of coated and laminated fabrics	67
4.5	Testing of coated and laminated fabrics	72
4.6	Environmental issues	76
4.7	Current applications	78
4.8	Future trends	80
4.9	Sources of further information and advice	81
4.10	References	82
5	The use of smart materials in cold weather apparel	84
	J. HU and MURUGESH BABU K., Hong Kong Polytechnic University, Hong Kong	
5.1	Introduction	84
5.2	Design requirements for cold weather clothing	85
5.3	Types of smart fibres and fabrics	89
5.4	The use of shape-memory materials	92
5.5	The use of phase-change materials	101
5.6	Future trends	107
5.7	References and further reading	109
6	Biomimetics and the design of outdoor clothing	113
	V. KAPSALI, University of the Arts London, UK	
6.1	Introduction	113
6.2	Inspiration from nature	114
6.3	Biological paradigms for outdoor clothing	118
6.4	Future trends	128
6.5	Sources of further information and advice	128
6.6	References	129

7	**Designing for ventilation in cold weather apparel**	**131**
	N. GHADDAR and K. GHALI, American University of Beirut, Lebanon	
7.1	Introduction: importance and function of ventilation in cold weather apparel	131
7.2	Water vapour transport through cold weather textiles at low temperatures	133
7.3	Layering cold weather clothing	135
7.4	Mechanism of ventilation in cold weather	136
7.5	Factors affecting ventilation	142
7.6	Recommendations and advice on clothing design for ventilation	147
7.7	Future trends	149
7.8	References	149
7.9	Nomenclature	151
8	**Factors affecting the design of cold weather performance clothing**	**152**
	J. BOUGOURD, University of the Arts London, UK and J. MCCANN, University of Wales, Newport, UK	
8.1	Introduction	152
8.2	Traditional design development processes	153
8.3	Stages in the process	155
8.4	Case studies: motorcycling and climbing	184
8.5	Future trends	190
8.6	Acknowledgements	191
8.7	Sources of further information and advice	191
8.8	References	191

Part II Evaluation and care of cold weather clothing

9	**Standards and legislation governing cold weather clothing**	**199**
	H. MÄKINEN, Finnish Institute of Occupational Health, Finland	
9.1	Introduction	199
9.2	Development of legislation and standards	200
9.3	Directives on personal protective equipment	201
9.4	European standards for cold protective clothing	203
9.5	Cold protective clothing standards outside Europe	212
9.6	Future trends	212
9.7	Sources of further information and advice	213
9.8	References	214

10 Laboratory assessment of cold weather clothing 217
G. HAVENITH, Loughborough University, UK

10.1	Introduction	217
10.2	Clothing properties relevant in cold	219
10.3	Material/fabric testing	220
10.4	Garment and ensemble testing: physical apparatus	223
10.5	Garment and ensemble testing: human subjects	229
10.6	Special applications	233
10.7	Future trends	239
10.8	References	240

11 Evaluation of cold weather clothing using manikins 244
E. A. McCULLOUGH, Kansas State University, USA

11.1	Introduction	244
11.2	Manikin tests vs. fabric tests	244
11.3	Thermal manikins	245
11.4	Measuring the thermal resistance of cold weather clothing systems	246
11.5	Measuring the evaporative resistance of cold weather clothing systems	249
11.6	Moving manikins	251
11.7	Using manikins under transient conditions	251
11.8	Temperature ratings	253
11.9	Conclusions	253
11.10	References	254

12 Human wear trials for cold weather protective clothing systems 256
I. HOLMÉR, Lund University, Sweden

12.1	Introduction	256
12.2	Types of human wear trials	257
12.3	Discussion	270
12.4	Sources of further information and advice	271
12.5	References	272

13 Care and maintenance of cold weather protective clothing 274
N. KERR, J. C. BATCHELLER and E. M. CROWN, The University of Alberta, Canada

13.1	Introduction	274
13.2	Home (domestic) laundering procedures	276
13.3	Professional textile care	281

13.4	Problem areas for maintenance of cold weather clothing	285
13.5	Care of cold weather clothing – case studies	289
13.6	New developments	295
13.7	Sources of further information and advice	296
13.8	References	298
	Appendix: Examples of home laundry detergents tailored for special purposes	301

Part III Cold weather clothing applications

14 Cold weather clothing for military applications 305
R. A. SCOTT, Colchester, UK

14.1	Introduction	305
14.2	History of military cold weather operations	306
14.3	General military clothing requirements	307
14.4	Incompatibilities in combat clothing systems	309
14.5	Biomedical aspects of protective combat clothing	310
14.6	Underwear materials	311
14.7	Thermal insulation materials	312
14.8	Waterproof/water vapour permeable materials	316
14.9	Materials for current UK combat clothing systems	320
14.10	Military hand- and footwear for cold climates	321
14.11	Research and development of future materials	323
14.12	References	326

15 Protective clothing for cold workplace environments 329
I. HOLMÉR, Lund University, Sweden

15.1	Introduction	329
15.2	Directives and standards	330
15.3	Protection requirements	331
15.4	Clothing for cold protection	336
15.5	Sources of further information and advice	340
15.6	References	340

16 Footwear for cold weather conditions 342
K. KUKLANE, Lund University, Sweden

16.1	Introduction	342
16.2	Criteria for cold protective footwear	343
16.3	Feet in cold	344
16.4	Foot and footwear related injuries in cold	348

16.5	Footwear insulation	350
16.6	The effect of moisture in the footwear	355
16.7	Design of cold protective footwear	361
16.8	Socks	367
16.9	References	370

17	**Gloves for protection from cold weather**	**374**
	P. I. DOLEZ and T. VU-KHANH, École de technologie supérieure, Canada	
17.1	Introduction: key issues of gloves in cold environments	374
17.2	Design, structure and materials used for hand protection in cold environments	376
17.3	Effect of cold temperatures on physical and mechanical properties of materials	381
17.4	Protection properties	383
17.5	Functionality and comfort	386
17.6	Applications/examples	388
17.7	Future trends	390
17.8	Sources of further information and advice	391
17.9	Acknowledgments	392
17.10	References	392

| *Index* | 399 |

Contributor contact details

(* = main contact)

Editor
Dr John Williams
Textile Engineering and Materials
 Research Group
De Montfort University
The Gateway
Leicester LE1 9BH
UK
E-mail: jtw@dmu.ac.uk

Chapter 1
Dr René Rossi
Empa Materials Science and
 Technology
Lerchenfeldstrasse 5
CH-9014 St. Gallen
Switzerland
E-mail: rene.rossi@empa.ch

Chapter 2
Dr Guowen Song
University of Alberta
Department of Human Ecology
302 Human Ecology Building
Edmonton
Alberta
Canada T6G 2N1
E-mail:
 Guowen.Song@afhe.ualberta.ca

Chapter 3
Professor Raechel Laing
Clothing and Textile Sciences
University of Otago
P O Box 56
Dunedin 9054
New Zealand
E-mail: r.laing@otago.ac.nz

Chapter 4
Dr Robert Lomax
Baxenden, a Chemtura Company
Union Lane
Droitwich WR9 9BB
UK
E-mail: robert.lomax@chemtura.com

Chapter 5
Professor Jinlian Hu* and Murugesh
 Babu K.
Institute of Textiles and Clothing
The Hong Kong Polytechnic
 University
Hung Hom
Kowloon
Hong Kong
E-mail: tchujl@inet.polyu.edu.hk

Chapter 6

Veronika Kapsali
20 John Princes Street
London College of Fashion
University of Arts
London W1G 0BJ
UK
E-mail: v.kapsali@fashion.arts.ac.uk

Chapter 7

Professor N. Ghaddar*
Department of Mechanical
  Engineering
Faculty of Engineering and
  Architecture
American University of Beirut
P.O. Box 11-236 – Riad El Solh
Beirut 1107 2020
Lebanon
E-mail: farah@aub.edu.lb

Associate Professor K. Ghali
Department of Mechanical
  Engineering
American University of Beirut
PO Box 11-236 Raid Elsolh
Beirut 11072020
Lebanon
E-mail: ka04@aub.edu.lb

Chapter 8

J. Bougourd*
Senior Research Fellow
London College of Fashion
University of the Arts London
20 John Princes Street
London W1M 0BJ
UK
E-mail:
  jenibougourd@blueyonder.co.uk

J. McCann
Director of Smart Clothes and
  Wearable Technology
Newport School of Art, Media and
  Design
University of Wales, Newport
Caerleon Campus
P O Box 179
Newport NP18 3YG
UK

Chapter 9

Dr Helena Mäkinen
Finnish Institute of Occupational
  Health (FIOH)
Work Environment Development
Protection and Product Safety
Topeliuksenkatu 41 aA
FI-00250 Helsinki
Finland
E-mail: Helena.Makinen@ttl.fi

Chapter 10

Professor George Havenith
Department of Human Sciences
Loughborough University
Ashby Road
Loughborough LE11 3TU
UK
E-mail: G.Havenith@lboro.ac.uk

Chapter 11

Professor Elizabeth A. McCullough
Institute for Environmental Research
Kansas State University
64 Seaton Hall
Manhattan
Kansas 66506
USA
E-mail: lizm@ksu.edu

Chapters 12 and 15
Professor Ingvar Holmér
Thermal Environment Laboratory
Department of Design Sciences
Faculty of Engineering
Lund University
SE-221 00 Lund
Sweden
E-mail: ingvar.holmer@design.lth.se

Chapter 13
Dr Nancy Kerr, Dr Jane Batcheller
   and Dr Elizabeth M. Crown*
Department of Human Ecology
302 Human Ecology Building
The University of Alberta
Edmonton
Canada T6G 2N1
E-mail: betty.crown@ualberta.ca

Chapter 14
Dr Richard A. Scott
'Mirabeau'
102 Abbots Road
Colchester CO2 8BG
UK
E-mail: dlo_rascott@hotmail.com

Chapter 16
Associate Professor Kalev Kuklane
Thermal Environment Laboratory
Department of Design Sciences
Faculty of Engineering
Lund University
Box 118
SE-221 00 Lund
Sweden
E-mail: Kalev.Kuklane@design.lth.se

Chapter 17
Dr P. I. Dolez and Professor T. Vu-
   Khanh*
Department of Mechanical
   Engineering
École de technologie supérieure
1100 rue Notre-Dame Ouest
Montréal
Québec
Canada H3C 1K3
E-mail: toan.vu-khanh@etsmtl.ca

Woodhead Publishing in Textiles

1. **Watson's textile design and colour Seventh edition**
   Edited by Z. Grosicki
2. **Watson's advanced textile design**
   Edited by Z. Grosicki
3. **Weaving Second edition**
   P. R. Lord and M. H. Mohamed
4. **Handbook of textile fibres Vol 1: Natural fibres**
   J. Gordon Cook
5. **Handbook of textile fibres Vol 2: Man-made fibres**
   J. Gordon Cook
6. **Recycling textile and plastic waste**
   Edited by A. R. Horrocks
7. **New fibers Second edition**
   T. Hongu and G. O. Phillips
8. **Atlas of fibre fracture and damage to textiles Second edition**
   J. W. S. Hearle, B. Lomas and W. D. Cooke
9. **Ecotextile '98**
   Edited by A. R. Horrocks
10. **Physical testing of textiles**
    B. P. Saville
11. **Geometric symmetry in patterns and tilings**
    C. E. Horne
12. **Handbook of technical textiles**
    Edited by A. R. Horrocks and S. C. Anand
13. **Textiles in automotive engineering**
    W. Fung and J. M. Hardcastle
14. **Handbook of textile design**
    J. Wilson
15. **High-performance fibres**
    Edited by J. W. S. Hearle
16. **Knitting technology Third edition**
    D. J. Spencer
17. **Medical textiles**
    Edited by S. C. Anand
18. **Regenerated cellulose fibres**
    Edited by C. Woodings

19 **Silk, mohair, cashmere and other luxury fibres**
    Edited by R. R. Franck
20 **Smart fibres, fabrics and clothing**
    Edited by X. M. Tao
21 **Yarn texturing technology**
    J. W. S. Hearle, L. Hollick and D. K. Wilson
22 **Encyclopedia of textile finishing**
    H-K. Rouette
23 **Coated and laminated textiles**
    W. Fung
24 **Fancy yarns**
    R. H. Gong and R. M. Wright
25 **Wool: Science and technology**
    Edited by W. S. Simpson and G. Crawshaw
26 **Dictionary of textile finishing**
    H-K. Rouette
27 **Environmental impact of textiles**
    K. Slater
28 **Handbook of yarn production**
    P. R. Lord
29 **Textile processing with enzymes**
    Edited by A. Cavaco-Paulo and G. Gübitz
30 **The China and Hong Kong denim industry**
    Y. Li, L. Yao and K. W. Yeung
31 **The World Trade Organization and international denim trading**
    Y. Li, Y. Shen, L. Yao and E. Newton
32 **Chemical finishing of textiles**
    W. D. Schindler and P. J. Hauser
33 **Clothing appearance and fit**
    J. Fan, W. Yu and L. Hunter
34 **Handbook of fibre rope technology**
    H. A. McKenna, J. W. S. Hearle and N. O'Hear
35 **Structure and mechanics of woven fabrics**
    J. Hu
36 **Synthetic fibres: nylon, polyester, acrylic, polyolefin**
    Edited by J. E. McIntyre
37 **Woollen and worsted woven fabric design**
    E. G. Gilligan
38 **Analytical electrochemistry in textiles**
    P. Westbroek, G. Priniotakis and P. Kiekens
39 **Bast and other plant fibres**
    R. R. Franck
40 **Chemical testing of textiles**
    Edited by Q. Fan
41 **Design and manufacture of textile composites**
    Edited by A. C. Long
42 **Effect of mechanical and physical properties on fabric hand**
    Edited by Hassan M. Behery

43 **New millennium fibers**
    T. Hongu, M. Takigami and G. O. Phillips
44 **Textiles for protection**
    Edited by R. A. Scott
45 **Textiles in sport**
    Edited by R. Shishoo
46 **Wearable electronics and photonics**
    Edited by X. M. Tao
47 **Biodegradable and sustainable fibres**
    Edited by R. S. Blackburn
48 **Medical textiles and biomaterials for healthcare**
    Edited by S. C. Anand, M. Miraftab, S. Rajendran and J. F. Kennedy
49 **Total colour management in textiles**
    Edited by J. Xin
50 **Recycling in textiles**
    Edited by Y. Wang
51 **Clothing biosensory engineering**
    Y. Li and A. S. W. Wong
52 **Biomechanical engineering of textiles and clothing**
    Edited by Y. Li and D. X-Q. Dai
53 **Digital printing of textiles**
    Edited by H. Ujiie
54 **Intelligent textiles and clothing**
    Edited by H. Mattila
55 **Innovation and technology of women's intimate apparel**
    W. Yu, J. Fan, S. C. Harlock and S. P. Ng
56 **Thermal and moisture transport in fibrous materials**
    Edited by N. Pan and P. Gibson
57 **Geosynthetics in civil engineering**
    Edited by R. W. Sarsby
58 **Handbook of nonwovens**
    Edited by S. Russell
59 **Cotton: Science and technology**
    Edited by S. Gordon and Y-L. Hsieh
60 **Ecotextiles**
    Edited by M. Miraftab and A. Horrocks
61 **Composite forming technologies**
    Edited by A. C. Long
62 **Plasma technology for textiles**
    Edited by R. Shishoo
63 **Smart textiles for medicine and healthcare**
    Edited by L. Van Langenhove
64 **Sizing in clothing**
    Edited by S. Ashdown
65 **Shape memory polymers and textiles**
    J. Hu
66 **Environmental aspects of textile dyeing**
    Edited by R. Christie

67 **Nanofibers and nanotechnology in textiles**
   Edited by P. Brown and K. Stevens
68 **Physical properties of textile fibres Fourth edition**
   W. E. Morton and J. W. S. Hearle
69 **Advances in apparel production**
   Edited by C. Fairhurst
70 **Advances in fire retardant materials**
   Edited by A. R. Horrocks and D. Price
71 **Polyesters and polyamides**
   Edited by B. L. Deopora, R. Alagirusamy, M. Joshi and B. S. Gupta
72 **Advances in wool technology**
   Edited by N. A. G. Johnson and I. Russell
73 **Military textiles**
   Edited by E. Wilusz
74 **3D fibrous assemblies: Properties, applications and modelling of three-dimensional textile structures**
   J. Hu
75 **Medical textiles 2007**
   Edited by J. Kennedy, A. Anand, M. Miraftab and S. Rajendran
76 **Fabric testing**
   Edited by J. Hu
77 **Biologically inspired textiles**
   Edited by A. Abbott and M. Ellison
78 **Friction in textiles**
   Edited by B. S. Gupta
79 **Textile advances in the automotive industry**
   Edited by R. Shishoo
80 **Structure and mechanics of textile fibre assemblies**
   Edited by P. Schwartz
81 **Engineering textiles: Integrating the design and manufacture of textile products**
   Edited by Y. E. El-Mogahzy
82 **Polyolefin fibres: Industrial and medical applications**
   Edited by S. C. O. Ugbolue
83 **Smart clothes and wearable technology**
   Edited by J. McCann and D. Bryson
84 **Identification of textile fibres**
   Edited by M. Houck
85 **Advanced textiles for wound care**
   Edited by S. Rajendran
86 **Fatigue failure of textile fibres**
   Edited by M. Miraftab
87 **Advances in carpet manufacture**
   Edited by K. K. Goswami
88 **Handbook of textile fibre structure**
   Edited by S. Eichhorn, J. W. S Hearle, M. Jaffe and T. Kikutani
89 **Advances in knitting technology**
   Edited by T. Dias

90 **Smart textile coatings and laminates**
    Edited by W. C. Smith
91 **Handbook of tensile properties of textile fibres**
    Edited by A. Bunsell
92 **Interior textiles: Design and developments**
    Edited by T. Rowe
93 **Textiles for cold weather apparel**
    Edited by J. Williams
94 **Modelling and predicting textile behaviour**
    Edited by X. Chen
95 **Textiles for construction**
    Edited by G. Pohl
96 **Engineering apparel fabrics and garments**
    J. Fan and L. Hunter
97 **Surface modification of textiles**
    Edited by Q. Wei
98 **Sustainable textiles**
    Edited by R. S. Blackburn

Introduction

Man has come a long way in the last 5000 years, from the concept of wearing leaves and animal skin clothing to creating fibres from oil. New discoveries, inventions and innovations have seen him explore the deepest seas, climb the highest mountains and reach out in to space. Protection from the elements is a key feature of clothing for comfort and survival and has been enhanced by a century of new materials, finishes and production methods. Clothing differs in requirement depending on application whether it be for fashion or performance. This book brings together an array of knowledge and experience to understand the principles behind cold weather protective clothing.

Comfort

Man's thermoregulatory system is adapted to tropical climatic conditions where the usual air temperatures associated with thermal comfort fall in the range 15 to 28 °C (Goldman, 1988). However, the range for physiological comfort is much narrower, i.e., the comfort range being where temperature regulation can be achieved without shivering or sweating, to the point where skin wetness exceeds 20% to obtain the evaporative cooling required. He further suggested that the human comfort zone for physiological regulation of the body, wearing shirt and trousers, is between 22.2 and 25.5 °C. For a nude man at 20 °C the heat loss from the skin is effected primarily by a dry heat flux caused by conduction, convection and radiation and is dependent on the difference between the skin and the ambient temperatures, although in normal circumstances heat loss by conduction is minimal. Some 10% of the heat produced by the body is lost by respiration and 90% lost from the skin.

Thermal comfort outside Goldman's zone of physiological regulation is provided by behavioural temperature regulation, by adding or removing clothing. White and Ronk (1984) reported that simply donning a shirt and trousers leads to a 40% reduction in heat loss due to insulation. Above this temperature range less heat is removed from the body than is produced and the excess heat is stored within the body. To maintain equilibrium, man's thermoregulation causes secretion of sweat and removes excess heat from the body by the evaporation of liquid sweat from the skin, i.e., an evaporative heat flux which depends on the humidity of the ambient air. It follows that any layer of clothing imparts a resistance to the migration of this water vapour away from the body and therefore impedes cooling.

Greenwood *et al.* (1970) described comfort as being determined by skin temperature and relative humidity (RH) prevailing close to the skin. Both these quantities tend to increase with increasing ambient air temperature and RH in the surrounding air and with increasing level of activity. The average male will be comfortable when the skin temperature is between 33 and 35 °C and they also suggested that there is little variation in subjective comfort when there is no deposition of liquid sweat on the skin, that is, the RH near the skin is below 100%. Umbach and Mecheels (1977) also used a measure of skin wetness as a discomfort factor to describe a subjective feeling of comfort. From simple subjective measurements they assessed that the average sweat wetted area of skin should be less than 30% for an average male to remain comfortable.

Heat balance

The human body is an energy generator continually producing metabolic heat and moisture at varying rates by digestion of food and by muscle activity. The heat produced by this activity will vary with the level of exercise.

In thermal equilibrium, man as a warm-blooded animal strives to keep his body core temperature constant at 37 °C. The heat flow to the environment must therefore be continually altered to balance with that produced, so as to maintain this equilibrium and thus thermo-physiological comfort (Spencer-Smith, 1976). Thermal energy can be lost to the environment by conduction, convection, radiation or evaporative cooling, as shown in a heat balance equation (White and Ronk, 1984).

This balance is achieved by means of the body's own temperature-regulating system where blood flow to the skin is increased and the temperature of the extremities raised and finally evaporative cooling begins. The evaporation of one litre of sweat from the skin in one hour causes a heat loss of approximately 670 Watts. Many workers have studied the effects of blood flow on human temperature regulation and have described mathematical models to determine head, limb and whole body temperature changes. Such physiological models are constantly being refined and will be discussed later.

It has been assumed that if a man is comfortable about a quarter of the heat produced in his body will be lost by evaporation and, to remain in heat balance, the remainder must be lost as sensible heat from the skin. If, for any reason, the heat loss exceeds the heat produced then the heat content of the body will decrease. Such a condition can be maintained only for certain tolerance periods before a dangerous state of hypothermia is reached. The converse condition of increased heat content of the body will lead to a state of hyperthermia. In addition to producing heat, the body is constantly producing perspiration which evaporates from the skin, known as insensible perspiration. In contrast, the liquid sweat that appears on the skin when the ambient temperature is high or when doing hard physical work is known as sensible perspiration.

The use of clothing

Wear comfort has been described by Mecheels (1977) as a measure of how well clothing assists the functioning of the body, or at least impairs it to a minimum. To be physiologically comfortable and maintain the state of comfort, clothing must be designed to allow the body's heat balance to be maintained under a wide range of environmental conditions and body activity. The important climatic considerations for comfort are temperature, humidity and air movement for both indoor and outdoor environments and additionally solar load and precipitation outdoors.

Clothing for hot/wet climates is normally loose fitting and may well be as little as modesty allows to promote free evaporation of sweat and gain full advantage of any convective cooling from the skin. A hot/dry climate, such as in a desert, with clear skies and intense direct and reflected solar radiation, low humidity, possible wind storms and a wide diurnal temperature variation would dictate that clothing should be loose fitting. This will allow evaporative sweat loss by day and retain body warmth at night. The solar load would also require that the skin should be covered and the clothes be light coloured to increase reflectance.

The cold environment

Man, by nature, strives to push the boundaries of his normal habitation by travel and exploration to the furthest parts of the globe where he may encounter harsh climatic conditions, ranging from mountain peaks to deserts. He may live in regions where average air temperatures remain below zero for several months of the year or work in artificial sub-zero conditions, for example, food storage facilities.

The body has its own inbuilt mechanism to reduce heat loss; blood flow to the skin and hence heat loss to the skin surface are reduced by vasoconstriction, equivalent in effect to putting on an extra clothing layer. As the body temperature falls there is a decrease in the temperature of the extremities which reduces the heat loss from those parts. The onset of shivering as involuntary muscle activity can produce compensatory energy production of up to $450\,W\,m^{-2}$ (Goldman, 1988).

Clothing for the cold environment

Clothing for cold climates should slow down the rate at which man loses heat to his environment to a level which can be balanced by that generated by metabolic processes. It should provide protection from wind penetration and afford thermal insulation without hindering agility due to bulk and also prevent any bellows action from removing warm air trapped within the clothing layers. The air layers between the fabrics should be narrow to further impair air movement. The clothing layers should be kept dry as the presence of water will lower the thermal insulation and a waterproof outer layer may also be necessary. The layers should allow water vapour migration from the body to the environment and ideally prevent condensation of sweat within the clothing which would again reduce its insulation.

In terms of the clothing worn it is therefore the two parameters of thermal insulation and water vapour resistance that are considered to be the major factors in controlling the body/environment interaction. These thermo-physiological aspects of clothing must be considered in respect of a particular wear situation to give a microclimate within the clothing which is comfortable. The four major environmental parameters affecting thermo-physiological comfort in the cold are air temperature, relative humidity, wind and radiant temperature since they directly affect heat transfer away from the body (Rees, 1971). Because these environmental conditions and work rates change, the clothing system must then be able to control the microclimate over a range of conditions.

Textile properties

The ability to alter the layers in a clothing system in a changing climate or as a result of a change in the level of work is therefore important. Textile materials are the basic building blocks with which protective clothing systems are built. Textile fibres can be either natural, for example, wool and cotton, or man made, for example, nylon and polyester. These fibres can be in short staple or long filament form, which are then spun into yarns and then woven or knitted together. Loose fibres can also be made into a non-woven matrix. Finishing processes, including coating and laminating, can be used to improve properties such as waterproofness, repellency, rot resistance, flame retardancy, etc. New technologies incorporating smart materials, microencapsulation and electronics are beginning to find application in cold weather clothing.

An understanding of the primary functions of clothing and the properties of the most suitable clothing for a specific application requires a detailed knowledge of the interactions of the body, clothing and the environment. In order that the textile and clothing industry can manufacture and supply materials to meet the requirements of comfort and protection for a given set of conditions, it is desirable to be able to compare actual wear characteristics with a set of laboratory measurements that can be used in the clothing development.

Book contents

This book is divided into three parts; Part I: Materials and design issues in cold weather clothing; Part II: Evaluation and care of cold weather clothing; Part III: Cold weather clothing applications. All contributing authors are experts in their particular fields, many from academia with industrial expertise.

Part I: Materials and design issues in cold weather clothing

The book opens by introducing the comfort and the thermoregulatory requirements of cold weather clothing. The concept of fabric layering, insulation and barrier properties then considers different fabric and fibre types to enhance warmth including the use of smart materials. This is followed by discussion of design principles to enhance protection from the cold environment and also the benefits of ventilation.

Part II: Evaluation and care of cold weather clothing

The current standards and legislation governing cold weather clothing introduces this section. The evaluation of individual textiles and garments as well as whole garment assemblies is discussed in detail ranging from the laboratory assessment of fabrics and the use of manikins to actual human wear trials. Consideration is also given to service life in terms of care and maintenance of clothing.

Part III: Cold weather clothing applications

Clothing in its broadest sense is discussed ranging from specialist items like gloves and footwear to whole body protection. Two specific examples of the latter are given, one for outdoor wearers beyond leisure use and secondly for cold workplace environments.

References

Goldman R F, 1988, 'Biomedical Effects of Clothing on Thermal Control and Strain', Chapter 2, *Handbook on Clothing*, NATO Research Study Group 7 on Biomedical Research Aspects of Military Protective Clothing.
Greenwood K, Rees W H, Lord J, 1970, 'Problems of Protection and Comfort in Modern Apparel Fabrics', *Studies in Modern Fabrics: Papers of the Diamond Jubilee Conference of the Textile Institute*, 197–218.
Mecheels J, 1977, 'Korper-Klima-Kleidung-Textil', *Melliand Textilberichte*, 58, 773–776, 857–886, 942–946.
Rees W H, 1971, 'Physical Factors Determining the Comfort Performance of Textiles', *3rd Shirley International Seminar*, Manchester, 15–17 June.

Spencer-Smith J L, 1976, 'The Physical Basis of Clothing Comfort, Part 1: General Review', *Clothing Research Journal*, 4, 126–138.

Umbach K H, Mecheels J, 1977, 'Thermophysiological Properties of Clothing Systems', *Melliand Textilberichte*, 57, 1976, 1029–1032 and 58, 73–81.

White M K, Ronk R, 1984, 'Chemical Protective Clothing and Heat Stress', *Professional Safety*, December, 34–38.

Part I

Materials and design issues in cold weather clothing

1
Comfort and thermoregulatory requirements in cold weather clothing

R. ROSSI, Empa Materials Science and Technology, Switzerland

Abstract: The human body has different thermoregulatory mechanisms to fight the cold, like the constriction of blood vessels or shivering. Wearing clothing is a behavioural means to prevent excessive heat loss from the body. In the cold, the function of the clothing is to reduce the heat transfer to the environment, especially by limiting the convective heat loss by air movements and the radiant heat loss. The body constantly releases humidity, either as insensible perspiration or sweat loss. At low temperature, this humidity may condense within the textile layers and negatively affect their thermal insulation and, as a consequence, the thermal and the tactile comfort of the wearer.

Key words: human thermoregulation, clothing comfort, thermal physiology, cold protection.

1.1 Introduction

Ideally, the heat production of the human body and heat dissipation to the environment should be balanced. By wearing clothing, humans can reduce the heat exchange between the body and the environment and can thus withstand extreme weather conditions. When metabolic heat production greatly varies (during alternating activities), heat dissipation also has to change in order to avoid excessive heat storage or a heat deficiency in the body. For this purpose, the human body has different thermoregulatory mechanisms. The function of clothing is to support this body thermoregulation as much as possible. In the cold, clothing should be designed to prevent either the whole body or local areas from being exposed to potentially harmful climatic conditions.

This chapter deals with human thermoregulation in the cold (Section 1.2) and shows the importance of clothing in maintaining the body core temperature. The interactions between the body and the clothing will be discussed (Section 1.3) with a special emphasis on the influence of body moisture production on the insulation properties of the garment. When the heat and especially the moisture transfer are reduced, thermal comfort will logically be affected, but other comfort types, such as the tactile may be altered as well (Section 1.4). In the last section (Section 1.5), possible solutions for optimal thermoregulatory textiles for cold protection are given.

1.2 Human thermoregulation in the cold

The human body is homeothermic and therefore has to maintain its core temperature within narrow limits around 37 °C. The body cells, especially in the organs and the muscles, produce heat that is partly released to the environment. This metabolic heat production can largely vary depending on the activity, from about 80 W at rest to over 1000 W during most strenuous efforts.

Heat transfer from the body to the environment occurs in several ways:

- dry heat transfer, either by conduction (heat transfer between two surfaces in contact with each other), convection (heat exchange between a surface and a surrounding fluid, e.g. air or water) and radiation (emission or absorption of electromagnetic waves)
- evaporation of sweat
- heat transfer by respiration.

In normal climatic conditions (about 20 °C and 50% relative humidity), radiant heat transfer is dominant (about 45%). Heat loss through respiration accounts for only about 10% of the total heat loss, but at low outside temperatures, it may increase to over 30% (Aschoff, Günther *et al.*, 1971).

As the body has to maintain a constant temperature, heat generation and heat loss should ideally be equal. This principle can be expressed in a heat balance equation (in W or W/m^2) for the human body:

$$M - W = E + R + C + K + S \qquad 1.1$$

where M is metabolic rate of the body, W is mechanical work, E is heat transfer by evaporation, R is heat transfer by radiation, C is heat transfer by convection, K is heat transfer by conduction, and S is heat storage.

When the body is in a thermally neutral state, S is equal to 0, but if the heat production is higher than the heat loss, S will be positive. On the other hand, if the heat loss is higher than heat production, S will be negative. The heat production term $M - W$ is obviously always positive, but R, C and K can be either positive if the body releases heat to the environment or negative if heat is gained from the environment. In principle, E could also be negative if there is a high relative humidity in the environment and water vapour condenses on the skin or in the textile layers near the skin, but this is rarely the case and therefore, E is usually positive.

1.2.1 Thermoregulatory functions of the body

In the cold, humans can consciously behave to avoid excessive heat loss and related cold stress problems and high discomfort. Adding clothing or increasing activity are two possible methods among others (like finding a shelter, exposure to sunlight, etc.). Apart from this behavioural thermoregulation, the human body has different autonomic thermoregulatory mechanisms to react to a cold environment. A thorough review of the human physiological responses to cold exposure can be found in Stocks, Taylor et al. (2004). In order to prevent an excessive heat loss, the body shell is capable of adapting its insulation to maintain the body core temperature constant without changing its heat production. The blood vessels constrict (vasoconstriction) and thus reduce the blood flow to the skin. The skin temperature will hence decrease and lead to a reduced heat transfer to the environment due to the smaller temperature gradient between body and environment. The insulation of the skin, especially the dermis, also reduces by a factor of three to four between vasodilated (thermal conductivity of about 0.9 W/mK in the dermis) and vasoconstricted state (about 0.25 W/mK) (Dittmar, Delhomme et al., 1999). In the limbs, superficial veins constrict in the cold and the cool blood returns along the veins close to the artery, leading to a heat exchange between veins and artery which will further reduce heat loss. Another autonomic means of the body to react against excessive heat loss is shivering, which is described as an asynchronous contraction of the muscles. The onset of shivering is dependent on both the skin and the core temperatures of the body.

Another vital thermoregulatory mechanism is the ability of the body to produce moisture that may evaporate and thus cool the body. In the heat, this mechanism is very often the dominant factor of heat loss. However, in the cold, the production of moisture may be counterproductive, as the presence of moisture in the clothing will affect its thermal insulation. There are two mechanisms of moisture release of the human skin: the insensible perspiration and the production of liquid sweat by the sweat glands. Insensible perspiration is defined as a diffusion of moisture through the skin and is dependent on the partial water vapour pressure gradient between the skin and the environment. Kerslake (1972) gave an empirical formula to calculate the moisture rate produced by insensible perspiration (in g/m^2h):

$$\dot{m} = 6.0 + 1.75(p_H - p_a) \qquad 1.2$$

where p_H is partial water vapour pressure near the skin (kPa), and p_a is partial water vapour pressure in the environment surrounding the body (kPa). For an average human, this will represent a moisture release of about 15 to 25 g/h in most conditions.

Sweating, like other thermoregulatory mechanisms as vasoconstriction or shivering, is much more affected by the core temperature than by the skin temperature (Wyss, Brengelmann et al., 1974; Bulcao, Frank et al., 2000). As the core temperature may rise during an activity in the cold, the human body will produce sweat, which may accumulate in the clothing layers. The sweating process is not simultaneous to the activity which means that the onset of sweating may be delayed from the beginning of the activity. Similarly, the production of sweat may continue for a while after the activity has stopped.

Insensible perspiration as well as this shift in sweat production represent a challenge for the clothing in the cold as the insulation of the textile layers should be as high as possible during a resting phase and therefore, thermal conductivity should be affected as little as possible by the presence of moisture.

1.2.2 Cold stress

People can experience different types of cold stress like exposure to cold air, immersion in water or through contact with cold surfaces. Prolonged exposure to cold, associated with insufficient physical activity or clothing insulation will result in a decrease in core temperature (hypothermia). However, this cooling might also be restricted to the extremities (head, hands and feet). Skin cooling can result from peripheral pain to freezing cold injuries. The degree of cold stress is dependent on several factors such as gender, age, health status or morphology. The adaptation to cold by repeated cold exposures can reduce cold stress responses (Rintamäki, 2001). The main cold adaptation responses are either the possibility for the body to drop the core temperature before heat production mechanisms are initiated (thermoregulatory response to hypothermia), the increase of body insulation (increase in subcutaneous fat or improved vasoconstriction) or the increase of heat production by shivering or non-shivering thermogenesis (Bittel, 1992). Nowadays, however, the most substantial part of cold adaptation is probably due to an increase in the thermal insulation of clothing or behavioural changes such as seeking a shelter.

1.3 Clothing and comfort

Comfort is a complex state of mind that depends on many physical, physiological and psychological factors. Four different types of comfort may be defined: thermal or thermophysiological comfort, sensorial comfort, garment fit and psychological comfort (aesthetics). Thermal comfort was defined as 'the condition of mind which expressed satisfaction with the thermal environment' (ISO 7730, 1984), which is the case when we are feeling neither too cold nor too warm, and when the humidity (sweat) produced by the body can be evacuated to the environment. The sensations of heat or cold, as well as that of skin wetness, determine thermal comfort. Human skin contains heat and cold receptors, as well as mechanical, tactile sensors. Humidity, however, cannot be directly detected by the skin. This sensation is determined as a mix of temperature perception (for instance, cooling through evaporation) and tactile sensors that perceive liquid sweat on the skin or altered touch properties of the wet textiles near to the skin. The factors affecting thermal comfort are the loss (or gain) of heat by radiation, conduction and convection, loss of heat by evaporation of sweat, the physical work being done by the person, and the environment (ambient temperature, air humidity and air movement). Fanger (1970) defined several conditions for a person to be in thermal comfort. In the cold, the most important factors are:

- the body must be in heat balance
- the mean skin temperature must be within comfort limits
- there should be no local thermal discomfort.

Sensorial comfort is the sensation of how the fabric feels when it is worn near to the skin. This feeling addresses properties of the fabric like prickling, itching, stiffness or smoothness. It can also be related to thermal comfort, as a fabric wetted through with sweat will change its properties and may, for instance, cling to the skin. Sensorial comfort is very difficult to predict as it involves a large number of different factors. Different studies have been performed mostly with human subjects (Garnsworthy, Gully et al., 1968, 1988; LaMotte, 1977; Demartino, Yoon et al., 1984; Elder, Fisher et al., 1984; Li, Keighley et al., 1988, 1991; Behmann, 1990; Matsudaira, Watt et al., 1990a,b; Sweeney and Branson, 1990a,b; Ajayi, 1992; Schneider, Holcombe et al., 1996; Naylor and Phillips, 1997; Wang, Zhang et al., 2003) to try to understand the relationship between fabric properties (protruding fibres, fibre and yarn diameters, fabric thickness, stiffness, etc.) and sensorial feelings on the skin. There are few objective methods to assess the sensorial properties of a textile. The most widely recognized and used around the world is probably the KES-F system developed by Kawabata and his co-workers to measure the fabric hand. The system consists of four different apparatuses (tensile and shear, bending, compression and surface friction/roughness) (Kawabata, 1980). An interesting new method has been used by the group of Bueno (Breugnot, Bueno et al., 2006; Praene, Breugnot et al., 2007): they placed microelectrodes percutaneously to record the activity of the nerve when stimulating the skin by putting it into contact with different fabrics with a defined pressure and movement. This method allowed them to determine parameters for the discrimination of the hairiness of fabrics. These parameters were used for the optimization of their 'vibrating thin plate tribometer', an apparatus used for the characterization of the surface topography of fabrics.

The garment fit considers not only the tightness of the garment, but also its weight and the overall freedom of movement of the wearer. The fourth type of comfort is psychological comfort, dealing with aesthetics (colour, garment construction, fashion, etc.) and the suitability of the clothing for the occasion. These four types of comfort are not independent of each other. The overall comfort of a person is the integration of all physical and physiological factors and their subjective perception by the wearer.

1.4 Thermal and tactile comfort in the cold

As explained before, thermal comfort is very dependent on heat and moisture transfer from the body through the clothing to the environment, as well as the heat and moisture buffering effect of the clothing. In the cold, one of the most important properties of clothing to maintain the thermal comfort of the body is to avoid condensation of insensible perspiration near the body, as skin wettedness is a good predictor of thermal discomfort, and to keep the thermal insulation as constant as possible, even when wet.

1.4.1 Environmental parameters

Heat and moisture transfer between the body and the environment depend on different external parameters:

Air temperature

As the convective heat transfer is dependent on the temperature gradient between skin and air, an elevation of air temperature leads to a reduced heat transfer. If the air temperature is higher than the skin temperature, the heat transfer will actually be reversed and the body will gain heat from the environment.

Radiant temperature

The mean surface temperature of all the objects surrounding the body determines the radiant heat exchange. In the cold, an exposure to sunlight is beneficial to decrease the radiant heat loss of the body.

Surface temperature

The temperature of surfaces in contact with the skin determines the conductive heat transfer. Different properties of the material in contact also influence the heat exchange and the temperature felt by the sensors in the skin. The most important property is the thermal inertia of the material I ($J/(m^2 K^1 s^{1/2})$), which is defined as the square root of the product of the material's bulk thermal conductivity k (W/mK), the density ρ (kg/m^3) and the specific heat capacity c (J/(kgK)):

$$I = \sqrt{k\rho c} \qquad 1.3$$

Thermal inertia is especially important for the first contact of the skin with an object but for longer contacts thermal conductivity becomes more and more important (Dittmar, Delhomme et al., 1996). A good overview of temperature limit values for touching cold surfaces with the fingertip is given by Geng, Holmér et al. (2006), showing that skin temperature decreases faster when touching aluminium or steel than when touching wood or nylon.

Relative humidity in the air

The amount of moisture in the air defines the water vapour partial pressure in the environment. The pressure difference (or the moisture concentration difference) between the skin and the environment determines the water vapour flow and the evaporative heat loss from the skin. The moisture content in the air is dependent on its temperature. The colder the air, the less moisture can be stored until saturation is reached. If the local partial water vapour pressure in the clothing is higher than the pressure of saturation, condensation occurs. The condensation pressure p_{sat} (mbar) in dependence of the temperature T (K) can by approximated by the integration of the Clausius–Clapeyron equation:

$$p_{sat} = 1.333 \cdot 10^{(-[2919.611/T] - 4.79518 \log T + 23.03733)} \qquad 1.4$$

Wind speed

Air movements affect the convective as well as the evaporative heat loss, which are logically higher with increasing wind speeds.

Precipitations

Although precipitations like rain or snow have no direct influence on the mechanisms of heat and mass transfer, they should be mentioned here, as the absorption of water by clothing can dramatically alter its thermal insulation.

1.4.2 Heat and mass transfer processes in clothing

From a thermoregulatory point of view, wearing clothing is a behavioural mechanism to prevent excessive body heat loss. Clothing impedes the transfer of heat and moisture; however, the insulation is only partly provided by the fabrics themselves, to a larger extent by the layers of air trapped between the layers of clothing. The fibre types do not have a large influence on the thermal insulation of a fabric, as it is mainly due to the air contained in them. Different mechanisms affect thermal and moisture transport through fabric layers:

- dry (conductive, convective and radiant) heat transfer between the body
- thermal energy stored within the clothing
- diffusion of water vapour molecules through the pores of the textile
- adsorption and migration of water vapour molecules and liquid water along the fibre surfaces, as well as transport of liquid through the capillaries between the fibres and the yarns
- absorption and desorption of water vapour, and transport of liquid water, in the interior of the fibres. This process is dependent on the hygroscopicity of the fibres

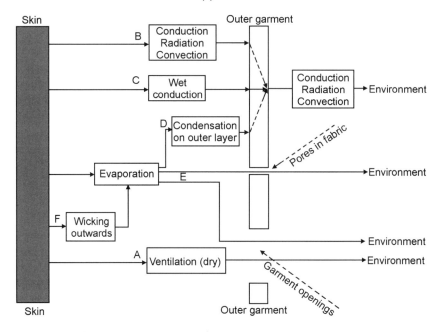

1.1 Heat and moisture transfer between the skin, the clothing and the environment.

- evaporation of liquid with thermal energy consumption or condensation with thermal energy release.

Havenith, Richards *et al.* (2008) have modelled these different heat and moisture transfer mechanisms between the skin, the clothing and the environment (Fig. 1.1). They depend on several factors such as the fibre type, the yarn and the fabric structure, the thickness and the porosity of the fabric.

Clothing represents a barrier to the free exchange of heat and moisture from the body to the environment. The thermal resistance of each clothing layer R_{ct} (m²K/W) is an intrinsic property of the textile and is measured by determining the dry heat flux through the layer for a defined temperature gradient across the textile and a fixed surface area:

$$R_{ct} = A \cdot \frac{(T_s - T_a)}{Q} \quad 1.5$$

where T_s is temperature on the inner side of the textile layer (K), T_a is temperature on the outer surface of the textile layer (K), A is surface area of the textile layer (m²), and Q is heat flux through the layer (W).

The most important dry heat transfer mechanism in clothing layers is conduction. Woo, Shalev *et al.* (1994) showed for nonwovens that only conduction is of importance as long as the fibre volume fraction is higher than 9%.

The convection in air layers between the fabrics can be neglected as long as the air layer is thinner than about 8 mm (Spencer-Smith, 1977a).

The resistance for the diffusion of water vapour through the pores of the textile (water vapour resistance R_{et} (m²Pa/W)) is dependent on the water vapour pressure difference across the layer. It is often determined by measuring the evaporative heat loss on the inner side of the layer in isothermal conditions:

$$R_{et} = A \cdot \frac{(p_s - p_a)}{Q} \qquad 1.6$$

where p_s is water vapour partial pressure on the inner side of the textile layer (K), p_a is water vapour partial pressure on the outer surface of the textile layer (K), A is surface area of the textile layer (m²), and Q is evaporative heat loss (W).

Heat and moisture transfer interact in the clothing, as water vapour may condense or be absorbed by the fabric, releasing heat of condensation or heat of sorption. Furthermore, wet fabric has different heat transfer properties from dry fabric. Dias and Delkumburewatte (2007) studied the influence of moisture content on the thermal conductivity of knitted structures and established a theoretical model of thermal conductivity k (W/mK) depending on the porosity of the material and the moisture content:

$$k = \frac{k_m k_a k_w}{(1-p)k_a k_w + (p-pw)k_m k_w + pw k_m k_a} \qquad 1.7$$

where k_m, k_a, k_w represent thermal conductivity of the material, air and water, respectively, p is porosity of the material ($p = 1 -$ [volume of the yarn in the unit cell]/[volume of the unit cell]), and w is water volume fraction.

The transport of liquid moisture is a complex mechanism dependent on the hydrophilic properties of the material (fibres), the inter- and intra-yarn capillaries, as well as the water absorption capacity (hygroscopicity) of the fibres. The spreading of liquid moisture can basically occur in two directions: spreading into the surface of the fabric (lateral wicking effect) or transfer of liquid from one side to the other (vertical wicking effect). Yao, Li et al. (2006) have developed a new test method, the moisture management tester, to investigate these wicking effects and divide the lateral wicking effect into two separate water spreading mechanisms for both sides of the fabric. This differentiation is important for many functional fabrics like double-face fabrics or denier-gradient fabrics. Van Langenhove and Kiekens (2001) have studied this lateral wicking effect and established an empirical formula for the water spreading in the surface:

$$s^2 = k_s \cdot t \qquad 1.8$$

where k_s is capillary transport constant and t is time (s).

Layer to layer wicking is possible only if in one layer a threshold amount of moisture is reached (Spencer-Smith, 1977b; Adler and Walsh, 1984; Crow and Osczevski, 1998) and depends on the kind of fabrics.

1.4.3 Condensation effects in the cold

Ideally, sweat produced by the body evaporates on the skin and is transmitted through the clothing to the environment. This occurs if the water vapour flow from the skin to the clothing is the same as from the clothing to the atmosphere. But if the local partial water vapour pressure is higher than the pressure of saturation, condensation occurs. The released heat can be related to the condensation rate with the evaporative heat:

$$q_{cond} = \varphi \cdot g_{cond} \qquad 1.9$$

where g_{cond} is condensation rate (kg/m^{2}s) and φ is evaporative heat of water (J/kg).

The heat produced will increase the thermal flow between textile and environment (as long as the outside temperature is lower than the inside temperature). Therefore, non-breathable, water-vapour-tight clothing can also allow some cooling due to sweat, even if no moisture can be released to the outside (Farnworth, 1986; Lotens, Vandelinde et al., 1995). Lotens (1983) has shown that the 'internal' water vapour transfer can lead to a 40% lower insulation.

The transfer of water vapour and the formation of condensation in fabrics at low temperatures were studied by different authors, generally showing a decrease of the permeability with decreasing temperature (Osczevski, 1996; Rossi, Gross et al., 2004; Kim, Yoo et al., 2006). Fukazawa, Kawamura et al. (2003) studied the combined influence of temperature and altitude and found a more important effect of the decreasing pressure at higher altitudes than temperature, causing a decrease in the water vapour resistance and an increase of condensation.

The accumulation of moisture can be a major problem in sleeping bags used in cold environments, as these are usually bulky and thick. Therefore, the risk that the dew point of water vapour lies within the sleeping bag is very high and thus condensation and moisture accumulation will take place. The amount of condensation is dependent on the water vapour permeability of the sleeping bag, and on the thermal resistance as the latter will determine the temperature gradient across the sleeping bag and thus the water vapour saturation pressure. Havenith (2002) and Havenith, den Hartog et al. (2004) studied this moisture accumulation for sleeping bags with different permeability at different subzero temperatures. The use of semi-permeable covers lead to acceptable moisture accumulation levels in mild cold ($-7\,°C$), but the moisture build-up increased substantially at $-20\,°C$. Apart from the moisture accumulation, the compression of the insulation layer on the lying area can also be problematic, as a large part of the body will lose heat by conduction to the ground. Camenzind, Weder et al. (2001) showed that the thickness of 10 commercially available sleeping bags was reduced to about 10 to 20% of the original value when compressed at 2 kPa, with a corresponding reduction of thermal resistance of 33% to 86%.

1.4.4 Textile/skin friction

The tribology of skin in contact with textiles is important in connection with the comfort of clothing, because the tactile properties of fabrics are closely related to their surface and frictional properties. It can also be a critical factor for skin injuries like irritations and blisters, which are caused by cyclic mechanical loads if contact pressures and shear forces are high or continue over long enough periods of time. It is generally agreed that skin hydration, lipid films and surface structure are important factors for the frictional properties of skin (Dowson, 1997; Sivamani, Goodman et al., 2003). It has been found that the presence of liquids on the skin can greatly increase the frictional resistance of the human skin (Elsner, Wilhelm et al., 1990). Kenins (1994) showed that moisture on the skin was more important than the fibre type or the fabric construction parameters in determining fabric-to-skin friction. For skin friction, factors of 1.5 to 7 have been reported between wet and dry conditions (Comaish and Bottoms, 1971; Highley, Coomey et al., 1977; Wolfram, 1983; Johnson, Gorman et al., 1993; Kenins, 1994; Adams, Briscoe et al., 2007). Figure 1.2 shows the typical influence of moisture on the friction of textiles (Derler, Schrade et al., 2007). The coefficient of friction largely increases with moisture content and stabilizes after a critical moisture regain has been reached.

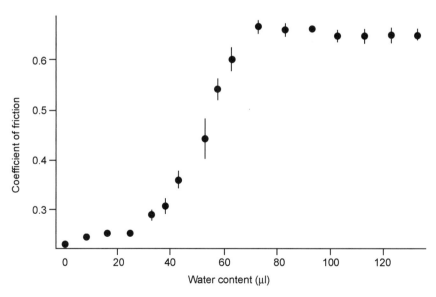

1.2 Influence of moisture on the friction of a wool fabric against the skin model Lorica (Derler, Schrade *et al.* 2007).

1.5 New trends in thermoregulatory textiles for cold protection

Persons exposed to severe cold as, for instance, during polar expeditions, cannot easily remove or add layers of clothing. For this reason, the clothing has to be adapted to different levels of activities and changing weather conditions. In order to minimize the well-known 'post-exercise chill' after an effort, when the clothing layers may be wet from the body sweat produced, participants in such expeditions know that they have to avoid any excessive activity to prevent the production of sensible (liquid) sweat. The development of textiles for cold protection has to consider the different thermoregulatory mechanisms of the human body. The measurements of the overall thermal insulation of clothing is usually made using thermal manikins (ISO 15831, 2003; ASTM F 1291, 1999). For the measurements, the temperature of the whole manikin is adjusted to a uniform temperature (34 °C) which should correspond to a mean skin temperature. However, especially in the cold, the differences in skin temperature can be quite high. As the temperatures of the extremities are usually cooler than 34 °C in the cold, the heat transfer in the sleeves of the clothing and in the trousers is probably overestimated in many cases.

The conductive heat transfer Q from a cylindrical body through a fabric is dependent on its radius:

$$Q = \frac{A \cdot \lambda (T_1 - T_2)}{r \cdot \ln(r_2/r_1)} \qquad 1.10$$

where A, r are area and radius of the cylindrical body (m), T_1, r_1 are temperature and radius on the inner surface of the fabric (K, m), and T_2, r_2 are temperature and radius on the outer surface of the fabric (K, m).

This means that smaller cylinders will transfer more heat for a given temperature gradient than larger ones, which will further overstate the importance of the extremities in overall thermal transfer. New developments in cold weather apparel should take into account this overestimation of the heat transfer of the legs and the arms and provide solutions with body-adapted thermal insulation. This body mapping approach would mean that the thermal insulation of the sleeves could be reduced in comparison to the trunk. However, the risk of local cold injuries, for instance, in the hands, has to be considered as well.

This concept of body mapping has already been used by different sports apparel manufacturers for improved moisture management, especially in the heat. As seen in this chapter, moisture management also has tremendous importance in the cold to avoid the storage of moisture in the textile layers that reduces their thermal insulation. Therefore, new products have to consider the production of insensible and sensible perspiration from the body. Different scenarios have to be established and advanced models of heat and mass transfer have to be developed to predict the transfer of moisture and the possible formation of condensation. At very low temperatures, this condensation can probably not be completely avoided and therefore new insulation layers have to be developed that will absorb this condensation. Fibres and fabrics have to be optimized to show a low increase of thermal conductivity when wet. Recent new developments of synthetic fibres, like the creation of hollow fibres to simulate the fur of polar bears, offer new possibilities in the field of biomimetics.

The development of e-textiles as well as powerful batteries with reduced dimensions and weight will open new possibilities for clothing with active heating elements and as they can already be found for instance in ski boots or cold protective gloves. The integration of electronics in textiles will also allow the development of fabrics with electrically switchable insulation. Apart from these actively adaptive materials, the development of passive adaptive materials, like new phase change materials for heat storage or breathable membranes with selective permeability will continue. In this field, as in many others, the advances in nanotechnology for the development of new polymers, nanocomposites and nanocoatings will give interesting opportunities that can be used for textile applications for improved well-being and optimized protection.

1.6　References

Adams, M. J., B. J. Briscoe, *et al.* (2007). 'Friction and lubrication of human skin'. *Tribology Letters* **26**(3): 239–253.

Adler, M. M. and W. K. Walsh (1984). 'Mechanisms of transient moisture transport between fabrics'. *Textile Research Journal* **54**(5): 334–343.

Ajayi, J. O. (1992). 'Fabric smoothness, friction, and handle'. *Textile Research Journal* **62**(1): 52–59.

Aschoff, J., B. Günther, *et al.* (1971). *Energiehaushalt und Temperaturregulation.* München, Germany, Urban & Schwarzenberg.

ASTM F 1291 (1999). Standard Test Method for Measuring the Thermal Insulation of Clothing Using a Heated Manikin. West Conshohocken, PA, ASTM.

Behmann, F. W. (1990). 'Versuche über die Rauhigkeit von Textiloberflächen'. *Melliand Textilberichte*: 438–440.

Bittel, J. H. (1992). 'The different types of general cold adaptation in man'. *International Journal of Sports Medicine* **13**: 172–176.

Breugnot, C., M. A. Bueno, *et al.* (2006). 'Mechanical discrimination of hairy fabrics from neurosensorial criteria'. *Textile Research Journal* **76**(11): 835–846.

Bulcao, C. F., S. M. Frank, *et al.* (2000). 'Relative contribution of core and skin temperatures to thermal comfort in humans'. *Journal of Thermal Biology* **25**(1–2): 147–150.

Camenzind, M., M. S. Weder, *et al.* (2001). 'Influence of body moisture on the thermal insulation of sleeping bags'. *Human factors and medicine panel symposium.* Dresden.

Comaish, S. and E. Bottoms (1971). 'The skin and friction: deviations from Amontons' laws, and the effects of hydration and lubrication'. *British Journal of Dermatology* **84**(1): 37–43.

Crow, R. M. and R. J. Osczevski (1998). 'The interaction of water with fabrics'. *Textile*

Research Journal **68**(4): 280–288.

Demartino, R. N., H. N. Yoon, *et al.* (1984). 'Improved comfort polyester. 5. Results from two subjective wearer trials and their correlation with laboratory tests'. *Textile Research Journal* **54**(9): 602–613.

Derler, S., U. Schrade, *et al.* (2007). 'Tribology of human skin and mechanical skin equivalents in contact with textiles'. *Wear* **263**: 1112–1116.

Dias, T. and G. B. Delkumburewatte (2007). 'The influence of moisture content on the thermal conductivity of a knitted structure'. *Measurement Science & Technology* **18**(5): 1304–1314.

Dittmar, A., G. Delhomme, *et al.* (1996). 'Le corps humain: un système thermique complexe'. *La thermique de l'homme et de son proche environnement*. Amsterdam, Elsevier.

Dittmar, A., G. Delhomme, *et al.* (1999). 'The human body: a complex thermal and controlled system'. *Thermal protection of man under hot and hazardous conditions*. Paris, Centre d'Etudes du Bouchet: 1–20.

Dowson, D. (1997). 'Tribology and the skin surface'. *Bioengineering of the skin: skin surface imaging and analysis*. K. P. Wilhelm, Elsner, P., Berardesca, E. and Maibach, H. I. Boca Raton, FL, CRC Press: 159–179.

Elder, H. M., S. Fisher, *et al.* (1984). 'Fabric softness, handle, and compression'. *Journal of the Textile Institute* **75**(1): 37–46.

Elsner, P., D. Wilhelm, *et al.* (1990). 'Frictional properties of human forearm and vulvar skin – Influence of age and correlation with transepidermal water loss and capacitance'. *Dermatologica* **181**(2): 88–91.

Fanger, P. O. (1970). *Thermal comfort*. Copenhagen, Danish Technical Press.

Farnworth, B. (1986). 'A numerical model of the combined diffusion of heat and water vapor through clothing'. *Textile Research Journal* **56**(11): 653–665.

Fukazawa, T., H. Kawamura, *et al.* (2003). 'Water vapor transport through textiles and condensation in clothes at high altitudes – combined influence of temperature and pressure simulating altitude'. *Textile Research Journal* **73**(8): 657–663.

Garnsworthy, R. K., R. L. Gully, *et al.* (1968). 'Understanding the causes of prickle and itch from the skin contact of fabrics'. *CSIRO – Textile and Fibre Technology*: 6.

Garnsworthy, R. K., R. L. Gully, *et al.* (1988). 'Identification of the physical stimulus and the neural basis of fabric-evoked prickle'. *Journal of Neurophysiology* **59**(4): 1083–1097.

Geng, Q., I. Holmér, *et al.* (2006). 'Temperature limit values for touching cold surfaces with the fingertip'. *Annals of Occupational Hygiene* **50**(8): 851–862.

Havenith, G. (2002). 'Moisture accumulation in sleeping bags at subzero temperatures – effect of semipermeable and impermeable covers'. *Textile Research Journal* **72**(4): 281–284.

Havenith, G., E. den Hartog, *et al.* (2004). 'Moisture accumulation in sleeping bags at −7 °C and −20 °C in relation to cover material and method of use'. *Ergonomics* **47**(13): 1424–1431.

Havenith, G., M. G. Richards, *et al.* (2008). 'Apparent latent heat of evaporation from clothing: attenuation and "heat pipe" effects'. *Journal of Applied Physiology* **104**: 142–149.

Highley, K. R., M. Coomey, *et al.* (1977). 'Frictional properties of skin'. *Journal of Investigative Dermatology* **69**(3): 303–305.

ISO 7730 (1984). *Moderate Thermal Environments – Determination of the PMV and PPD indices and specification of the conditions for thermal comfort*. Geneva, Switzerland, International Organization for Standardization.

ISO 15831 (2003). *Clothing – Physiological effects – Measurement of thermal insulation by means of a thermal manikin.* Geneva, Switzerland, International Organization for Standardization.
Johnson, S. A., D. M. Gorman, *et al.* (1993). 'The friction and lubrication of human stratum corneum'. *Thin Films in Tribology.* D. Dowson. Amsterdam, Elsevier Science Publishers B.V.: 663–672.
Kawabata, S. (1980). *The standardization and analysis of hand evaluation.* Osaka, The Textile Machinery Society of Japan.
Kenins, P. (1994). 'Influence of fiber type and moisture on measured fabric-to-skin friction'. *Textile Research Journal* **64**(12): 722–728.
Kerslake, D. M. (1972). *The stress of hot environments.* Cambridge, Cambridge University Press.
Kim, E., S. J. Yoo, *et al.* (2006). 'Performance of selected clothing systems under subzero conditions: determination of performance by a human–clothing–environment simulator'. *Textile Research Journal* **76**(4): 301–308.
LaMotte, R. H. (1977). 'Psychophysical and neurophysiological studies of tactile sensibility'. *Clothing comfort: interaction of thermal, ventilation, construction and assessment factors.* N. H. R. Goldman. Ann Arbor, MI, Science Publishers.
Li, Y., J. H. Keighley, *et al.* (1988). 'Physiological responses and psychological sensations in wearer trials with knitted sportswear'. *Ergonomics* **31**(11): 1709–1721.
Li, Y., J. H. Keighley, *et al.* (1991). 'Predictability between objective physical factors of fabrics and subjective preference votes for derived garments'. *Journal of the Textile Institute* **82**(3): 277–284.
Lotens, W. A. (1983). *Clothing, physical load and military performance.* Aspects médicaux et biophysiques des vêtements de protection, Lyon, France, Centre de Recherches du Service de Santé des Armées.
Lotens, W. A., F. J. G. Vandelinde, *et al.* (1995). 'Effects of condensation in clothing on heat transfer'. *Ergonomics* **38**(6): 1114–1131.
Matsudaira, M., J. D. Watt, *et al.* (1990a). 'Measurement of the surface prickle of fabrics. 1. The evaluation of potential objective methods'. *Journal of the Textile Institute* **81**(3): 288–299.
Matsudaira, M., J. D. Watt, *et al.* (1990b). 'Measurement of the surface prickle of fabrics. 2. Some effects of finishing on fabric prickle'. *Journal of the Textile Institute* **81**(3): 300–309.
Naylor, G. R. S. and D. G. Phillips (1997). 'Fabric-evoked prickle in worsted spun single jersey fabrics. 3. Wear trial studies of absolute fabric acceptability'. *Textile Research Journal* **67**(6): 413–416.
Osczevski, R. J. (1996). 'Water vapor transfer through a hydrophilic film at subzero temperatures'. *Textile Research Journal* **66**(1): 24–29.
Praene, J. M., C. Breugnot, *et al.* (2007). 'Mechano-acoustical discrimination of hairy fabrics from neurosensorial criteria'. *Textile Research Journal* **77**: 462–470.
Rintamäki, H. (2001). 'Human cold acclimatization and acclimation'. *International Journal of Circumpolar Health* **60**: 422–429.
Rossi, R. M., R. Gross, *et al.* (2004). 'Water vapor transfer and condensation effects in multilayer textile combinations'. *Textile Research Journal* **74**(1): 1–6.
Schneider, A. M., B. V. Holcombe, *et al.* (1996). 'Enhancement of coolness to the touch by hygroscopic fibers. 1. Subjective trials'. *Textile Research Journal* **66**(8): 515–520.
Sivamani, R. K., J. Goodman, *et al.* (2003). 'Coefficient of friction: tribological studies in man – an overview'. *Skin Research and Technology* **9**(3): 227–234.

Spencer-Smith, J. L. (1977a). 'The physical basis of clothing comfort, Part 2: Heat transfer through dry clothing assemblies'. *Clothing Res. J.* **5**(1): 3–17.

Spencer-Smith, J. L. (1977b). 'The physical basis of clothing comfort, Part 4: The passage of heat and water through damp clothing assemblies'. *Clothing Res. J.* **5**(3): 116–128.

Stocks, J. M., N. A. S. Taylor, *et al.* (2004). 'Human physiological responses to cold exposure'. *Aviation Space and Environmental Medicine* **75**(5): 444–457.

Sweeney, M. M. and D. H. Branson (1990a). 'Sensory comfort. 1. A psychophysical method for assessing moisture sensation in clothing'. *Textile Research Journal* **60**(7): 371–377.

Sweeney, M. M. and D. H. Branson (1990b). 'Sensory comfort. 2. A magnitude estimation approach for assessing moisture sensation'. *Textile Research Journal* **60**(8): 447–452.

Van Langenhove, L. and P. Kiekens (2001). 'Textiles and the transport of moisture'. *Textile Asia*: 32–34.

Wang, G., W. Zhang, *et al.* (2003). 'Evaluating wool shirt comfort with wear trials and the forearm test'. *Textile Research Journal* **73**(2): 113–119.

Wolfram, L. J. (1983). 'Friction of skin'. *Journal of the Society of Cosmetic Chemists* **34**(8): 465–476.

Woo, S. S., I. Shalev, *et al.* (1994). 'Heat and moisture transfer through nonwoven fabrics. 1. Heat transfer'. *Textile Research Journal* **64**(3): 149–162.

Wyss, C. R., G L Brengelmann, *et al.* (1974). 'Control of skin blood flow, sweating, and heart rate – role of skin vs core temperature'. *Journal of Applied Physiology* **36**(6): 726–733.

Yao, B. G., Y. Li, *et al.* (2006). 'An improved test method for characterizing the dynamic liquid moisture transfer in porous polymeric materials'. *Polymer Testing* **25**(5): 677–689.

2
Thermal insulation properties of textiles and clothing

G. SONG, University of Alberta, Canada

Abstract: This chapter reviews how textiles contribute to the thermal comfort of the wearer. It discusses mechanisms of heat transfer and moisture transport in fabrics before going on to describe how fibre properties, yarn and fabric structure affect thermal insulation properties. It concludes by assessing ways of predicting heat and moisture transfer in fabrics.

Key words: thermal insulation, comfort, thickness, heat transfer, moisture transfer.

2.1 Introduction

Clothing serves both as a barrier to the outside environment and a transporter of heat and moisture from the body to the surrounding environment. Protective clothing protects the wearer from environmental hazards such as cold while providing the heat and moisture balance the body needs. The interaction between the human body and clothing system includes four basic processes: physical processes in the clothing and surrounding environment, physiological processes in the body, neurophysiological and psychological processes (Li, 2001). The physical process between the body, clothing and environment involves heat and moisture transfer. The heat and moisture balance between the body and environment determines the level of thermal comfort for the human body.

Human comfort can be defined as a neutral state that is free from pain and discomfort. Slater (1985) describes comfort as a pleasant state of physiological, psychological and physical harmony between a human being and the environment. Hatch (1993) defines physiological comfort as including thermal comfort, sensorial comfort and body movement comfort. Thermal comfort refers to the thermal balance between the human body and the environment, and the proper balance between body heat production and heat loss. It involves transport of heat and moisture through the body-clothing-environment system. Sensorial comfort refers to neural sensations when a fabric or garment comes into contact with human skin. It includes the warmth/coolness, prickliness, surface roughness and electrical properties (e.g. static) of fabric against skin. Body movement comfort refers to the ability of a textile to allow freedom of movement. It includes qualities such as fabric stretch and weight. Among these, thermal comfort is critical in achieving physiological comfort, particularly in cold conditions.

2.2 Thermal comfort

Thermal comfort is defined in the ISO 7730 Standard 'Moderate Thermal Environments – Determination of the PMV and PPD Indices and Specification of the Conditions for Thermal Comfort' as being 'That condition of mind which expresses satisfaction with the thermal environment'. It is a simple definition but it implies many parameters which include body heat generation (metabolic heat), heat and moisture transfer from the body to the clothing micro-environment (the area between the skin and clothing), and then heat and moisture transfer from clothing to the environment. Metabolic heat production can be described as

$$H = M - W \qquad 2.1$$

where H = the total heat, M = the metabolic heat generated by the human body, and W = the total external work performed by the body.

Human heat balance (eqn 2.1) shows how the body can maintain its core temperature at 37 °C in terms of heat generation and heat exchange with the environment. Normally this is not a steady state but a dynamic equilibrium. As environmental conditions continually change, the body needs to respond to maintain a relatively constant core temperature:

$$S = M - W - E - R - C - K \qquad 2.2$$

where S = heat storage in the body, E = evaporative heat flow, R = radiative heat flow, K = conductive heat flow, and C = heat exchange through clothing.

When the human body is in a state of thermal comfort, a heat balance is obtained (i.e. constant temperature) and the rate of heat storage is zero ($S = 0$). If S is positive, heat is stored in the body and the body temperature will rise. If S is negative, heat is being lost and body temperature will fall.

A classic analysis of conditions for thermal comfort was presented by Fanger (1970). He defined three conditions for a person to be in (whole-body) thermal comfort:

1. the body is in heat balance
2. sweat rate is within comfort limits
3. mean skin temperature is within comfort limits.

The schematic diagram of heat and moisture transfer between the body, clothing and environment is illustrated in Fig. 2.1. He describes conceptual heat balance using the following heat balance equation:

$$M - E_{\text{dif}} - E_{\text{sw}} - E_{\text{res}} - C_{\text{res}} = R + C \qquad 2.3$$

Thermal insulation properties of textiles and clothing 21

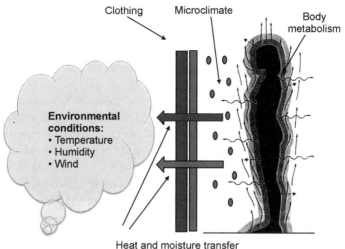

2.1 Schematic diagram of human body, clothing and environment.

where M = metabolic heat production, E_{dif} = heat loss by vapour diffusion through skin, E_{sw} = heat loss by evaporation of sweat, E_{res} = latent respiration heat loss, C_{res} = dry respiration heat loss, R = heat transfer by radiation from clothing surface, and C = heat transfer by convection form clothing surface.

The human body regulates its heat balance to a large degree by the evaporation of perspiration. The body dissipates moisture continuously. This can be insensible perspiration, as in a resting state, or sensible perspiration, when the body starts to sweat, as in a high level of activity. Sweat is primarily in liquid form. When the body temperature rises, sweat is secreted over the body to allow cooling by evaporation. Cold weather clothing must keep the sweat rate within comfort limits, particularly within the clothing microclimate, and manage the transfer of moisture from the skin as well as maintain the body in overall heat balance.

2.3   Heat transfer in fabrics

Heat transfer refers to the rate of energy that transfers from a high-temperature medium to a low-temperature medium. Heat transfer will continue until the two media are the same temperature and have reached equilibrium. The rate of the energy transferred depends on temperature difference and the degree of resistance between the two media. Heat can be transferred in fabrics by conduction through air and fibres, radiation from fibre to fibre, and convection of the air within the fabric structure.

Conduction is the transfer of heat by physical contact of two surfaces with temperature differences. The larger the difference in temperature, the faster the heat flow between the two surfaces. The conduction process occurs at the molecular level and involves the transfer of energy from the more energetic molecules to those with a lower energy level. When wearing clothing, conduction can take place between two contact fabrics or between fabrics and human skin.

Radiation involves heat transfer through space. Heat is transported by electromagnetic waves from an object with higher temperature to the one with lower temperature. In conduction, the transfer of energy requires the presence of a material medium, while in radiation it does not. Heat transfer by radiation can be significant in low-density textile materials. Fabrics with smooth, flat surfaces can be used as linings to insulate against radiant heat by reflecting the heat back. In some fabrics finished metallic surfaces are designed to reflect radiant heat in thermal protective clothing.

Convection is heat transfer by a fluid or gas caused by molecular motion. Air in contact with a warm surface absorbs heat and becomes less dense. The difference in density produced by the temperature differences forces warmed air to rise and natural convection occurs. Wind can strongly affect convection and speed up heat exchange – forced convection. In clothing, convective heat transfer is caused by air movement in the fabric which depends on the openness of the fabric. Radiation is not affected by the fabric thickness. Heat flow by conduction depends on the thickness of fabric. As indicated in eqn 2.2, the effective thermal insulation of fabric is controlled by the conductivity of the air trapped in the fabric structure. With the increase in fabric thickness, more air is trapped in the structure and, as a result, thermal insulation increases.

With the presence of moisture or water (sweating), heat and mass transfer in fabrics become more complicated and are normally treated as coupled heat and moisture transfer (Henry, 1948). Generally heat transfer in textiles takes place in three different forms (Greenwood *et al.*, 1970). The first form, which takes place in dry conditions, involves conduction, radiation and convection. The second form involves diffusion of insensible perspiration (moisture) and the last form involves the diffusion of liquid perspiration. Total heat transfer consists of a dry heat transfer component, a moisture and a liquid transfer component. The measurements of these properties for fabric are outlined in ASTM F 1868, 'Standard Test Method for Thermal and Evaporative Resistance of Clothing Materials using a Sweating Hot Plate'. The total heat loss (THL) contributed by dry heat transfer and moisture transfer is defined in the standard as the amount of heat transferred through a material or a composite by the combined dry and evaporative heat exchanges under specified conditions expressed in watts per square metre. The THL is calculated from eqn 2.4 (ASTM, 2002):

$$Qt = \frac{10\,°C}{Rcf + 0.04} + \frac{3.57\,kPa}{Ref^a + 0.0035} \qquad 2.4$$

where Qt = total heat loss (W/m^2), Rcf = average intrinsic thermal resistance of the sample (K.m²/W), and Ref^a = average apparent intrinsic evaporative resistance of the sample (kPa.m²/W).

This test specifically measures the ability of the garment to allow heat to pass away from the body through the multi-layers that make up the clothing systems by dry and wet heat exchange. Generally, the higher the THL, the more likely the system will be able to dissipate excess body heat. Higher THL values are created by lighter thermal liners and shell fabrics, but most effectively by high-performance breathable moisture barriers. Garments with non-breathable moisture barriers or heavier thermal liners will inhibit total heat loss and carry the risk of elevating the body's core temperature to extreme levels. The NFPA requirement states that a three-layer ensemble must provide a minimum total heat loss of 130 W/m².

2.4 Moisture transport in fabrics

Diffusion and convection are common modes of moisture transport in textile materials. Diffusion involves transport of molecules from a region of higher concentration to one of lower concentration by random molecular motion. Moisture can diffuse in fabrics through the air spaces between fibres or yarns. Moisture diffusion is affected by yarn and fabric structure as well as the size and number of fabric interstices that are formed between fibres within yarns. Parameters such as fabric count, yarn linear density, and yarn twist, affect the size and number of fabric interstices. Fabrics with larger interstices normally allow rapid diffusion of water.

Moisture transfer also involves adsorption, absorption or desorption between the fibres and the surrounding air as well as the movement of condensed liquid water as a result of external forces, such as capillary pressure and gravity. Adsorption occurs when water molecules are attracted to the surface of a solid. A larger fibre surface area within a fabric can increase the amount of water adsorbed. In absorption, molecular moisture diffuses through the material. Desorption is the process of moisture release from adsorbed or absorbed water. The process of moisture absorption or desorption within textile materials absorbs or releases heat, which further complicates the heat transfer process.

Fibres differ substantially in the amounts of water they can retain. Fabrics made of hydrophilic fibres allow water molecules to penetrate deeper into the fine fibre structure. Liquid water also can be transported by wicking through the capillary interstices between fibres and yarns. Capillary action may occur due to the interstices in the fabric. These phenomena depend on the liquid surface tension, the size of the interstices and wettability of the fibre surface. The capillaries (interstices) in the fabric must form a continuous channel with the proper size. Sweat forming on the skin can be transported from the skin surface to the outer surface of the fabric by fabric wicking, where it evaporates to the environment and keeps the skin dry.

Moisture regain in the fibre affects wicking performance. Fibre with larger moisture regain tends to decrease the wicking effect. Polyester fabric with very low moisture regain can be developed to create a high wicking fabric. Significant research effort is being made to develop fabrics with an enhanced ability to wick moisture. Zhang *et al.* (2006) compared the capillary effect of six fibres with different cross-section shapes. He found the radius of capillaries a key factor in deciding capillary effects. He developed an improved model (Zhang *et al.*, 2007) to investigate the effect of different fibre profiles on wicking performance. The model concluded that the best wicking effect was obtained from the convave cross-sectional shape.

2.5 Fibre properties and thermal insulation

Fibre is the key component of fabrics. Normally fibres are formed into yarns, and these yarns are interlaced (woven) or interloped (knitted) or formed by other means into a porous fabric structure. Nonwoven fabrics are formed directly from fibres or filaments without the need to produce yarns (Collier *et al.*, 2008). One of the most important factors for fabrics that resist heat flow, i.e. fabric thermal insulation, is the trapped still air in the fabric structure. The thermal resistance of air is eight times higher than that of fibres. The effective thermal conductivity of the fabric system can be written in terms of the air volume and fibre fractions of the system using the equation developed by Farnworth (1983):

$$k_{sys} = (1 - P_\varphi)k_A + P_\varphi k_F \qquad 2.5$$

where P_φ is packing factor, k_A is the thermal conductivity of air volume fraction, and k_F is the thermal conductivity of the fibrous component.

From eqn 2.5, the more trapped (still) air in the fabric structure, the better the thermal insulation. Fibres in the fabric structure serve two main functions in providing thermal insulation. Firstly, they develop air spaces and prevent air movement. Secondly, the fibre provides a shield to heat loss from radiation. The efficiency of the thermal insulation of a fabric depends on fibre physical properties, such as fineness and shape, as well as the structure of the fabric.

The boundary air layer existing on the fabric surface contributes significantly to fabric thermal insulation. This can be explained by boundary-layer theory. When moving air comes into contact with a solid, a thin region will be formed which is known as the boundary layer. The air immediately adjacent to the fibre or fabric surface area is at rest relative to the fibre or fabric. Therefore the existing boundary air layer, along with the existing still air, increases thermal insulation and provides considerable resistance to heat transfer through the fabric.

The physical structure of fibre includes its fineness, crimp and length. The finer the fibre, the higher surface-to-volume ratio. As a result, many small spaces for still air between fibres can be formed. Conduction and convection through the still air are limited, achieving a higher thermal insulation. Fibrefill® is a good example of using fine microfibres to achieve high thermal insulation. Microfibres have been used in garments in recent years to improve garment function, particularly in sportswear, sleeping bags and tents. Microfibre yarns possess larger surface area and twice the bulk of the normal fibre. Additionally radiant heat transfer in textiles is related to fibre diameter where a fabric with finer fibres exhibits a lower heat transfer by radiation.

Some fibres have physical characteristics that enhance this effect of air insulation. Fibre crimp develops more trapped air by creating interstices in the yarn structure that effectively increase the fabric thermal insulation and also resist its movement. Man-made fibres can be produced with a degree of crimp or surface irregularity that increases thermal resistance. Wool fibres possess natural crimp and maintain a high volume of still air, explaining its traditional use in cold weather clothing. High shrinkage fibres can also improve thermal insulation because of their ability to trap air.

Experiments performed by Bogaty *et al.* (1957) suggest fabric thermal insulation is closely related to fibre arrangement and fabric thickness. Fibre arrangements with fibres lying parallel and perpendicular to the fabric surface show improved thermal insulation. A study by Kong *et al.* (2001) found that fibre configuration and its porosity have a significant effect on heat and moisture transfer. They found different levels of heat and mass transfer in three configurations of the fibre: hollow, solid and a mixture of the two (multilayer). Fabrics or garments made from hollow fibres can provide better thermal insulation values due to the larger trapped air volume provided. Teijin Ltd and Du Pont have similar products with their own brand name such as Aerocapsule®. These hollow fibres were developed with different cross-sectional shapes and even some voids on the fibre surface, or more convolutions created along the fibre. The hollow centre holds air, resulting in additional thermal insulation as well as 20% reduction in fibre weight. Due to these modifications, a capillary phenomenon can be developed and moisture management can be improved. Hollow fibres can be applied to swimming costumes to provide lift for the body in the water and prevent it from being cooled. The blending of hollow fibres and microfibres can significantly improve the fabric thermal property. Hollow fibres that inherently entrap air are produced specifically for use in cold weather apparel.

2.6 Yarn/fabric structure and thermal insulation

Differences in yarn structure affect the thermal insulation capacity of fabrics. The yarn structure together with the fibre geometry and fabric construction may influence radiation, moisture transfer and conduction in fabrics. A spun yarn or textured filament yarn can hold more trapped air than a flat filament yarn and, therefore, provide more thermal insulation. The degree of yarn twist also affects heat and moisture transfer as higher twist yarns are more compact, providing less air volume and possibly improving the capillary effect in moisture transfer.

A number of studies indicate that fabric thickness is the single most important variable in determining thermal insulation (Holcombe and Hoschke, 1983; Bandyopadhyay et al., 1987). A thicker fabric provides more air space and, therefore, more entrapped air in the fabric. In some cases fabric thermal insulation exhibits a linear relation with fabric thickness. Multilayer fabrics provide even better insulation than a single thick fabric. For multilayer fabrics, however, the thermal resistance and thickness show a non-linear relation (Wilson et al., 1999).

Different fabric structures have different levels of porosity, hence different amounts of the entrapped air in the fabric. Knitted fabrics usually entrap more air than woven fabrics, although the tightness of the weave or knit is a factor as well. Fabrics with pile or napped constructions can improve their thermal insulation as the yarns or fibres perpendicular to the surface provide numerous spaces for dead air. These fabrics provide optimized insulation when the fabric is used as an inner layer next to the skin.

Fabric weight is an important factor that affects fabric thermal insulation performance. For cold weather clothing it is important to balance high thermal insulation with low weight to improve comfort. Lightweight fabrics can be achieved by using very low-density fibre assemblies or laminates using polyurethane foam to bond knitted fabric to woven fabrics. The fabric-foam-fabric structure creates high insulation with light weight due to the low density of polyurethane foam. Low-density fabrics are particularly vulnerable to heat loss by radiation, and this must be taken into account in design.

Frydrych et al. (2002) compared fabrics made from regular cotton and man-made cellulose fibres. They found that thermal insulation differed noticeably both with different types of fibre and different weaves using each fibre. They also reported a difference in the thermal absorption of the fabrics when different finishes were applied. Similar studies were also performed on different clothing materials using a sweating cylinder by Celcar et al. (2008). In the study, eight different material combinations, arranged in a four-layer clothing system, were investigated and the results showed significant differences in thermal insulation among these combinations.

Multiple fabric layers add to the overall thickness of the garment which contributes to the total insulation provided by the clothing system. The multilayer principle normally consists of three layers, the inner layer, the middle layer, and the outer layer. The inner layer forms the microclimate with the skin. Moisture and heat transfer starts directly from this layer. The middle layer normally provides most of the thermal insulation. This layer tends to be made of materials that possess a large amount of trapped air. The outer layer's main function is to provide protection to maintain the garment's integrity. The outer layer must be constructed from materials that display excellent mechanical properties. Normally a chemical finish can be applied to the outer layer in order to achieve some desirable properties such as water repellency.

In addition to fibre, yarn, and fabric properties, applying chemicals to a fabric can change its thermal properties. Chemical finishes have been developed that can store heat and then release it when environmental conditions change. There have been a number of studies on treating fabrics by incorporating phase change materials (PCMs) (Shim *et al.*, 2001; Bo-an *et al.*, 2004). When phase change materials are incorporated into textiles, they absorb heat energy as they change from a solid to a liquid state and release heat as they return to a solid state. It has been suggested that garments made from PCM-treated fabric can regulate body temperature automatically in changing conditions. In a study by Shim *et al.* (2001), for example, a small, temporary heating/cooling effect from clothing using PCMs was shown. The application of PCMs can be incorporated into the spinning dope of manufactured fibres or incorporated into the structure of fabrics by coating, printing or laminating on a foam with PCMs.

An additional development is the use of inflatable tubings sewn into fabrics. The mechanism allows people to simply blow into a tube to inflate the tubes in the fabric with more still air, increasing the amount of insulation. The use of electrical heating incorporated into the structure of a garment is another new development. Electrical heating involves integrating wires into the garment and providing electrical power through a battery pack to heat the wires up. The method is very effective in producing heat but requires higher capacity portable batteries and a durable but lightweight and flexible wiring system.

2.7 Predicting heat and moisture transfer in fabrics

There has been a significant amount of research into ways of predicting the thermal insulation properties of textile materials, including understanding the mechanisms involved in heat and moisture transfer (Morse *et al.*, 1973; Torvi, 1997; Williams and Curry, 1992; Futschik and White, 1994). This includes conduction in solid fibres and air, thermal radiation between fibres and convection of the air in the fabric structure. Morse *et al.* (1973) have calculated fabric thermal conductivity using the flowing equation:

$$K = x(V_f k_f + V_a k_a) + y \frac{k_f k_a}{V_a k_f + V_f k_a}, \qquad 2.6$$

where V_f, V_a = volume fraction of fibre and air, and k_f, k_a = conductivity of fibre and air. In this equation, $x + y = 1$ and $V_a + V_f = 1$.

Torvi (1997) used a simplified model to weight the contributions from the solid fibres and the air, as well as the contribution of radiation heat transfer between fibres such that:

$$k_{\text{eff}} = (k_{\text{gas}} + k_{\text{solid}}) + k_{\text{rad}}. \qquad 2.7$$

Since heat transfer in fibrous materials is a combination of conduction/convection in the air between fibres, conduction in solid fibres, and radiation heat transfer between fibres, effective conduction can be represented as:

$$k_{\text{eff}}(T) = (\nu_{\text{air}}k_{\text{air}}(T) + (1 - \nu_{\text{air}})k_{\text{fibre}}(T)) + k_{\text{rad}}(T), \qquad 2.8$$

where ν_{air} is the volume fraction of air in the fibrous material.

Applied, for example, to Nomex® fibre (Williams and Curry, 1992; Futschik and Witte, 1994),

$$k_{\text{fibre}}(T) = 0.13 + 0.0018(T(K) - 300k) \qquad T \leq 700k$$
$$= 1.0 \qquad T > 700k$$

The thermal conductivity of air has been represented by a linear relationship

$$k_{\text{air}}(T) = 0.026 + 0.000068(T(K) - 300K) \qquad T \leq 700k$$
$$= 0.053 + 0.000054(T(K) - 700K) \qquad T > 700k$$

In these models, the fibres in the fabric are assumed to function as infinite plates acting as radiation shields. Hence, the portion of the thermal conductivity due to thermal radiation between the fibres is assumed to be equal. The thermal conductivity of fabrics can therefore be calculated as follows:

$$k_{\text{rad}} = \frac{\sigma \varepsilon_{\text{fibre}} \Delta x (T_1^2 + T_2^2)(T_1 + T_2)}{2 - \varepsilon_{\text{fibre}}}, \qquad 2.9$$

where $\varepsilon_{\text{fibre}}$ = emissivity of the fibres, and Δx = width of the particular finite element.

The radiation portion of thermal conductivity is known to be very small. For a 100K temperature difference across a finite fabric element in the fabric, the contribution due to thermal radiation is about 5% of the total thermal conductivity (Torvi, 1997). This model can be used to calculate thermal conductivity of fabrics across a wide range of temperatures. Figure 2.2 shows the relationship between thermal conductivity and temperature predicted for Nomex® IIIA. An improved model (Stuart and Holcombe, 1984) separates the process of conduction through air and fibre and heat exchange by radiation.

The coupled heat and moisture transfer in textile materials that affects thermal insulation has been widely recognized and studied extensively (Henry, 1939; Crank, 1975; Ogniewicz and Tien, 1981). Farnworth (1986) has developed a dynamic model of coupled heat and moisture transfer with sorption and condensation in clothing. In this model the mass of absorbed water was assumed to be directly proportional to the relative humidity. This model was simplified with the assumption that the temperature and moisture content in each clothing layer were uniform. When the fabric's water content is low, the exchange of water can be treated as a sorption and desorption process. An experimental relationship has been established between the rate of change of water content of

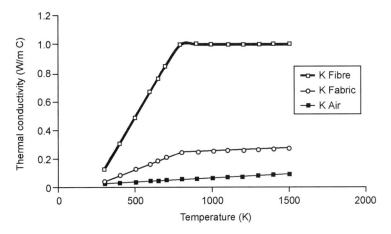

2.2 Thermal conductivity vs. temperature.

the fibres and the absolute difference between the relative humidity of the air and fibre (David and Nordon, 1969).

Dynamic moisture diffusion into different fabrics made from hydrophobic and hydrophilic fibre has been investigated using different mathematical models (Li and Holcombe, 1992; Li and Luo, 1999). The two-stage sorption model (Li and Holcombe, 1992) provides a good description of the water vapour transfer process in hydrophilic fabrics such as wool and cotton. Fick's diffusion model describes the water transfer in hydrophobic fabrics such as acrylic and polypropylene. The results demonstrate the different mechanism involved in dynamic moisture transfer between hydrophilic fabrics and hydrophobic fabrics. Hydrophilic fabrics show greater mass and energy exchange with the environment than hydrophobic fabrics.

More recent models of heat and moisture transfer in textile materials (Li et al., 2002; Li and Zhu, 2003) take into account that liquid water moves in the composite network of capillaries in porous textiles and is carried to the surface by capillary action. These models have been validated with experimental results on fabrics with varying hygroscopicity, thickness and surface properties and can be used to examine the effect of key structural parameters of fabric. Model predictions and experimental results indicate that fabric thickness and porosity significantly impacts moisture transport processes. Other models take into account moisture absorption, phase change, moisture bulk flow, radiative heat transfer and mobile condensates (Fan et al., 2000, Fan and Wen, 2002). They have been used to study the effectiveness of clothing consisting of fibrous battings sandwiched by two layers of thin fabrics. They show that inner fibrous battings having higher fibre content, finer fibre, greater fibre emissivity, higher air permeability, lower disperse coefficient of surface free water, and lower moisture absorption rate, cause less condensation and moisture absorption, which provides better thermal insulation during and after exercising in cold weather conditions. The results also demonstrate that covering fabrics have a significant effect on the water content within the fibrous batting.

Another model has simulated steady state heat conduction and moisture diffusion from skin to environment through a fabric system including a microclimate layer and a single fabric layer (Min *et al.*, 2007). This model showed that the microclimate (air layer) plays a significant role in the heat and moisture transfer from skin to environment, within which radiation and conduction contributed approximately 20% each of the total heat flux, and moisture diffusion contributed about 60% of the total heat flux. Variation in fabric thickness was found less important than variation in microclimate size in affecting heat transfer. Modelling was also used to compare a one-dimensional horizontal system (for fibre alignment) with a vertical system with a two-dimensional flow. The model showed the total heat flux was 10% greater in the vertical system than in the horizontal system.

2.8   Conclusions

In summary, the factors contributing to thermal fabric properties are:

- The thermal properties of the polymer of which the fibre is composed, such as thermal conductivity, specific heat, etc. In general existing polymers used for common fibres do not vary much in thermal conductivity and specific heat.
- Fibre fineness and geometry. Fibre diameter and the cross-section shape are important to the amount of air contained in yarn and the fabric structure. Hollow and crimped fibres trap more dead air and significantly increase fabric thermal insulation. A smaller diameter of fibre increases the surface area to which the air can adhere. Boundary layers can develop on the fabric surface as well as at fibre surfaces.
- Fabric thickness.
- Bulk density of a fabric which indicates the number, sizes and distribution of spaces within the fabric.
- Moisture transfer properties of fibres and fabrics.
- Fabric surface properties such as smoothness, surface emissivity, etc.
- Chemical finishes applied to fibre and fabrics.

2.9   References

ASTM F 1868 (2002). *Standard Test Method for Thermal and Evaporative Resistance of clothing Materials Using a Sweating Hot Plate,* 100 Barr Harbor Drive, PO Box C700, West Conshohocken, PA 19428-2959, United States.

Bandyopadhyay, S. K., Ghose, P. K., Bose, S. K. and Mukhopadhyay, U. (1987). *Journal of Textile Institute*, 78, 255.

Bo-an, Y., Yi-Lin, K., Yi, L., Yeung, C. and Song, Q. (2004). Thermal regulating functional performance of PCM garments. *International Journal of Clothing Science and Technology*, 16 (1/2), 84–96.

Bogaty, H., Hollies, N. R. S. and Harris, M. (1957). Some thermal properties of fabrics: Part I: The effect of fiber arrangement. *Textile Research Journal*, 27, 445–449.

Celcar, D., Meinander, H. and Gersak, J. (2008). A study of the influence of different clothing materials on heat and moisture transmission through clothing materials, evaluate using a sweating cylinder. *International Journal of Clothing Science and Technology*, 20 (2), 119–130.

Collier, B. J., Bide, M. and Tortora, P. G. (2008). *Understanding Textiles*, Pearson Prentice Hall.

Crank, J. (1975). *The Mathematics of Diffusion*, Clarendon Press, Oxford.

David, H. G. and Nordon, P. (1969). Case studies of coupled heat and moisture diffusion in wool beds. *Textile Research Journal*, 39, 166–172.

Fan, J. and Wen, X. (2002). Modelling heat and moisture transfer through fibrous insulation with phase change and mobile condensates. *Int. J. Heat Mass Transfer*, 45, 4045–4055.

Fan, J., Luo, Z. and Li, Y. (2000). Heat and moisture transfer with sorption and condensation in porous clothing assemblies and numerical simulation. *Int. J. Heat Mass Transfer*, 43 (12), 2989–3000.

Fanger, P. O. (1970). *Thermal Comfort*, Copenhagen: Danish Technical Press.

Farnworth, B. (1983). Mechanisms of heat flow through clothing insulation. *Textile Research Journal*, 53, 717–725.

Farnworth B. (1986). A numerical model of the combined diffusion of heat and water vapor through clothing. *Textile Research Journal*, 56 (11), 653–665.

Frydrych, I., Dziworska, G. and Bilska, J. (2002). Comparative analysis of the thermal insulation properties of fabrics made of natural and man-made cellulose fibres. *Fibers & Textiles in Eastern Europe*, 10–12.

Futschik, M. W. and Witte, L. C. (1994). Effective thermal conductivity of fibrous materials, General Papers in Heat and Mass Transfer, Insulation, and Turbon machinery, HTD-Vol. 271, *American Society of Mechanical Engineers*, New York, pp. 123–134.

Greenwood, K., Rees, W. H. and Lord, J. (1970). *Studies in Modern Fabrics*, edited by P. W. Harrison, The Textile Institute, Manchester, p. 197.

Hatch, K. L. (1993). *Textile Science*, West Publishing Company, New York.

Henry, P. S. H. (1939). Diffusion in absorbing media. *Proceedings of the Royal Society of London*, Series A, 171, 215–241.

Henry, P. S. H. (1948). The diffusion of moisture and heat through textiles. *Discussions of the Faraday Soc.*, 3, 243–257.

Holcombe, B. V. and Hoschke, B. N. (1983). *Textile Research Journal*, 53, 368.

Kong, L. X., Li, Y., She, F. H., Gao, W. M. and Wong, B. Z. (2001). Effect of fiber geometry and porosity on heat and moisture transfer in textiles. *Proceedings of the Textile Institute 81st World Conference – An Odyssey in Fibers and Spaces*, Melbourne, Australia.

Li, Y. (2001). The science of clothing comfort. *Textile Progress*, 31.

Li, Y. and Holcombe, B. V. (1992). A two-stage sorption model of the coupled diffusion of moisture and heat in wool fabrics. *Textile Research Journal*, 62, 211–217.

Li, Y. and Luo, Z. (1999). An improved mathematical simulation of the coupled diffusion of moisture and heat in wool fabric. *Textile Research Journal*, 69 (10), 760–768.

Li, Y. and Zhu, Q. (2003). Simultaneous heat and moisture transfer with moisture sorption, condensation, and capillary liquid diffusion in porous textiles. *Textile Research Journal*, 73 (6), 515–524.

Li, Y., Zhu, Q. and Yeung, K. W. (2002). Influence of thickness and porosity on coupled heat and liquid moisture transfer in porous textiles. *Textile Research Journal*, 72 (5), 435–446.

Min, K., Son, Y., Kim, C., Lee, Y. and Hong, K. (2007). Heat and moisture transfer from skin to environment through fabrics: a mathematical model. *International Journal of Heat and Mass Transfer.*

Morse, H. L., Thompson, J. G., Clark, K. J., Green, K. A. and Moyer, C. B. (1973). Analysis of the Thermal Response of Protective Fabrics, Technical report, AFML-RT-73-17, Air Force Materials Information Service, Springfield, VA.

Ogniewicz, Y. and Tien, C. L. (1981). Analysis of condensation in porous insulation. *J. Heat Mass Transfer*, 24 (4), 421–429.

Shim, H. E., McCullough, A. and Jones, B. W. (2001). Using phase change materials in clothing. *Textile Res. J.*, 71 (6), 495–502.

Slater, K. (1985). *Human Comfort*, Thomas Springfield, USA.

Stuart, I. M. and Holcombe, B.V. (1984). Heat transfer through fiber beds by radiation with shading and conduction. *Textile Research Journal*, 54 (3), 149–157.

Torvi, D. A. (1997). Heat Transfer in Thin Fibrous Materials under High Heat Flux Conditions, Doctoral Thesis, University of Alberta, Canada.

Williams, S. D. and Curry, D. M. (1992). Thermal Protection Materials: Thermophysical Property Data, NASA Reference Publication 1289, National Aeronautics and Space Administration, Scientific and Technical Information Program.

Wilson, C. A., Niven, B. E. and Laing, R. M. (1999). Estimating thermal resistance of the bedding assembly from thickness of materials. *International Journal of Clothing Science and Technology*, 11 (5), 262–267.

Zhang, Y., Wang, H. P. and Chen, Y. H. (2006). Capillary effect of hydrophobic polyester fiber bundles with noncircular cross section. *Journal of Applied Polymer Science*, 102 (2), 1405–1412.

Zhang, Y., Wang, H., Zhang, C. and Chen, Y. (2007). Modeling of capillary flow in shaped polymer fiber bundles. *Journal of Materials Science*, 42 (19), 8035–8039.

3
Assessing fabrics for cold weather apparel: the case of wool

R. M. LAING, University of Otago, New Zealand

Abstract: Although wool use for cold weather apparel has a long history, common in this application until the mid-20th century, it accounts for a small proportion of fibres now used. Developments and demonstration of the efficacy of wool apparel are examined – test methods, thermal and buffering effects, water and water vapour, odour and odour retention.

Key words: absorption, buffering, perception, odour retention.

3.1 Introduction

Wool use for cold weather apparel has a long history, common in this application until the mid-20th century. Since then, wool has accounted for a decreasing proportion of the world total fibre production/demand (all purposes, including apparel) (i.e. 1950 11.3%, 1960 9.8%, 1980 5.4%, 1990 5.2%, 2000 2.8% (Simpson, 2006, 2007; Simpson and Madkaikar, 2007). This relative decline, coupled with intense competition among fibre producers/marketers, resulted in applications for wool in apparel being more targeted, including 'high performance' applications.

From a wearer's perspective, requirements for cold weather clothing differ little, irrespective of the fibre(s) from which it is manufactured: to maintain thermoneutrality (or at least minimise thermoregulatory changes and the rate of any change), to minimise skin wettness and condensation within the garment microclimate, to minimise odour retained in the fabric/garment, to maintain acceptable tactile properties, and to minimise any change in mass with wetting. Priorities depend on the particular combination of activities undertaken and conditions of the cold environment. Cold conditions are an integral part of some sporting codes and working environments (e.g. snow skiing, mountaineering, utilities supply, exploration (http://www.antarcticanz.govt.nz, 2008)). Such conditions may be common in other codes and activities (e.g. ocean yacht racing, tramping/hiking/trekking), often with long periods of time during which clothing cannot be changed, thus necessitating assemblies which can accommodate effects of exercise, sweat/body odour, and intermittent wetting/drying (e.g. MAPP survey of sports participants, Champion, personal communication, 2003) (Laing *et al.*, 2007a).

34 Textiles for cold weather apparel

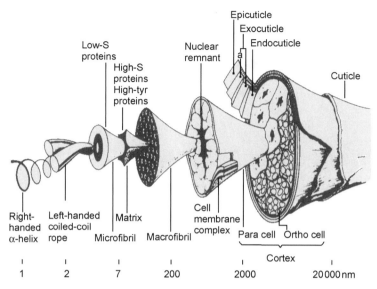

3.1 Structure of wool fibre (drawn by Robert C Marshall, CSIRO) (Hearle, 2002).

In this chapter, wool is defined as '1. The fibrous covering of a sheep (*Ovis aries*). 2. Appertaining to wool generally' (Denton and Daniels, 2002)[1] and the structure of a wool fibre given in Fig. 3.1.

Issues in which investigations involving wool apparel and its performance seem to provide points of difference from apparel manufactured from fabrics in other fibres include dealing with transient environmental conditions, minimising the rate of physiological change in the wearer to these transient conditions, and dealing with the limited scope for refurbishment/change of the apparel. The chapter is structured around a review of test methods, thermal and buffering effects, water and water vapour, and odour retention (Sections 3.2.1, 3.2.2, 3.2.3 and 3.2.4, respectively). Each section includes comment on early understanding of the issue along with evidence from investigations on fabric layers and/or human performance. The focus is on wool apparel rather than on the fibre itself and fibre processing.

---

1. And hair (excluded from the chapter) defined as 'Animal fibre other than sheep's wool or silk.' (M. J. Denton and P. N. Daniels eds. 2002. *Textile terms and definitions*. 11th edn. Manchester, UK: The Textile Institute).

3.2 Developments and demonstration of efficacy of wool apparel

3.2.1 Test methods

Developments in understanding connections between the garment/clothing system and the human body span many centuries and many notions, with emphasis on natural fibres such as wool and cotton from the 18th century until the more common use of synthetic polymer fibres by the mid-20th century. Early experiments included the following:

- Test methods for fabrics. In 1907, a book of methods for determining several fabric properties relevant to clothing for cold conditions (thickness, compressibility, thermal resistance, air permeability) was published. Rubner, physiologist and Director of the Hygiene Institute, Berlin, reported on fabrics manufactured from wool, linen, cotton, and several other fibres, noting their thermal resistance increased approximately in proportion to thickness; and that primary properties of fibres needed to be separated from secondary properties of fabrics (Renbourn, 1972).
- Garments/human trials. Early in the 20th century, Rubner measured skin/garment temperature gradients, relating physiological techniques to clothing (including wool clothing) (Renbourn, 1972).

During the early to mid-20th century, the demand to better understand performance of clothing was driven by military rather than civilian requirements, and by the mid-20th century by exploration in space. However, most research during the first part of the 20th century was on fibres and yarns rather than fabric properties, the latter considered more relevant to a garment ensemble (Fourt and Harris, 1949), although still unlikely to represent a garment ensemble. Some elements critical to the thermal resistance of clothing assemblies and to the wearer's thermal condition were partially understood (e.g. the relevance of air contained both in fabric and between layers of fabric, introduction of measures of thermal insulation (resistance) of a garment ensemble (e.g. 'clo' in 1949) (Newburgh, 1968 reprint); the cooling effect of garment ventilation (Forbes, 1949); and effects of sweat accumulation in a garment ensemble during exercise in the cold (Belding, 1949). As for human trials, the distinction between experiments carried out on fabrics and those carried out on garments worn by people under normal conditions of use, recognised by Rubner in the early 20th century was largely ignored for 30 years (Forbes, 1949). By the 1960s, many gaps in knowledge were apparent. For example, although synthetic polymer fibres and fabrics had been developed, most research on apparel systems focused on those manufactured from the principal fibres/products of that time (i.e. cotton, wool); accessing knowledge of experimental design, data collection and analysis was generally difficult; understanding sensory stimuli, how to measure these and relate them to physiological indicators was unclear; and the external demands for human survival under more extreme ambient conditions needed resolution. Trial and error was no longer an option.

Manikins, controlled human laboratory trials, and test methods relevant to clothing requirements for cold conditions, became better understood during the early part of the 21st century (e.g. Subzero Project) (Meinander *et al.*, 2004). The water vapour resistance of cold protective clothing measured using the Skin Model (for fabric) and both a thermal manikin and human trials (for garment assemblies) (Bartels and Umbach, 2003) were compared; and a comparison of thermal manikins of different body shapes and sizes showed lesser effects when garments were 'tight fitting' than when insulation was unevenly distributed (Kuklane *et al.*, 2004). Three factors relating to use of sweating manikins for determining clothing for protection against cold were identified. First, their use was for qualitative and comparative purposes rather than measurement of evaporative resistance of an assembly (Holmér, 2004b). Second, while reproducibility of results obtained using manikins was reasonable, different garment fit (i.e. manikins differed in 'body' dimensions and shapes) confounded results. And third, while the thermal insulation values defined with the thermal manikin corresponded with those of the related wear trial at moderately cold temperatures (i.e. 0 °C, −10 °C), at the lower temperature of −25 °C, a correction factor was required to account for greater sweating than predicted in the human trial (Meinander *et al.*, 2004). (However, note that wool was not among fibres included in the garments of the Subzero Project.) These, and other investigations and reviews of published work, formed the basis of several international (ISO) and *de facto* international (EN) Standards relevant to clothing in cold environments:

- ISO 7730: 2005 (E) Ergonomics of the thermal environment – analytical determination and interpretation of thermal comfort using calculation of the PMV and PPD indices and local thermal comfort criteria (International Organization for Standardization, 2005b).
- ISO 9920: 2007 (E) Ergonomics of the thermal environment – estimation of thermal insulation and water vapour resistance of a clothing ensemble (International Organization for Standardization, 2007a).
- ISO 11079: 2007 (E) Ergonomics of the thermal environment – determination and interpretation of cold stress when using required clothing insulation (IREQ) and local cooling effects (International Organization for Standardization, 2007b).
- ISO 15831:2004 (E) Clothing – physiological effects – measurement of thermal insulation by means of a thermal manikin (International Organization for Standardization, 2004).
- EN 342: 2004 (E) Protective clothing – ensembles and garments for protection against cold (European Committee for Standardization, 2004).

(See chapters on Standards, including those on laboratory assessment, assessment using manikins, and human wear trials.)

Effects of wind, mass of a garment assembly, friction between garment layers, fabric/garment stiffness have not been taken into account in the Standards, as noted in each introductory statement e.g.

> The thermal characteristics determined in this International Standard are values for steady-state conditions. Properties like 'buffering', adsorption of water and similar are not dealt with.
> ...
> This International Standard does not deal with the local thermal insulation on different body parts, nor the discomfort due to a non-uniform distribution of the clothing on the body.
> pvi (International Organization for Standardization, 2007a)

> Wind chill is commonly encountered in cold climates, but it is low temperatures that first of all endanger body heat balance. By proper adjustment of clothing, human beings can often control and regulate body heat loss, to balance a change in the ambient climate.
> pv (International Organization for Standardization, 2007b)

While understandable omissions, these variables reflect real, often transient, ambient conditions and potentially important consequential interactions between the clothing assembly and the human body. Over many years, desirable properties of wool and wool/blend apparel fabrics and apparel have been demonstrated under transient conditions and/or periods of extended use. One high profile example is the late Sir Peter Blake, KBE, Skipper of ENZA and 1994 winner of the Jules Verne Trophy for non-stop around the world sailing, whose diary entry, in relation to prototype next-to-skin wool products (from Icebreaker^{TM} a New Zealand apparel company), reads

> Icebreaker is superior in every way to anything I've ever worn. I wore it for 40 days and 40 nights and it didn't itch or get whiffy.

> The Icebreaker garments are fantastic. They have performed way beyond anyone's expectations. I think there were those in the crew who might have thought I was extolling the virtues of Icebreaker too much at the beginning of the trip but no longer. They are now all converts. I wear my Icebreaker under my wet weather gear, under my dry suit when diving, round the boat, and to bed at night.

> When you are among the icebergs, driving hard for a world record, the last thing you need is to have to strip to change your grundies. I didn't change my Icebreakers for 40 days and 40 nights. They didn't itch at all, were comfortable at all times, were very warm, didn't get whiffy (as most polyprop does after a few days) and dried quickly when damp. Icebreaker is a real breakthrough for sporting and outdoors people who want minimum bulk and maximum protection against the elements.
> (http://www.icebreaker.com/site/aboutus/reviews, 2008)

Also worth noting is that in moderately cold conditions (i.e. +10 °C), humans can feel cold but comfortable, with the body perceived as warm and the peripheral regions as cold (Holmér, 2004a).

3.2.2 Thermal and buffering effects

Early understanding of warmth and thermal resistance of clothing date back at least to the late 18th century experiments on thermal resistance of fibres used in clothing by Sir Benjamin Thompson; and later (1858) experiments with fabrics arranged in cylindrical form (wool, hemp, linen, cotton), Coulier, Professor of Hygiene at the Military Hospital, Paris, stated 'I have shown by a very simple experiment, that a cloth is considerably heated by the hygroscopic water ...' (Renbourn, 1972). Maintenance of a thermo-neutral state when in cool and cold conditions is the principal objective in textile/apparel selection for this situation, although other concurrent requirements for clothing are more common than not (e.g. resistance to burning, resistance to wind penetration, minimum garment mass). That the thermal conductivity of different fibres used in apparel is comparable has been known for many years (e.g. fibre type and conductivity in W/(m.K): aramid 0.130, cellulose acetate 0.226, cotton 0.461, polyester 0.141, polyamide 0.243, polypropylene 0.117, viscose rayon 0.289, wool 0.193) (Holcombe, 1984; see also Morton and Hearle, 1993). It is the air (conductivity 0.026 W/(m.K)) contained in the fabric/garment which dominates thermal resistance (assuming a similar covered surface area of the body).

Given all other variables are constant, air contained in wool yarns, fabrics, garments may be expected to be greater than that in yarns, fabric, and garments made from most other fibres because of the effect the physical structure of the wool fibre has on processing. However, the important question is whether any differences can be detected by wearers. Do differences in the properties of fibres and fabrics yield detectable differences in the thermal state of wearers when in cold conditions?

Few clothing-related human trials conducted in cold conditions during the latter part of the 20th century and early part of the 21st century included wool garments. Nevertheless, three studies are of interest; two do not include wool, and one does. In terms of physiological responses to exercise in the cold (exercise and rest (60 minutes each) repeated four times over an eight-hour period, -10 °C, $n = 6$ participants) wearing chemical protective clothing (no wool) resulted in a continuous decrease in skin temperature of the extremities during and following exercise and rest up to the third session, but a gradual increase in skin temperature of the torso to the fourth session (Rissanen and Rintamäki, 1998). In a similar study, effects of nuclear, biological and chemical protective clothing (again no wool) on cold and heat strain ($n = 11$ participants) during and following moderate and heavy activity levels (60 minutes) -33 °C to 0 °C showed rectal temperature was affected by changes in metabolic intensity rather than skin temperature (Rissanen and Rintamäki, 2007). Even at −33 °C ambient temperature, rectal temperature increased above 38 °C during heavy activity; with an ambient temperature of −25 °C, mean skin temperature decreased to tolerance level (25 °C) with moderate exercise; and at −15 °C or cooler ambient temperature, the temperature of fingers decreased below 15 °C (degraded performance) (i.e. the risk of concurrent heat and cold strain was demonstrated) (Rissanen and Rintamäki, 2007).

In the third study, effects of three fibre/fabric types two of which were wool, were included in a trial of long-sleeved garments for the upper body worn by well-trained participants ($n = 10$) exercising under conditions varying from cold (8 ± 2 °C, $40 \pm 2\%$ R.H.) to hot (32 ± 2 °C, $20 \pm 2\%$ R.H.) with a simulated wind, matched to the relevant running and walking speeds of each participant, applied during the 30-min run (~11 km/h) and the 10-min walk (~6 km/h) (Laing et al., 2008). Differences in human responses to the fabrics/garments included heart rate; core temperature during run (cold and hot), rest (hot only), and walk (cold only); heat content of the body; humidity under garments during rest and run; and time to the onset of sweating. No such differences were identified for change in body mass; core temperature during walk (hot only) and rest (cold only); skin temperature; temperature of skin covered by the garment; humidity under the garments during walk; or for any perceptions (thermal sensations, thermal comfort of torso, exertion, wetness). The garment manufactured from wool single jersey resulted in the least compromise to the thermoregulatory system during exercise in both the cold and hot conditions (i.e. time to the onset of sweating, a smaller increase in heat content of the body, smaller changes in core temperature) (see Fig. 3.2).

In cold conditions, separate garments may be worn one on top of another, or alternatively, one or more garments may be constructed of multiple layers of fabric/membranes, some or all or which are integrated. Air spaces between the textile layers may lead to either increased or decreased thermal resistance, sometimes the reverse of expectation with a smaller air space being a 'warmer' assembly even though the basic thermal resistance values of different clothing assemblies as determined on a manikin are the same (human and manikin trials) (Nielsen et al., 1989). The thermal resistance of multiple fabric layers under wind assault has been shown to increase with the number of textile layers (as would be expected), with less additional resistance per layer achieved with the change from two to three layers (Babus'Haq et al., 1996). Permeability of the outer layer to wind apparently has a dominant effect, different ensembles with the same outer layer not necessarily behaving in the same way (Babus'Haq et al., 1996).

The order of layers in a clothing assembly from skin layer to outer layer is another variable. Human trials have been conducted to better understand layering, where layers in an assembly comprise different fibres/fabric structures. Two such clothing systems, each comprising four garments (nine fabric layers), were

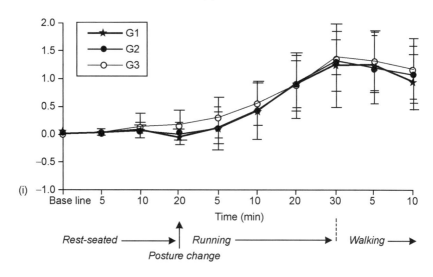

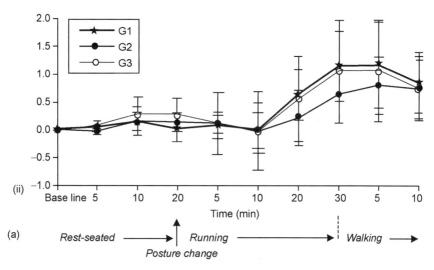

3.2 Selected physiological responses to three garments during exercise (G1 = polyester interlock, G2 = plated (wool and polyester), G3 = wool single jersey): (a) core temperature in (i) hot and (ii) cold conditions, (b) skin (covered) temperature in (i) hot and (ii) cold conditions, (c) relative humidity under garment in (i) hot and (ii) cold conditions (Laing *et al.*, 2008).

worn by young males ($n = 11$) during a protocol of 90 minutes (sitting, walking, sitting) in cold conditions ($-15 \pm 0.5\,°C$, air velocity <0.1 m/s) (Wang *et al.*, 2007). For both assemblies, the layer against the skin was a wool/cotton blend knit, and the outer layer a woven polyamide (although not stated, apparently the same fabrics for each assembly). The remaining layers of the assemblies differed

Assessing fabrics for cold weather apparel: the case of wool 41

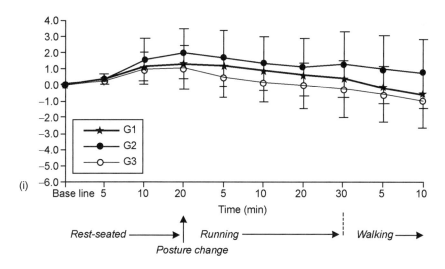

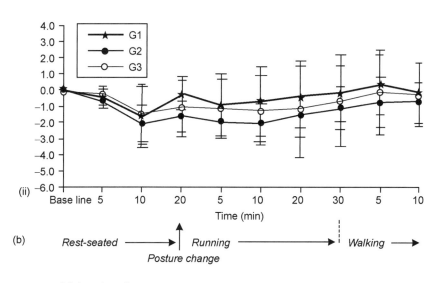

3.2 (continued)

slightly (A – woven, non-woven, mesh knit structures of polyester and polyamide; B – woven, non-woven, knit structures of wool/cotton blend, polyester, polyamide) (Wang *et al.*, 2007). With assembly B, both absolute humidity and relative humidity at the skin surface were lower, no liquid appeared on the skin surface, and the humidity outside the coat (but inside the outer jacket) was higher indicating more moisture was transferred from the skin surface to the outer layers than was the case with assembly A (Wang *et al.*, 2007).

42   Textiles for cold weather apparel

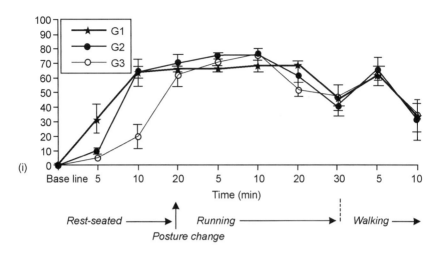

(i)

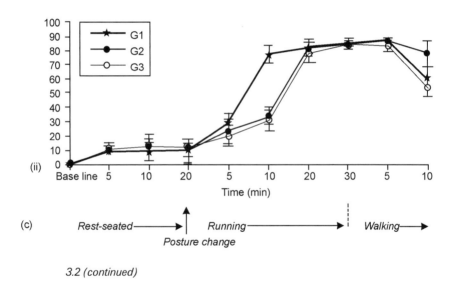

(ii)

(c)

3.2 (continued)

3.2.3 Water and water vapour

Early understanding of water and water vapour relationships with fibres/fabrics and clothing date to the first half of the 19th century, when military hygienists examined the relative merits of linen, wool and cotton for use in tropical conditions. Results from quantitative tests for permeability of fabrics to water vapour (and air) were first published in 1865 by von Pettenkofer, Professor of Hygiene, University of Munich (Renbourn, 1972). Wool and wool fabrics are inherently hydrophilic compared to fabrics from many other fibres, and under cold conditions, this may be either an advantage (leading to absorption of liquid sweat from the body and perception of a drier skin) or a disadvantage (potentially absorbing a large volume of liquid from the environment over time; although see (Laing *et al.*, 2007b)). This inherent hydrophilic property can be increased using a low pressure plasma treatment, demonstrated on a range of wool and wool/polyester blend fabrics with the claimed potential of enhanced physiological 'comfort' (Nunes *et al.*, 2005). Whether or not that potential has been reached is unclear. Interest in water vapour sorption and desorption of wool has been protracted, with debate over how water molecules are bonded with the fibre. Some suggest strongly or loosely bonded network water, while others consider water is physically sorbed and located in fibre capillaries. The mobility of water molecules in the fibre reportedly increases with increasing the amount of water in the system, as demonstrated with woven wool (21 μm) fabrics using Thermogravimetric Analysis (TGA), and Differential Scanning Calorimetry (DSC) (Popescu and Wortmann, 2000).

Fabric/garment layering with respect to permeability/transfer of water vapour and condensation was the subject of several investigations during the latter part of the 20th and early part of the 21st centuries. Layers of fabrics typical of outdoor clothing (three-layer combinations, no air space between) were shown to affect each other, the base liner fabric dominating water vapour permeability of the whole system (no temperature gradient), and condensation forming on all base liners (Ruckman, 2005). The water vapour transfer of a three-layered fabric seemed to be able to be improved and condensation reduced either by decreasing the thickness of the waterproof membrane and outer layer, or by increasing the average diffusion coefficient of the outer layer and membrane (Ren and Ruckman, 2004). What effect different spacing between the layers in the experimental set-up might have had on results of the 2005 study is not clear.

In another study, four layers typical of clothing for cold conditions (underwear, two garment layers, an outer layer of three-ply laminate) examined on a sleeve/sweating arm, showed the permeability of the fabrics and their condensation rates depended strongly on the outside climate and the hydrophilicity of the outer layers (Rossi *et al.*, 2004). Interestingly, when tested at the Standard conditions for textile testing (i.e. $20 \pm 2\,°C$, $65 \pm 4\%$ R.H. (International Organization for Standardization, 2005a), the water vapour resistances of the ensembles were similar, but differences became more apparent as the ambient temperature was reduced (Rossi *et al.*, 2004). The least condensation was observed when the hydrophilic membrane was laminated to the hydrophilic inner side. Hydrophilic layers under the outer shell generally absorbed more moisture than similar hydrophobic layers, suggesting liquid moisture moves from the outer shell to the inner layers (Rossi *et al.*, 2004).

A third study (also fabric layers rather than garments on the body), the position of wool batting in two multi-layered arrangements representing outer garments (Gore-tex® inner, multiple-ply wool batting, multiple-ply polyester batting, Gore-tex® outer; Gore-tex® inner, multiple-ply polyester batting, multiple-ply wool batting, Gore-tex® outer) measured under cold ($-20 \pm 1\,°C$) and hot ($25 \pm 1\,°C$) conditions, showed positioning the wool battings closer to the inner and the polyester closer to the outer could reduce moisture accumulation within (Wu and Fan, 2008). Total heat loss through the assemblies was reduced and this was attributed to wool releasing heat when absorbing moisture (Wu and Fan, 2008). Unlike many studies, Wu and Fan (2008) examined moisture (and heat) diffusion over times reflecting some typical times of wear (i.e. 24 hours, 8 hours), but whether these findings have been extended to human trials is unclear.

Sorption of water vapour by wool was the topic of several investigations during the late 20th century; the heat of absorption being substantially greater for wool than for any other fibre (Stuart et al., 1989) (Fig. 3.3). The liberation of heat of sorption is a transient effect, continuing until equilibrium is reached and, unlike many other fabrics, reaching equilibrium can take several hours with wool fabrics. This slow change is desirable as the rate of temperature change determines

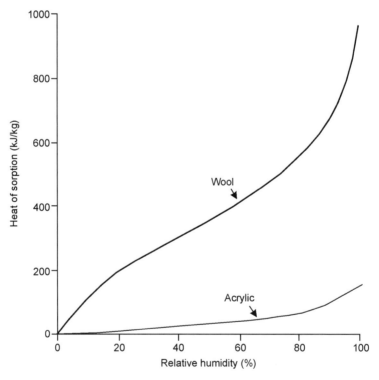

3.3 Heat of sorption generated when 1 kg of wool or acrylic absorbs water vapour to come to equilibrium with relative humidities between 0% and 100% (Stuart et al., 1989).

whether an actual change in temperature can be detected by the wearer; humans adapt continuously to their immediate environment. When cold, humans are reportedly more aware of added warmth and reject cold more strongly (Hensel, 1981). Using mittens in human trials ($n = 12$ participants) in cold conditions (7 °C, 80% R.H.), Stuart *et al.* (1989) showed: (i) with hands cold and wet, dry wool mittens were rated warmer than acrylic or conditioned mittens (i.e. warmth due to heat of sorption can be perceived in the dry wool garments); and (ii) with hands cold and wet, dry wool mittens were rated drier than others, because the wool removed moisture from the hands of wearers (Stuart *et al.*, 1989).

Skin wettedness, arising either from non-absorbed sweat on the skin surface or from externally saturated garments (wetted by rain) affects the wearer's perceptions of discomfort. The extent of skin wettedness which exceeds a certain threshold results in perceptions of damp and discomfort (e.g. Havenith *et al.*, 2003), the threshold suggested was around $w > 0.3$ (e.g. Goldman, 1988; Nishi and Gagge, 1970), or some other value related to the wearer's metabolic rate. For any given sweat production, the ambient vapour pressure and the vapour resistance of the clothing assembly determine the skin wettedness. At higher sweat rates, diffusion of sweat through fabric is too slow to ensure skin wettedness is low (and the person is 'comfortable'), and therefore garment ventilation is critical (Havenith *et al.*, 2003). Under cold conditions, sweat production tends to be rather low (e.g. -10 °C, 60 minutes exercise, 60 minutes rest for 8 hours wearing impermeable and semi-permeable protective clothing; Rissanen and Rintamäki, 1997) but obviously many factors determine this. However, if moisture does accumulate in the clothing system, under cold conditions this may freeze, leading to discomfort. Alternatively, sweat may saturate the clothing, decreasing thermal resistance of that clothing and thus more rapid cooling of the body when at rest.

Under conditions of cold (5 °C, 70% R.H., air velocity $1.5 \, \text{m.s}^{-1}$) and using a moveable sweating manikin, heat and water vapour transfer of three types of protective clothing were examined, and although wool was not included, of interest was the fact that while the mean weighted and regional skin wettedness did not depend on the fabric combinations when the exercise intensity was low, this was not the case when the exercise intensity was high (Fukazawa *et al.*, 2004). In this latter case, skin wettedness was lowest with the garment ensemble which did not include a laminated film (Fukazawa *et al.*, 2004). This result suggests use of the breathable film under conditions of high exercise intensity needs scrutiny.

Perception of dampness during wear influences product acceptability. Perceived dampness tends to decrease with increasing hygroscopicity of fabric, and the difference in perceived dampness between fabrics increases as the ambient relative humidity and moisture content of the fabric decrease (Plante *et al.*, 1995). Fabrics from fibres which are strongly hygroscopic, such as wool, feel drier over a wider range of moisture content than fabrics which are made from less hygroscopic fibres (Li *et al.*, 1995). Unfortunately, this experimental work (wool, cotton, polyester fabrics) was carried out under warm conditions only (25 °C, 25% R.H.). To what extent would perceptions of dampness of these fabrics differ under cold conditions? How do perceptions of dampness alter with the cold, both cold ambient conditions and a cold body? Perceptions of temperature, moisture and comfort in clothing are known to be influenced by environmental transient conditions (Li, 2005). Sweaters made from wool and from acrylic, matched in all major properties were worn in a series of wear trials with varying conditions, including simulated moderate rain. The perception of warmth was positively related to temperature of both the skin and fabric, while the perception of dampness was negatively related to skin temperature and positively (but non-linearly) related to humidity in the clothing microclimate (Li, 2005). The perception of comfort was negatively related to the perception of dampness and negatively related to humidity in the clothing microclimate (Li, 2005) (Fig. 3.4). Wool absorbs more moisture, more heat is produced and the body remains warmer, and because the changes occur more slowly, adaptation occurs.

3.2.4 Odour and odour retention

Concern with both biological warfare and the spread of infectious disease during the latter part of the 20th and early part of the 21st centuries prompted several organisations to confer infection-resistance to fibres/yarns/fabrics (e.g. resistance to five bacterial and one fungal species conferred on wool and wool blend fabrics, the treated 100% wool having a slightly higher zone of inhibition than the blends (El-Ola, 2007); chitosan-treated wool fabric, although superior in antimicrobial and deodorant properties, exhibited adverse effects on some mechanical properties (Jeong *et al.*, 2002); deodorant properties of wool fabrics were developed through acid mordant dyeing (Kobayashi *et al.*, 2002)). Enhanced infection resistance to textiles for wider applications in medicine and surgery has generally not included wool/wool blends. Anti-microbial/anti-odour treatments for fabrics/garments have practical relevance in situations involving long periods of wear/use without opportunity for cleaning, drying, and/or replacement, especially those undertaken in cold environments (e.g. ocean sailing, tramping/hiking/trekking, utilities supply/restoration, scientific investigations in Antarctica (http://www.antarcticanz.govt.nz, 2008)). (See also the review by Gao and Cranston (2008) which includes treatments of wool, e.g. silver nanoparticles and chitosan.)

Resistance to retention of body odour is a property of apparel fabrics destined for long-duration wear in cold weather apparel, since although the sweat rate is typically lower under cold than under warm conditions, sweat production continues. Concurrent heat and cold stress at different parts of the body when working under cold ambient conditions may also occur (Rissanen and

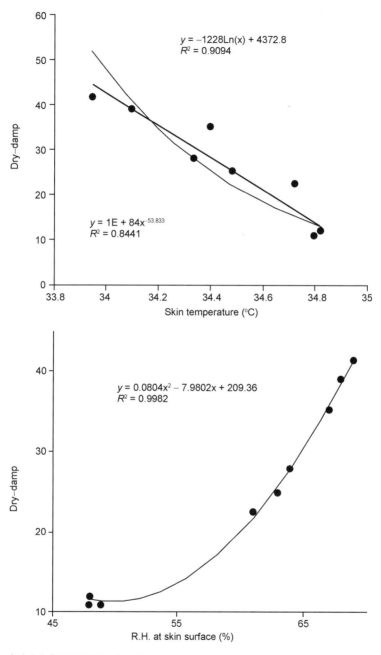

3.4 (a) Perception of moisture and skin temperature, (b) perception of moisture and clothing microclimate humidity (Li, 2005).

Rintamäki, 2007), presumably with some sweat absorbed by clothing covering the trunk region even while peripheral regions of the body may be at temperatures likely to impair performance, and with negligible sweat. An outer 'wind proof' or 'water proof' layer worn to minimise heat loss may not facilitate vaporisation of sweat, even when described as 'breathable' (see e.g. Fukazawa *et al.*, 2004). Thus under cold conditions, when sweat is produced and moves to clothing layers, only part may be vaporised at the garment surface. Skin debris and secretions are retained in the fabric layers, leading to greater or less intense odour.

Anecdotal reports of wool garments being worn satisfactorily for long periods of time without cleaning (see Section 3.2.1) have some scientific support (McQueen *et al.*, 2007, 2008). Features of studies by McQueen and colleagues were that structural properties and finishing of the fabrics were standardised other than for the fibre content (cotton, wool, polyester) and fabric structures (single jersey, 1 × 1 rib, interlock). Axillary secretions and microorganisms on the fabric were obtained during wear by human participants, and the odour intensity was detected using both trained humans and instrumental methods (likely compounds from volatiles identified using proton transfer reaction-mass spectrometry PTR-MS, taking into account the characteristic smell sometimes emanating from wool). The wool apparel fabrics retained less body odour than either the cotton or polyester fabrics. Some fabric structural effects were evident, the single jersey typically retaining less odour than either the interlock or the 1 × 1 rib structures (Fig. 3.5). Thus, based on results of these studies, under conditions in which options for cleaning/drying apparel are limited (both recreational and work-related), wool apparel seemed to perform well.

3.3 Summary and future trends

There is no doubt that selecting apparel for use in cold weather involves compromise in the extent to which various concurrent conditions/hazards can be met. Sufficient thermal insulation is relatively simple to achieve, principally by layering. What is less simple is ensuring all other conditions are met – sufficient flexibility of the garment assembly to allow body movement for the wearer, minimising the mass of a garment assembly, minimising the risk of thermal strain which may arise during prolonged periods of physical activity when wearing thick/heavy clothing, and minimising odours retained on garments when little opportunity exists for garment refurbishment. Although there is evidence of some human adaptation to a cold environment, and evidence that humans can feel cold but still function satisfactorily, selection of various forms of clothing and textiles are the typical response to cold in an effort to maintain thermoneutrality. The objective of this chapter has been to highlight aspects of clothing systems in which wool, a minor fibre in global production and apparel end uses, seems to provide some advantage. These included thermal/moisture

Assessing fabrics for cold weather apparel: the case of wool 49

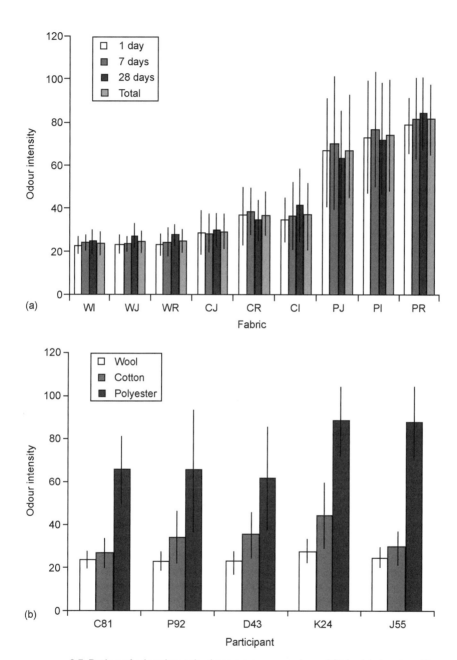

3.5 Rating of odour intensity (mean ± sem, panel *n* = 13) for fabrics stored for different times (a) and for fibre groups for five experts (b) (McQueen *et al.*, 2007).

buffering effects which lead to smaller, slower physiological changes in response to a transient environment, and the tendency for wool apparel fabrics worn next to the skin to be comparatively free of body odour.

Insufficient fine apparel-type wool is likely to be available to meet future demand because of the effects of climate change on the main grower countries (especially Australia), changed land use (e.g. 'retirement' of tracts of New Zealand high country through the land tenure review process), and increased interest in renewable, natural fibres which are also biodegradable. Therefore, further modifications in processing (yarn and fabric structure, fabric finishing) to use mid-micron wool successfully for apparel fabrics is anticipated.

Finer and lighter fabrics will continue to be sought for apparel worn next to the skin. This, too, can be achieved through modified yarn and fabric structures, different blends, and finishing treatments, e.g. fine micron (17–23 μm) fibre used for apparel fabrics and apparel; a modified wrap-spun yarn allows a perceptual reduction in the sense of 'prickle' of approximately 1 μm for a 25 μm mean fibre diameter wool, 2 μm for a 29 μm mean fibre diameter wool, 3 μm for a 31 μm mean fibre diameter wool (Miao et al., 2005).

Conversely, negative perceptions of wool apparel, perhaps related to early experiences or information (e.g. prickly, heavy, dimensionally unstable), or to perceived problems in various stages of processing, e.g., mulesing (the surgical removal of wool-bearing wrinkled skin from around the breech of a sheep), which is scheduled to cease in Australia by 2010 (http://www.wool.com.au, 2008) and chlorine-based shrink-resist treatments may suppress demand.

General environment-related issues (personal and global) assumed greater importance during the latter part of the 20th and early part of the 21st centuries (De Coster, 2007). Issues surrounding clothing and human health/performance generally are expected to be highlighted. For example, textile tests and Standards to determine specified harmful substances on apparel fabrics began in Austria during the late 1980s (oeko-tex) (see http://www.oekotex.com, 2008), and the related Standards and labelling were used in several parts of western Europe. Other aspects of quality (e.g. ethical and environmental issues) were being addressed at the time of publication including by several wool-related organisations. Developments in environmental and supply issues with respect to wool seem likely to be both positive (natural fibre, disposal, likely traceability from farm to disposal), and negative (effects of climate change on fibre production, e.g., Australian drought (Harle et al., 2007; Hatcher et al., 2003); animal care/handling (e.g. mulesing); demands in wool processing (Poole and Cord-Ruwisch, 2004) including the need for substitution of the chlorine-based treatments for shrink proofing (Chen et al., 2004; Das and Ramaswamy, 2006); environment-related effects (Chen and Burns, 2006); e.g., over and under supply in relation to demand (Cardellino, 2003).

Processing and finishing treatments will be required to meet increasingly stringent international standards of environmental acceptability. Pressure will be

Assessing fabrics for cold weather apparel: the case of wool   51

placed on manufacturers and distributors to provide evidence to support all claims (e.g. accountability, traceability, product performance particularly those related to enhanced human safety and performance).

3.4 Sources of further information and advice

Key references

Effects of clothing on human performance – Clothing, textiles and human performance, *Textile Progress* 32(2): 1–132 (Laing and Sleivert, 2002).

Moisture buffering, wetting and wicking – publications by Li, Holcombe and colleagues, e.g. Li *et al.* (1992, 1995), early investigations on the physical properties of fibres, e.g. Morton and Hearle (1993), and on wetting and wicking, e.g. Patnaik *et al.* (2006).

Odour retention and intensity – publications by McQueen and colleagues, e.g. McQueen *et al.* (2007, 2008).

Wool: science and technology (Simpson and Crawshaw, 2002).

Research and industry groups

During the period 2007–9, structural changes took place to research organisations dealing with wool and wool products in both Australia and New Zealand, i.e., the principal wool-producing countries.

Australia

In 2008, Australia's Commonwealth Scientific and Industrial Organisation (CSIRO) Fibre and Textiles Division (undertaking research on both wool and cotton) was merged into the Materials Science and Engineering sector of CSIRO (http://www.csiro.au/org/TextilesOverview.html, 2008). Aspects of the Australian wool industry as a whole are dealt with by Australian Wool Innovation (http://www.wool.com.au, 2008).

New Zealand

In January 2007 AgResearch purchased the assets of Canesis, formerly the Wool Research Organisation of New Zealand Inc (WRONZ), thus joining expertise in wool production (AgResearch) with that of wool processing and manufacturing (principally early processing and carpet technologies, at Canesis). A Textiles Group was formed (http://www.agresearch.co.nz/science/textiles.asp, 2008). Wool fibre ≤ 21 μm has been the marketing and research focus of the NZ Merino Co Ltd (http://www.nzmerino.co.nz/, 2008), with extension to the lower end of the mid-micron range during 2008 (i.e. ~23 μm).

Quality

oeko-tex (http://www.oekotex.com, 2008).

3.5 References

Babus'Haq, R. F., Hiasat, M. A. A. and Probert, S. D. 1996. Thermally insulating behaviour of single and multiple layers of textiles under wind assault. *Applied Energy* 54 (4): 375–391.

Bartels, V. and Umbach, K.-H. 2003. Measurement of the vapour resistance of cold protective clothing by means of the Skin Model and a thermal manikin. In *2nd European conference on protective clothing – Challenges for protective clothing*, edited by ECPC. Montreux, Switzerland: European Society of Protective Clothing.

Belding, H. S. 1949. Protection against dry cold. In *Physiology of heat regulation and the science of clothing*, edited by Newburgh, L. H., 351–367. New York: Hafner Publishing Co. Inc.

Cardellino, R. C. 2003. Perspectives and challenges in the production and use of mid-micron wools. *Wool Technology and Sheep Breeding* 51 (2): 192–201.

Chen, H.-L. and Burns, L. D. 2006. Environmental analysis of textile products. *Clothing and Textiles Research Journal* 24 (3): 248–261.

Chen, Q. H., Au, K. F., Yuen, C. W. M. and Yeung, K. W. 2004. An analysis of the felting shrinkage of plain knitted wool fabrics. *Textile Research Journal* 74 (5): 399–404.

Das, T. and Ramaswamy, G. N. 2006. Enzyme treatment of wool and specialty hair fibers. *Textile Research Journal* 76 (2): 126–133.

De Coster, J. 2007. Green textiles and apparel: environmental impact and strategies for improvement. *Textile Outlook International* 132: 143–164.

Denton, M. J. and Daniels, P. N. eds. 2002. *Textile terms and definitions*. 11th edn. Manchester, UK: The Textile Institute.

El-Ola, S. M. A. 2007. New approach for imparting antimicrobial properties for polyamide and wool containing fabrics. *Polymer-Plastics Technology and Engineering* 46 (9): 831–839.

European Committee for Standardization, 2004. *EN 342: 2004 (E) Protective clothing – ensembles and garments for protection against cold*. Brussels: European Committee for Standardization.

Forbes, W. H. 1949. Laboratory and field trials. In *Physiology of heat regulation and the science of clothing*, edited by Newburgh, L. H., 320–388. New York: Hafner Publishing Co. Inc.

Fourt, L. and Harris, M. 1949. Physical properties of clothing fabrics. In *Physiology of heat regulation and the science of clothing*, edited by Newburgh, L. H., 291–319. New York: Hafner Publishing Co. Inc.

Fukazawa, T., Lee, G., Matsuoka, T., Kano, K. and Tochihara, Y. 2004. Heat and water vapour transfer of protective clothing systems in a cold environment, measured with a newly developed sweating thermal manikin. *European Journal of Applied Physiology* 92 (6): 645–648.

Gao, Y. and Cranston, R. 2008. Recent advances in antimicrobial treatments of textiles. *Textile Research Journal* 78 (1): 60–72.

Goldman, R. 1988. *The handbook on clothing, biomedical effects of military clothing and equipment systems*. Brussels: Report NATO Research Study Group 7.

Harle, K. J., Howden, S. M., Hunt, L. P. and Dunlop, M. 2007. The potential impact of climate change on the Australian wool industry by 2030. *Agricultural Systems* 93 (1–3): 61–89.

Hatcher, S., Atkins, K. D. and Thornberry, K. J. 2003. Wool buyers do not adversely discount fine wool grown in 'non-traditional' environments. *Wool Technology and Sheep Breeding* 51 (2): 163–175.

Havenith, G., Ueda, H., Sari, H. and Inoue, Y. 2003. Required clothing ventilation for different body regions in relation to local sweat rates. In *2nd European conference on protective clothing*, edited by ECPC. Montreux, Switzerland: European Society of Protective Clothing.

Hearle, J. W. S. 2002. Physical properties of wool. In *Wool: science and technology*, edited by Simpson, W. S. and Crawshaw, G. H., 80–129. Cambridge, UK: Woodhead Publishing Company Limited.

Hensel, H. 1981. Thermoreception and temperature regulation. *Monographs of the Physiological Society* 38 (12.1).

Holcombe, B. 1984. The thermal insulation performance of textile fabrics. *Wool Science Review* 60 (April): 12–22.

Holmér, I. 2004a. Cold but comfortable? Application of comfort criteria to cold environments. *Indoor Air* 14 (Supplement 7): 27–31.

Holmér, I. 2004b. Thermal manikin history and applications. *European Journal of Applied Physiology* 92 (6): 614–618.

International Organization for Standardization, 2004. *ISO 15831: 2004 (E) Clothing – physiological effects – measurement of thermal insulation by means of a thermal manikin*. Geneva, Switzerland: International Organization for Standardization.

International Organization for Standardization, 2005a. *ISO 139: 2005 Textiles – standard atmospheres for conditioning and testing*. Geneva, Switzerland: International Organization for Standardization.

International Organization for Standardization, 2005b. *ISO 7730: 2005 (E) Ergonomics of the thermal environment – analytical determination and interpretation of thermal comfort using calculation of the PMV and PPD indices and local thermal comfort criteria*. Geneva, Switzerland: International Organization for Standardization.

International Organization for Standardization, 2007a. *ISO 9920: 2007 (E) Ergonomics of the thermal environment – estimation of thermal insulation and water vapour resistance of a clothing ensemble*. Geneva, Switzerland: International Organization for Standardization.

International Organization for Standardization, 2007b. *ISO 11079: 2007 (E) Ergonomics of the thermal environment – determination and interpretation of cold stress when using required clothing insulation (IREQ) and local cooling effects*. Geneva, Switzerland: International Organization for Standardization.

Jeong, Y. J., Cha, S. Y., Yu, W. R. and Park, W. H. 2002. Changes in mechanical properties of chitosan-treated wool fabric. *Textile Research Journal* 72 (1): 70–76.

Kobayashi, Y., Nakanishi, T. and Komiyama, J. 2002. Deodorant properties of wool fabrics dyed with acid mordant dyes and a copper salt. *Textile Research Journal* 72 (2): 125–131.

Kuklane, K., Sandsund, M., Reinersten, R. E., Tochihara, Y., Fukazawa, T. and Holmér, I. 2004. Comparison of thermal manikins of different body shapes and size. *European Journal of Applied Physiology* 92 (6): 683–688.

Laing, R. M. and Sleivert, G. G. 2002. Clothing, textiles, and human performance. *Textile Progress* 32 (2): 1–132.

Laing, R. M., Wilson, C. A., Gore, S. E., Niven, B. E. and Carr, D. J. 2007a. Determining the drying time of apparel fabrics. *Textile Research Journal* 77 (8): 583–590.

Laing, R. M., Niven, B. E., Barker, R. L. and Porter, J. 2007b. Response of wool knit apparel fabrics to water vapor and water. *Textile Research Journal* 77 (3): 165–171.

Laing, R. M., Sims, S. T., Wilson, C. A., Niven, B. E. and Cruthers, N. M. 2008. Differences in wearer response to garments for outdoor activity. *Ergonomics* 51 (4): 492–510.

Li, Y. 2005. Perceptions of temperature, moisture and comfort in clothing during environmental transients *Ergonomics* 48 (3): 234–248.

Li, Y., Holcombe, B. V. and Apcar, F. 1992. Moisture buffering behavior of hygroscopic fabric during wear. *Textile Research Journal* 62 (11): 619–627.

Li, Y., Plante, A. M. and Holcombe, B. V. 1995. Fiber hygroscopicity and perceptions of dampness: Part II: Physical mechanisms. *Textile Research Journal* 65 (6): 316–324.

McQueen, R. H., Laing, R. M., Brooks, H. J. L. and Niven, B. E. 2007. Odor intensity in apparel fabrics and the link with bacterial populations. *Textile Research Journal* 77 (7): 449–456.

McQueen, R. H., Laing, R. M., Delahunty, C. M., Brooks, H. J. L. and Niven, B. E. 2008. Retention of axillary odour on apparel fabrics. *The Journal of The Textile Institute* first published on line 28 May 2008.

Meinander, H., Anttonen, H., Bartels, V., Holmér, I., Reinersten, R. E., Soltynski, K. and Varieras, S. 2004. Manikin measurements versus wear trials of cold protective clothing (Subzero project). *European Journal of Applied Physiology* 92 (6): 619–621.

Miao, M., Collie, S. R., Watt, J. D. and Glassey, H. E. 2005. Prickle and pilling reduction by modified yarn structure. In *11th International Wool Textile Research Conference*, 150 YF Vol. Edited by Byrne, K., Duffield, P., Myers, P., Scouller, S. and Swift, J. A. 1–10. Leeds, UK: University of Leeds.

Morton, W. E. and Hearle, J. W. S. 1993. *Physical properties of textile fibres*. 3rd edn. Manchester, UK: The Textile Institute.

Newburgh, L. H. 1968 reprint. *Physiology of heat regulation and the science of clothing*. New York: Hafner Publishing Co. Inc.

Nielsen, R., Gavhed, D. C. E. and Nilsson, H. 1989. Thermal function of clothing ensemble during work: dependency on inner clothing fit. *Ergonomics* 32 (12): 1581–1594.

Nishi, Y. and Gagge, A. P. 1970. Moisture permeation of clothing – a factor governing thermal equilibrium and comfort. *ASHRAE Transactions* 76: 137–145.

Nunes, M. F., Geraldes, M. J. O., Belino, N. J. R. and Gouveia, I. C. 2005. Increase of hydrophily of fabrics made of wool and wool/polyester blends, using low pressure plasma techniques. In *11th International Wool Textile Research Conference*, CCF 35 Vol. Edited by Byrne, K., Duffield, P., Myers, P., Scouller, S. and Swift, J. A. 1–6. Leeds, UK: University of Leeds.

Patnaik, A., Rengasamy, R. S., Kothari, V. K. and Ghosh, A. 2006. Wetting and wicking in fibrous materials. *Textile Progress* 38 (1): 1–105.

Plante, A. M., Holcombe, B. V. and Stephens, L. G. 1995. Fiber hygroscopicity and perceptions of dampness Part 1: Subjective trials. *Textile Research Journal* 65 (5): 293–298.

Poole, A. J. and Cord-Ruwisch, R. 2004. Treatment of strongflow wool scouring effluent by biological emulsion destabilisation. *Water Research* 38 (6): 1419–1426.

Popescu, C. and Wortmann, F.-J. 2000. Water vapour sorption and desorption. In *10th International Wool Textile Research Conference*, ST-6 Vol. Edited by Hocker, H. and Kupper, B. 1–10. Aachen, Germany: Deutsches Wollforschunginstitut.

Ren, Y. J. and Ruckman, J. E. 2004. Condensation in three-layer waterproof breathable fabrics for clothing. *International Journal of Clothing Science and Technology* 16 (3): 335–347.

Renbourn, E. T. 1972. *Materials and clothing in health and disease* London: H K Lewis and Co Ltd.

Rissanen, S. and Rintamäki, H. 1997. Thermal responses and physiological strain in men wearing impermeable and semipermeable protective clothing in the cold. *Ergonomics* 40 (2): 141–150.

Rissanen, S. and Rintamäki, H. 1998. Effects of repeated exercise/rest sessions at $-10\,°C$ on skin and rectal temperatures in men wearing chemical protective clothing. *European Journal of Applied Physiology* 78 (6): 560–564.

Rissanen, S. and Rintamäki, H. 2007. Cold and heat strain during cold-weather field training with nuclear, biological, and chemical protective clothing. *Military Medicine* 172 (2): 128–132.

Rossi, R. M., Gross, R. and May, H. 2004. Water vapor transfer and condensation effects in multilayer textile combinations. *Textile Research Journal* 74 (1): 1–6.

Ruckman, J. E. 2005. The application of a layered system to the marketing of outdoor clothing. *Journal of Fashion Marketing and Management* 9 (1): 122–129.

Simpson, P. 2006. Global trends in fibre prices, production and consumption. *Textile Outlook International* 125: 82–106.

Simpson, P. 2007. Global trends in fibre prices, production and consumption. *Textile Outlook International* 131: 87–112.

Simpson, W. S. and Crawshaw, G. H. eds. 2002. *Wool: science and technology.* Cambridge, UK: Woodhead Publishing Company Limited.

Simpson, P. and Madkaikar, N. 2007. Global trends in fibre production, consumption, and prices *Textile Outlook International* 128: 119–141.

Stuart, I. M., Schneider, A. M. and Turner, T. R. 1989. Perception of the heat of sorption of wool. *Textile Research Journal* 59 (6): 324–329.

Wang, S. X., Li, Y., Tokura, H., Hu, J. Y., Han, Y. X., Kwok, Y. L. and Au, R. W. 2007. Effect of moisture management on functional performance of cold protective clothing. *Textile Research Journal* 77 (12): 968–980.

Wu, H. and Fan, J. 2008. Study of heat and moisture transfer within multi-layer clothing assemblies consisting of different types of battings. *International Journal of Thermal Sciences* 47 (5): 641–647.

4
Coating and laminating fabrics for cold weather apparel

R. LOMAX, Baxenden, a Chemtura Company, UK

Abstract: This chapter illustrates how continuous polymer membranes are used to impart weatherproof properties in clothing materials. It provides a historical background and describes the main technologies used to achieve the best possible balance between weather protection and wearer comfort under foul-weather or particularly cold conditions. Manufacturing techniques, fundamental properties and test methods for these products are also highlighted. Some developments that lessen the environmental impact of coated and laminated textiles, together with thoughts on future trends are mentioned at the end of the chapter.

Key words: coated fabrics, waterproof breathable fabrics, foul-weather clothing, hydrophilic membranes, microporous membranes.

4.1 Introduction

Garments incorporating polymer membranes (as films and unbroken coatings) provide exceptional weatherproof properties (Holmes, 2000). In particular, they are highly resistant to penetration by wind, spindrift and every type of atmospheric precipitation, which can make all the difference in some life-threatening situations. The downside is that they can restrict perspiration escaping away from the body, leading to an undesirable build-up of water vapour inside the clothing and ultimately, condensation. Sodden clothing loses much of its insulation value, so considerable effort has been spent in developing new polymer systems which add breathability to their repertoire (Träubel, 1999). Tackling this paradox of protection versus comfort is, of course, only one of many challenges for scientists and designers of cold weather apparel.

The technology for manufacturing textiles coated with, or laminated to continuous polymer layers is surprisingly diverse (Fung, 2002). It has proved one of the most active R&D areas for functional clothing during the past 30 years, with plenty of new fabrics introduced through a steady blend of innovation and evolution (Fung, 2005). One school of thought holds that they are more suitable for temperate foul weather, ranging say from +10 to −5°C than for much colder, drier conditions that might be encountered, e.g., in Arctic/ Antarctic climates and high-altitude mountaineering (Holmér, 2005).

Nonetheless, coated and laminated fabrics are routinely used in garment assemblies based on layering and moisture-management principles, and worn by sports enthusiasts and professionals in all extremes.

4.2 Historical aspects and evolution of the modern industry

The origins of the modern coating industry are not known for certain, but documentary evidence suggests that they pre-date 1600 AD. Long before then, man had used hides, furs and even the innards of large mammals to provide warmth and shelter against bad weather. Animal epidermis, suitably scraped and treated to limit putrefaction can be regarded as the prototype membrane for imparting wind and rain resistance. Caribou and seal skin clothing worn by Inuit people, for instance, still compares favourably with modern synthetic ensembles for working and travelling in harsh Arctic winters (Oakes *et al.*, 1995).

4.2.1 Proofing with rubber and other bio-renewable resins

Spanish conquistadors reported that inhabitants of the Central and South American rain forests were producing weatherproof cloth by painting sap and resin oozing from certain trees onto linen fabrics. The most suitable material was a milky white fluid, which could be tapped from the bark and cambium of the species *Hevea brasiliensis*. European traders were keen to exploit this new craft in their own countries, but were continually thwarted; unless the fluid was used immediately, it coagulated into a sticky mass which could no longer be spread onto fabric. It remained a curiosity until the mid-1700s, when small pieces of smoke-dried resin were found to be capable of erasing pencil marks, and they then sold for incredibly high prices. Soon afterwards, the material became known as rubber, or more correctly 'India rubber' since it emanated from the *East* Indies colonies (Lomax, 1985b).

The next significant advance came in 1791 when a British inventor, Samuel Peel, took out a patent for dissolving rubber in turpentine, coating or impregnating fabrics with the viscous solution, and finally drying off the solvent. This technique produced a fairly stiff, woven cloth with the interstices between yarns completely filled by impervious polymer. In 1823, Charles Macintosh improved Peel's process, first by casting rubber solution onto a marble slab and then by sewing the thin, dried rubber sheet onto cotton or silk fabric with needle and thread. These early rubberised fabrics were not very serviceable. In hot weather they became too soft and sticky, whereas in cold weather they became brittle, cracked and lost their waterproof properties. Over the next two decades, however, some remaining technical issues were overcome and the foundation for the rubber-proofing industry was firmly laid. Key innovations included the first spreading machine (1827), mastication, i.e., the mechanical kneading of rubber into a more pliable, soluble and consistent form, and most importantly, vulcanisation which was patented independently by Thomas Hancock and Charles Goodyear in 1843. The latter process involves adding sulphur to the coating compound and heat treating the finished article. Vulcanisation initiates a crosslinking mechanism that converts natural rubber from a tacky gum into a durable, flexible polymer with much wider temperature latitude.

As the proofing industry flourished during the latter half of the 19th century, vast areas of land in Ceylon, Malaya, Indonesia and Borneo were planted with imported *Hevea* seedlings and other local species. By this time, applications included ship's sails, storage bags, hoses, drive belts, tarpaulins and inflatables, as well as the original waterproof clothing (Lomax, 1985a). Records show that consumption of natural rubber increased dramatically, from about 2,000 tonnes in 1853 to more than 1,000,000 tonnes per year in the 1930s when the Far East plantations had fully matured. Other treatments derived from natural products include nitrocellulose (introduced in the 1880s) and linseed oil, which after repeated impregnation into fabric, oxidises and hardens to give the original 'oilskins'. These materials have been gradually displaced by synthetic membranes with greatly improved balance of properties.

4.2.2 Developments in man-made coatings and fabrics, 1900–1945

Stepwise advances in organic chemistry during the first part of the 20th century led to a new range of monomers capable of being converted into fibres, fabrics and membranes. The driving force was, and still is, the availability and low cost of carbon reserves, initially coal and nowadays petroleum and natural gas. These resources are mainly used for fuels and heat generation. Only a relatively small proportion (ca. 6%) is refined into the basic feedstock chemicals that now support a vast industry for plastics, synthetic fibres and other polymer items, see Fig. 4.1.

One of the first pioneers was Wallace Carothers, who undertook fundamental research at the E.I. du Pont de Nemours laboratories in Wilmington, Delaware. His team of chemists was responsible for at least two outstanding discoveries in the early 1930s. These were fibre-forming polyamides (later introduced as Nylon 6,6) and Neoprene rubber (polychloroprene), which had better chemical resistance and adhesive properties than the natural material. Carothers had also been tantalisingly close to discovering the greatest synthetic textile polymer, polyester, or more specifically poly(ethylene terephthalate), PET – but this new field of science was so wide open that he did not persevere, and simply moved on to the next chemical challenge. After Carother's death in 1937, his work on generic polyesters was reinvestigated by Rex Whinfield and James Dickson then employed by the Calico Printers Association in Accrington. Their 1940s patents were eventually licensed to ICI (Imperial Chemical Industries) and, perhaps

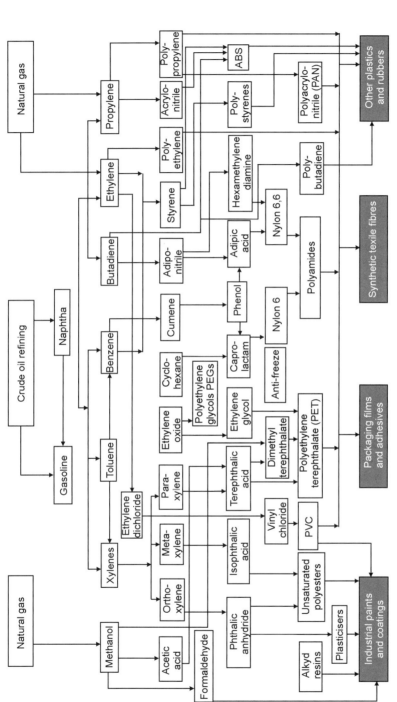

4.1 Main chemical feedstocks derived from petroleum and natural gas, and some synthetic routes to polymer fibres, plastics, coatings and rubbers.

ironically, the US rights were bought by du Pont. The new PET fibres Terylene and Dacron were first presented in the early 1950s for easy-care/non-iron fabrics and blending with cotton.

Other potential coating materials such as SBR (styrene-butadiene rubber), PVC (polyvinyl chloride, in plasticised form) and nitrile rubber (butadiene-acrylonitrile copolymers) had been developed in the 1920s–30s, but they were nowhere near as widely used as natural rubber. This situation changed abruptly with the outbreak of the Second World War (WWII). The Japanese occupation of Malaya in 1942 effectively curtailed supplies of rubber to the US, Britain, their European allies and foes, and an intense development programme was instigated to find a practical alternative. This was needed in large quantities by both sides for maintaining supplies of protective clothing, tyres, inflatable craft and other military items for the war effort. Almost simultaneously, SBRs were introduced in the US and Germany as a superior replacement for the natural product. Japan's earlier foray into mainland China had also wrecked the supply of silk to the west. Fortunately, it only hastened the commercialisation of Carothers's fibre. By 1939, Nylon 6,6 was already being used as a replacement for ponchos, flak jackets, ropes, tyre cord, parachute fabric, other military supplies and by no means less importantly, ladies stockings.

The first recognised waterproof, breathable fabric, Ventile was introduced in 1943 specifically to protect British aircrew patrolling North Atlantic supply convoys. It was made into coveralls, rubber sealed at the neck, wrists and feet to keep the wearer dry if he ditched in cold seas. Ventiles are in fact a series of cotton Oxford fabrics developed in the weaving department at Shirley Institute (now British Textile Technology Group, BTTG), Manchester originally as a replacement for flax used in waterproof hoses and fire buckets. They are not based on membranes but are constructed from specially treated cotton yarns with a controlled, high swelling capacity. In the dry state, a Ventile fabric is air-permeable, breathable and highly comfortable, but when it is suddenly wetted out the fibres swell and the interstitial pores become constricted, which imparts a good level of waterproofness. The new Ventile coveralls improved the survival rate for ditched aircrew from almost zero to more than 80%, and they are still used in similar applications today.

4.2.3 Significant introductions from 1945 onwards

After WWII hostilities ended, a further intense period of collaborative research followed through into the 1950s and early 1960s. One major introduction during this era was the Ziegler–Natta catalysts for stereoselective polymerisation, which later earned the co-inventors a Nobel Prize for Chemistry (1963). These are organometallic compounds, based on titanium and aluminium that mimic the behaviour of enzymes. At the molecular level, they provide a unique template platform for monomers and by doing so control the size and especially the shape of a growing polymer molecule. They are used in the manufacture of polyethylene (PE), poly(vinyl alcohol) and most importantly, polypropylene (PP) fibres and films.

The post-war decades also saw the scale-up and commercialisation of a number of coating materials and elastomers that had been studied in the earlier golden period of polymer research, but often only from an academic point of view. These included Teflon or PTFE (polytetrafluoroethylene, discovered by accident in 1938), fluoroelastomers, butyl rubber (isobutene–isoprene copolymers, 1937), acrylics, Hypalon (a vulcanisable plastic, formed by treating polyethylene with chlorine and sulphur dioxide) and silicone rubbers (Lomax, 1985b). Machines and techniques used to apply these compounds to fabric remained virtually unchanged from 19th-century rubber spreading. A vast class of polymers, collectively known as PUs (polyurethanes) was first discovered in 1937 by Otto Bayer. This PU technology has been further developed for textile coatings, films, Spandex and Lycra fibres, flexible foam layers, shoe soles, moulded boots and plastics, as well as many other applications where low-temperature performance is important.

By the end of the 1960s, therefore, the basic set of polymer materials (coatings, films and fibres) that still constitute modern weatherproof clothing, and processing machinery were already in place (Sen, 2007). Some of the later innovations and background technology are outlined in the following sections.

4.3 Breathable membranes

The trend towards lighter and more easily stowed waterproofs in the 1960s–70s aided the development of so-called breathable membranes, based on microporous and hydrophilic polymers, see Fig. 4.2. These materials retain excellent windproof and waterproof properties, but have the capability to transmit moisture vapour at substantially faster rates than normal textile coatings. Under some climatic conditions and workrates, they can provide a real physiological benefit even at $-20\,°C$ (Bartels and Umbach, 2002). According to Fick's Laws of Diffusion, which govern the flow of gases and vapours through a barrier layer, the rate of water vapour transmission (WVTR) is directly proportional to the pressure difference across the membrane surfaces (Δp) and inversely proportional to its thickness (t), i.e.

$$\text{WVTR} = k.\Delta p/t \qquad 4.1$$

where k is a proportionality constant. Logically therefore, the thinner the membrane the better, but high breathability has to be offset against flimsiness in terms of handling, other production issues and a suite of durability properties for use. Typical breathable layers are 15–30 microns thick, but laminated films down to 5–10 microns have intermittently appeared on the market. This contrasts sharply with earlier reasoning, where thicker, hydrophobic membranes were preferred for their perceived higher levels of protection.

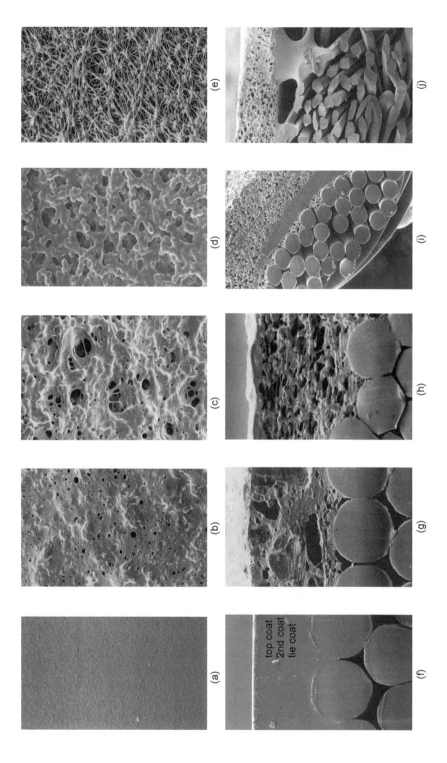

4.3.1 Microporous membranes

These materials have a vast, permanent labyrinth of holes and passageways that connect the inner and outer surfaces of a textile membrane. Overall, the structure is air permeable and therefore allows transfer of water vapour, which is an intimate constituent of air (0–4% w/w, depending on ambient relative humidity and temperature), under a suitable pressure gradient. The best known microporous membrane is expanded (or e) PTFE used, e.g., in the Gore-Tex laminates first introduced in 1976, see Fig. 4.2(e). Typically, a 30 micron ePTFE membrane comprises 20% polymer and 80% air by volume, and has about 'nine billion pores per square inch, with the largest pore being 0.2 microns' (Tanner, 1979).

Alternative microporous coatings and laminated films are currently made from a wide range of polymers including PU, acrylic, PE, and PP. They normally have a dull, white appearance, caused by diffraction of light through the surface cavities and other optical effects. Endless variations in porous structure are possible, in terms of mean, minimum and maximum pore size (surface and core), their shape and distribution, % pore volume, tortuosity, etc., all of which can be easily discerned under the SEM. Some of the more interesting manufacturing techniques are mentioned by Lomax (1985a):

- Mechanical fibrillation: the uniaxial or biaxial stretching of a film, which if carefully controlled produces a series of microscopic rips and tears through the membrane. The films often contain micron-size particulate fillers, e.g. 40–70% w/w of calcium carbonate that provides seeding points for tear initiation. Key products include ePTFE (from many different manufacturers), PP films such as Celguard, and PU, e.g., the now discontinued Triplepoint Ceramic.
- Wet coagulation: a process where the polymer is dissolved in a water-miscible solvent such as dimethylformamide (DMF), and the viscous solution applied to fabric or release paper. The porous structure is formed as the fabric is passed through a coagulation bath, where DMF is displaced by the non-

4.2 (opposite) Some features of waterproof, breathable fabrics (WBFs).
Top row: surfaces of membranes (a) solid, hydrophilic PU coating; (b) microporous PU coating by wet coagulation; (c) microporous PU coating by phase separation; (d) coating from mechanically foamed acrylic dispersion; (e) ePTFE film.
Bottom row: sections through fabrics (polyamide substrates unless otherwise stated) (f) solid, hydrophilic PU showing successive coating layers; (g) microporous PU coating by wet coagulation; (h) microporous PU coating by phase separation, with very thin pore sealing topcoat; (i) mechanically foamed acrylic coating with the first layer crushed; (j) microporous PU film made by salt extraction, laminated to polyester fabric with discontinuous (solid dot) adhesive.

solvent, water. This technique tends to produce membranes with an asymmetric pore structure, Fig. 4.2(g) and is widely used in the Far East, especially Japan (e.g. Entrant).

- Phase separation: the coating polymer (normally PU) is applied to fabric from a mixture of volatile solvent and relatively higher boiling non-solvent. As the true solvent evaporates faster during drying, the concentration of non-solvent remaining in the coating builds up, and the polymer eventually precipitates out. The process is controlled with special surfactants to ensure that the pore size of the resultant coating is suitable for purpose, see Fig. 4.2(c),(h). The best-known product is the Ucecoat 2000 coating system (UCB, Belgium now part of Cytec Industries), and derived fabrics such as Cyclone.
- Solvent extraction: finely divided, water-soluble salts or other compounds can be added to the polymer formulation, and then washed out of the dried membrane. Products formed using micronised sodium chloride such as the Porvair and Permair films and foils are typical examples, see Fig. 4.2(j).

4.3.2 Hydrophilic membranes

In comparison, hydrophilic membranes are naturally glossy, transparent, and clearly have a solid (aka monolithic or compact) structure, see Fig. 4.2(a),(f). Solidity confers an even higher level of wind and waterproofness, whilst the breathability element is built into the chemistry of the polymer. Most hydrophilics are segmented copolymers which give nanophase-separated structures as they consolidate down, Fig. 4.3(a)–(c).

Poly(ethylene oxide), PEO is the normal soft segment component with water vapour transfer properties, and the constraining hard segments can be based on ester (e.g. Sympatex film), amide (e.g. Pebax elastomers) or most commonly, urethane residues. The science behind hydrophilic PU membranes has recently been reviewed (Lomax, 2007), and gives an insight into some of the more unusual properties of PEO-copolymers. The former materials were first developed at Shirley Institute in the late 1970s, after project funding by the UK Ministry of Defence to improve the comfort of military PU-coated nylon waterproofs. This technology was subsequently licensed and formed the basis for the Witcoflex coating systems, introduced in about 1982. Since then, a number of the major European PU suppliers and increasingly, small manufacturers in the Far East have developed products of this type for direct coating, transfer coating, film production and other applications such as adhesives and pore sealants.

Breathability in hydrophilics is an absorption–diffusion–desorption process, accelerated by hydrogen bonding between individual water molecules and complementary functional groups built into the molecular chains. It can be likened to providing an infinite number of 'molecular stepping stones', correctly spaced apart so that water molecules transfer easily from a membrane surface of

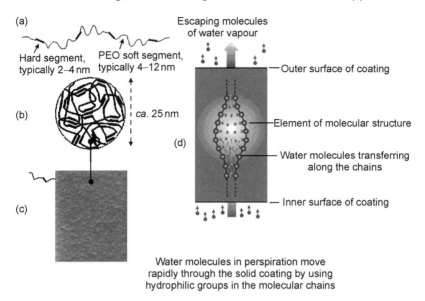

4.3 Highly schematic representation of hydrophilic polymer features: (a) 'stick and string' model of molecular chain in segmented copolymer; (b) phase separation initiated by hydrogen bonding between adjacent hard segments as the membrane consolidates; (c) SEM photomicrograph of hydrophilic PU surface etched with DMF vapour, showing hard and soft contours; (d) original pictogram for hydrophilic water vapour transfer, widely reproduced in technical brochures and swing tickets.

high concentration, to one of lower concentration, see Fig. 4.3(d). For a number of technical reasons (World Pat 2006075144, 2006), ether oxygen atoms (–O–) are preferred stepping stones compared with alternative hydrogen-bonding groups such as hydroxyl (–OH), amine (–NH_2) or carboxyl (–CO_2H). Under optimum breathability conditions, each stepping stone can attract up to three water molecules, as they transfer through the soft segment nanophase. The imbibed water causes the membrane to swell, and potential water uptakes of 40–100% w/w are quite normal. For an average anorak coated with hydrophilic polyurethane, this means that ca. 25 g of water can be absorbed under these conditions. Swelling is a double-edged sword, and has to be very carefully controlled (Lomax, 2001). On the one hand, it is a genuine mechanism for breathability in hydrophilics, but on the other it can jeopardise the waterproof properties as the swollen membrane is more susceptible to physical damage. Water-vapour-permeable, solid membranes with low swell properties (<5% w/w water uptake) became a new target for chemists working in this field, and this has been achieved to a modest extent (World Pat 2006075144, 2006). It opens up a different market for coatings worn on the outside, e.g., for wipe clean workwear, shoes and other applications where the liquid water sensitivity of normal hydrophilics precludes their use.

4.3.3 Bicomponent membranes

The surface pores of a true microporous membrane can become blocked by particulate matter, e.g., dirt, dried salts from seawater or sweat, or contaminated with oils, greases or surfactants present in skin exudates, cosmetics, barrier creams, insect repellents, washing powders and dry-cleaning fluids. This is a potential weakness that can affect breathability, waterproofness or both properties. For example, surfactants can either reduce the surface tension of a penetrating liquid and/or lower the contact angle that it makes with the pore walls. Both effects will automatically reduce the waterproofness of the membrane. In an extreme case, the porous structure can become completely flooded with water, and the white membrane turns transparent because of the major change in its refractive index.

Nearly all top-flight fabrics incorporating porous membranes therefore include a very thin, solid polymer layer (or layers) to seal the surface pores and prevent contamination. It penetrates a few microns into the surface cavities to achieve good adhesion, and forms a continuous coating a few microns above the surface for added protection. This part of the bicomponent membrane is no longer air permeable and the molecular diffusion mechanism takes over for breathability purposes. The measure was originally taken for second-generation Gore-Tex fabrics, launched in 1979. It was in direct response to criticism by some end-users that the prototype PTFE laminates could indeed be contaminated by salts and surfactants, and deteriorate in use. The sealing layer is sometimes described as oleophobic, i.e., implying that it repels body oils, grease and oil-based creams, but it also has to be hydrophilic so as to minimise the loss in overall breathability (US Pat 4194041, 1980). PEO-PU copolymers are therefore commonly used in this application.

4.3.4 General comments

It is fair to say that the technical, advertising and customer support effort that W.L. Gore and Associates put into their PTFE laminates catapulted the textile sub-genre of waterproof, breathable fabrics (WBFs) into prominence for sports, fashion and workwear markets. The 1980s saw a succession of competitive WBFs based on alternative membrane technologies. They were aimed first at the same top-end of the waterproofs market and then increasingly for lower cost, although not necessarily lower expectation garments. It also heralded a bewildering number of new trade names for coating systems, films and WBFs – many with suffixes or prefixes such as flex, tex, air, hydro, breathe, vent, acti(ve), dri(y) or similar – as each manufacturer strove to create their own identity and legacy. Many succeeded, but of course others failed. For a long period, confusion and controversy reigned as different producers, in the absence of agreed standards and test methods, made claim and counterclaim over the relative merits of their respective products. Breathability assessment was particularly contentious as certain test methods favoured the air permeability mechanism over assisted molecular diffusion, or *vice versa*. The situation has never been fully resolved, and probably never will be.

WBF membranes also provide some fascinating examples of technology transfer from other areas. For instance, use of PTFE in textiles derives from NASA research on heat-resistant polymers and, more specifically, astronaut suits for the Gemini Earth orbital/docking missions and Apollo Moon landings (Braddock and O'Mahony, 1999). They were improvements over earlier, non-pressurised spacesuits that contained an inner layer of Neoprene-coated nylon fabric under an aluminised outer, as used in the Mercury manned flights (McCann, 2005). On the other hand, microporous PU films and coatings are based on 'poromerics', such as Corfam, Porvair and Clarino leathercloths manufactured for shoe uppers in the mid-1960s. These are composites with microcellular PU or PVC foam layers that simulate the collagen structure of natural leather (Payne, 1970). Shirley Institute developments with hydrophilic polyurethanes relate back to an obscure paper on designing reverse osmosis membranes for an artificial kidney (Lyman and Loo, 1967). The synthetic method proposed by the Stanford investigators was completely impractical from an industrial viewpoint, but their ideas for polymer composition and water vapour transport proved to be perfectly valid.

4.4 Manufacture and properties of coated and laminated fabrics

The location of the membrane within a garment assembly is important, not just for weatherproofness but for comfort and aesthetic qualities, such as tactility, handle, drape and visual appearance. Some typical arrangements are shown in Fig. 4.4. In simple unlined garments, the membrane forms the outermost, or most usually the next inner layer, see Fig. 4.4(a) or (b). Early lightweight garments of this type often lacked subtlety and finesse. The weatherproof fabric tended to be stiff and crackly, stitched seams were visible and prone to leakage, and polymer membranes could come into direct contact with the skin.

Since the 1980s, however, and especially with the advent of breathable systems, positioning of membranes in fabrics and garments has evolved dramatically. Inner linings help wicking of perspiration, protect vulnerable membranes from scuff damage and take away the 'wet cling' sensation of membranes against skin. Constructions incorporating drop liners, such as Fig. 4.4(d),(e) allow a much wider choice of external fabrics and aesthetics are minimally impaired. As another option, the exposed membrane surface can be bonded to, e.g., lightweight tricot or insulating materials to give 3-ply laminates, see Fig. 4.4(g). These products were first developed for expensive sports and fashion wear (e.g. Gannex), and are now routinely used in high-visibility workwear.

68 Textiles for cold weather apparel

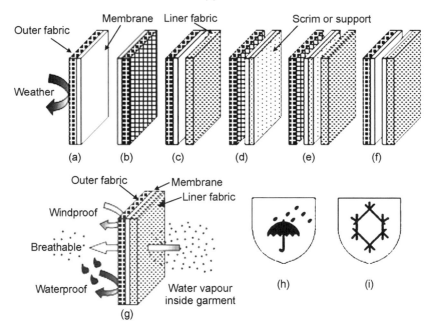

4.4 Location of membranes in outerwear garments and pictograms associated with their use: (a) basic 2-layer fabric with coating on inside, water repellent-treated outer surface; (b) 2-layer, coating on outside; (c) 2-layer plus loose liner fabric; (d) loose outer fabric plus scrim-supported membrane (drop liner); (e) drop liner insert; (f) double breathable membrane system, e.g., Dual Protection; (g) pictogram showing breathability and weatherproofness of a 3-layer laminate; (h) pictogram for EN343:2003, foul weather; (i) pictogram for EN342:2003, cold weather.

Coated and laminated fabrics are manufactured in continuous rolls, typically 1.5–2.0 m wide and 100–5000 m long by the direct coating method, transfer coating using release papers or belts, and various film lamination processes (Lomax, 1992). Only brief details of materials and manufacturing procedures are given here, and interested readers are directed towards several in-depth reviews (Fung, 2002 and 2005; Sen, 2007).

4.4.1 Base fabrics and polymer compounds

The original linen and cotton substrates have largely been replaced by synthetic materials, which offer higher strength-to-weight ratios and are less prone to microbial attack. Plain-weave polyamide fabrics made from continuous filament yarns were materials of choice in the 1960s–70s, since they present a perfectly flat, smooth surface ideal for direct coating with, e.g., thin PU layers. Typical qualities simply referred to as two or four ounce (oz) nylon (*ca.* 60 or 120 g/m^2) were used in most lightweight waterproofs of that era. Fabrics with higher tear strength could also be woven with a 'rip-stop' construction, i.e., noticeably stronger yarns included every 5 mm or so in both warp and weft directions, mainly for ultra-lightweight garments and tents. Even though coated nylon afforded much better functional properties and was easy to clean, it lacked the pleasant handle, drape and touch characteristics of natural fibre materials.

By the early 1980s, direct coaters had started to experiment with polyester/cotton blends, and more importantly with the latest 100% synthetics, e.g., Tactel, Taslan, heavier Cordura and microfibre fabrics woven from false twist and air textured yarns. These substrates have loose ends or fibre loops protruding from their surfaces, which more closely resemble natural staple fabrics and give improved mechanical adhesion to coating compounds. On the negative side, heavier or thicker coatings are required to ensure that all fibre protrusions are adequately 'flattened' and covered, so that waterproofness is maintained.

In common with ordinary textiles, polyester has now overtaken polyamide as the main coating substrate. It is generally cheaper and has better dimensional stability, shrink resistance and extensibility for manufacturing operations. Tightly woven fabrics (square weave, some twills and Oxfords) are still selected exclusively for direct coating because of the high machine tensions involved, but less dimensionally stable textiles such as scrims, knitted fabrics and nonwovens can be attached to membranes using alternative bonding techniques. Stretch fabrics incorporating Lycra or other PU fibres are increasingly used in tight-fitting weatherproof clothing for cycling, skiing and other 'speed' applications where aerodynamic drag is considered important.

Another consideration is cleanliness of textile surfaces. Most yarns and fabrics go through a series of finishing operations before they are ready for coating or lamination, and traces of anti-bonding material such as waxes, oils, silicones and residual dyestuffs may still be present. Surface contaminants invariably have a negative effect on adhesion between the polymer layer and substrate. The impact may be relatively minor or it could lead to more serious problems such as delamination and subsequent breakdown of membrane, e.g., at major flex points in garments or on laundering. Synthetic fabrics should therefore be scoured and thoroughly rinsed before use, but this step is sometimes overlooked or missed out deliberately for cost saving purposes. This is false economy if a major complaint involving garment failure follows a few months down the line.

Film and coating precursors may be applied to fabrics as viscous solutions, thickened aqueous dispersions, pastes or even as 100% solids in reactive or molten form. The membrane itself usually forms during a subsequent drying and curing process, which drives off residual volatile material, consolidates the molecular network and completes any chain extension or crosslinking reactions. Recipes for coating and laminating compounds can be complex, and are normally devised following laboratory trials. They invariably consist of a film-forming polymer together with a selection of additives that (i) assist with processing, (ii) increase the durability of the membrane in use, e.g., crosslinkers or (iii) impart specific properties, e.g., colour, flame retardancy, dulling, and heat reflectance.

The choice of polymer is influenced by the properties required, membrane thickness and ultimately, cost. Some rubber compounds contain more than twenty ingredients, but normally additives are kept to a minimum so that the final integrity of the membrane is not compromised. Increasingly stringent legislation on volatile organic compound (VOC) emissions has already had a dramatic effect on the direct/transfer coating industry. Many smaller European producers either closed down or merged in the 1990s, whilst larger manufacturers had to invest heavily in solvent recovery (mainly for DMF) or incineration facilities. A few accepted the restrictions and adapted their processes for aqueous dispersions, high solids and other low VOC compounds. It also prompted a shift of machinery, formulations and expertise to less well-regulated countries.

4.4.2 Application techniques

The main direct coating technique evolved progressively from rubber spreading. Basic equipment consists of a coating head, an associated drying/curing oven and a series of geared rolls including unwinding and rewinding stations, whose purpose is to pull a continuous roll of tightly woven fabric through the machine under controlled tension. Coating heads have many variations, but in their simplest form is a precision-ground knife, known as a doctor blade that is pressed taut against the fabric and aligned across its width. Direct coating compounds (typically 30–50% solids) have viscosity and flow characteristics similar to those of golden syrup or treacle. A bank of polymer compound is placed in front of the doctor blade, and the forward motion of the fabric causes the polymer to be scraped very thinly and uniformly over its complete surface. Cheeks are attached at each end of the doctor blade, a centimetre or so inside the fabric selvedge to prevent the compound from spilling out sideways. The tacky coated fabric then enters the oven through a narrow horizontal slit, solvent is evaporated and the dried coated fabric is finally wound up. If two or more polymer layers are required, the roll of fabric is returned to the front of the machine and rethreaded through it. The coated surface now forms the substrate for the second layer of polymer solution, applied and dried in precisely the same manner.

The main essence of the process is absolute control over polymer add-on across the full length and width of the fabric. Stenter frames can be incorporated into the machine to maintain fabric width, avoid creases and minimise shrinkage and relaxation effects during the coating cycle. The first coat or tiecoat is very important in defining the final properties of the fabric. It must penetrate far enough into the weave to achieve good mechanical adhesion, but not so far as to make the coated fabric unduly stiff. Penetration can be controlled by a skilled machine operator by adjusting compound viscosity, varying the blade edge profile (pointed, round, shoe), altering the angle that the blade makes with the fabric, tension and throughput speed and other factors. The tiecoat is usually heavier than subsequently applied layers, and must be carefully dried down to avoid skinning, craters and pinholes all of which can ultimately lower waterproof properties.

Building up relatively thick coatings on a single-head machine is slow, labour intensive and thus expensive. For example, twelve or more passes through the machine are often required to achieve an acceptable 'rubberised' textile. Even thin PU coatings (20–30 microns) are usually built up in several layers, each of which might have a different chemical composition and function. During the 1980s, many UK and European coaters invested in new or refurbished coating lines that placed two, three or even four coating heads and their ovens in series. Such coating lines could be up to 100 metres long and had to be aligned perfectly to avoid fabric tension problems. They enable a typical PU coating to be applied in a single pass at relatively high speed (20–30+ metres/min), rather than three separate passes and winding/rewinding operations, and give substantial savings on unit costs. In modern coating heads, the rolling bank of polymer compound is constantly topped up from a traversing reservoir, and the wet coating thickness can be monitored by β-ray gauges and automatically adjusted. The ovens are heated in electronically controlled zones providing an accurate temperature gradient that minimises flash off of solvents and ensures full curing in the final stage.

Transfer coating differs from the direct method because the first layer, which is actually the topcoat of the finished product, is laid down on release paper instead of fabric. The release paper takes all the applied machine tensions, which is usually a two- or three-head line. As the preformed film emerges from the oven, an adhesive tiecoat is applied and whilst this is still tacky the fabric substrate is joined under a nip roll. The adhesive is dried in the second oven and the coating layers fully cured, leaving the composite still firmly attached to release paper. Paper and fabric are finally stripped apart and both are wound up on separate stations. The siliconised paper can be reused a number of times before it deteriorates and then has to be scrapped. Paper costs are relatively high, so transfer coating tend to be a more expensive operation than direct coating. Then again, the transfer process, in common with film lamination, ensures that at least one layer is a uniform, complete membrane for weatherproofness. These techniques are particularly suitable for difficult substrates, e.g., 'hairy' woven fabrics, knits, nonwovens, fleeces, etc.

4.4.3 Main properties

For coated and laminated fabrics used in cold weather climates, the main requirements are water-shedding exterior surface, weatherproofness, durability in all uses, laundering and of course adequate low-temperature performance. Weather elements include wind, wind-driven solids including dust, grit and ice particles, precipitation of all types (rain, hail, sleet, and snow), droplet sizes and velocities, cold with wind chill factors, and in some extreme climates, intense direct sunlight and reflected solar glare. Off-shore activities will also need protection against seaspray, and possible immersion in icy waters. Some pursuits and occupations will expose the wearer to potential contact with microbes and/or chemicals (e.g. fuels, fish oils and other natural fluids) and assessment of breakthrough times and permeation rates might be appropriate (ISO 6529:2001).

4.5 Testing of coated and laminated fabrics

A useful protocol for evaluating fabrics, garments and clothing assemblies worn in cold weather is shown in Fig. 4.5. It comprises five levels of testing (Umbach, 1986), with the predictive capability improving and relative costs escalating as each successive level is gained. Levels 1–3 involve tests in a controlled laboratory environment that give numerical data for comparison with Standards and Specifications, previous results, etc. The higher levels yield subjective information on fitness for purpose, which can then be fed back down the pyramid, e.g. to gauge the usefulness of laboratory procedures and pass/fail criteria. Only the first level testing of fabrics will be considered here. Levels 2–5 relate more specifically to garment construction and design, seams, accessories such as zips, interactions with internal clothing layers and this is dealt with in other chapters.

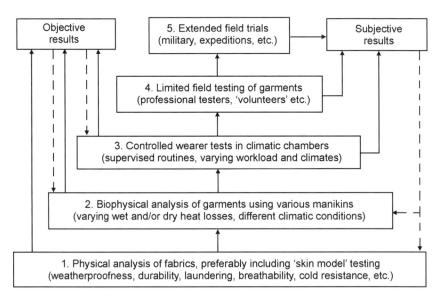

4.5 Suggested test protocol for weatherproof garments based on coated and laminated fabrics.

4.5.1 Most relevant tests

Wind resistance can be assessed by measuring the air permeability of a fabric (e.g. ISO 9237:1995), normally under a pressure drop of 100, 125 or 200 Pa depending on the Standard procedure. Solid membranes and most coated and laminated fabrics record zero values in these tests and are therefore considered windproof.

The water-shedding ability of the outer fabric surface is usually assessed by spray rating (e.g. ISO 4920:1981). A fabric specimen is held at an angle of 45°, and a measured amount of water is 'sprayed' onto its surface using a perforated nozzle to simulate rain. The degree of wetting out of the fabric surface is rated from 1 (nearly 100% covered by water) to 5 (few adhering droplets). A factory-applied fluorocarbon finish will normally generate a spray rating of 5, but its permanence to, e.g., washing and abrasion depends on add-on and level of crosslinking, if any. Water repellency can sometimes be restored by ironing the fabric surface, or there are a number of proprietary reproofing agents available.

Water penetration resistance can be measured under dynamic or static conditions. Dynamic methods such as the Bundesmann (ISO 9865:1991), WIRA shower, and horizontal water spray test (ISO 22958:2005) aim to simulate rainfall, usually of short duration under laboratory-controlled conditions. Some test conditions produce water droplets with far greater kinetic energy and penetrating power than would be experienced in the heaviest, natural precipitation. Once again, coated and laminated fabrics designed for foul-weather clothing should not be penetrated by simulated rain. In contrast, static tests (such as ISO 811:1981) provide information on waterproofness under high applied pressure, more characteristic of point contact, e.g., when kneeling or crawling on wet ground, or sitting on a wet boat seat.

Hydrostatic head tests are widely used by this industry, but several versions exist and correlating results from different sources requires caution. In its original design, the test apparatus literally measured the height of a column of water that the fabric could withstand before being penetrated, and this is why common units are inches, mm, cm, or m. These machines were often placed in stairwells, since they need sufficient head space for the glass tube (typically 3–5 metres tall) containing the water column, as well as the pulley system to raise the reservoir at constant speed. There are two common test variations: (i) hold the sample at a given hydrostatic pressure, and note the time taken for penetration, if any, to occur, and (ii) subject the fabric to a constantly increasing pressure and note the pressure at which penetration first occurs. Most testers are now semi-automated, bench-top models, using a pump to increase the water pressure at a regulated rate below the fabric sample, which is held in place and tightly sealed by an annular clamp and electronics to record the progressively applied pressure. In a type (ii) test, the upper surface, normally the membrane in

2-layer fabrics is closely observed and the end point is recorded as the applied pressure when, e.g., one droplet bursts through the fabric.

The numerical result is influenced by the diameter of the fabric specimen, and particularly the rate of pressure increase. Measures should be taken to stop the specimen from distorting or ballooning up as the applied pressure increases and to prevent leakage around the edges. Many Standards use hydrostatic head tests to assess membrane integrity before and after a series of accelerated durability tests (abrasion, flexing, contamination, washing, dry cleaning, etc.). Some Standards state that a value of 100 cm hydrostatic head is sufficient for waterproofness, whereas others contest that much higher values are required to cover all circumstances.

Numerous test methods exist for assessing breathability of textiles, and this has been a topic of great debate and dispute over the past 25 years (McCullough et al., 2003; Gretton et al., 1998; Ruckman, 2005). It is discussed more fully in other chapters. There are three totally different methods in recent ISO Standards that apply to coated and laminated fabrics, yet each set of test parameters produces water vapour pressure gradients (Δp values) that bear little semblance to real foul-weather environs, especially cold temperatures, see Table 4.1. All tests are carried out under isothermal conditions that deliberately avoid condensation, even though this is an accepted phenomenon in use and dealing with it is part of the *modus operandi* of layering systems and components such as wicking fibres and hydrophilic membranes (Gretton, 1999). Another point is that the water vapour permeability of polymers at low temperature often differs significantly from that in typical laboratory test conditions, because of changes around the glass transition temperature or unusual hydrogen bonding effects.

Table 4.1 Some nominal values for the applied water vapour pressure gradient (Δp) across a membrane, as a function of temperature (t) and relative humidity (RH), under typical laboratory test and realistic climatic conditions.

Laboratory test or climate	High-pressure side		Low-pressure side		Δp
	t (°C)	RH (%)	t (°C)	RH (%)	(Pa)
ISO 11092:1993	35	100	35	40	3093
ISO 15496:2004	23	100	23	23	2168
ISO 8096:2005/BS 7209:1990	20	100	20	65	818
ASTM E96:1999 method B	23	100	23	50	1404
JIS L1099 Method A1	40	90	40	0	6638
Temperate foul weather	7	100	5	80	304
Maximum Δp at t (°C)					
+35	35	100	35	0	5623
+20	20	100	20	0	2338
+05	5	100	5	0	872
−05	−5	100	−5	0	201
−20	−20	100	−20	0	52
−40	−40	100	−40	0	6

4.5.2 Main standards for coated and laminated fabrics

In principle, International Standards (ISO) should override regional (e.g. Europäische Norm, EN) and in turn national or industry-specific standards (e.g. BS, ASTM, DIN or JIS, military specifications), but the situation regarding fabrics incorporating membranes is still perplexing. The primary Standard is ISO 8096:2005 which specifies requirements for WVP and non-WVP coated fabrics for 'water penetration resistant clothing'. It covers a range of polymer types by inference (PU, silicone elastomers and their blends with PU, natural rubber, Neoprene, NBR, Hypalon and PVC) and defines single-face (normal 2-layer), double-face (i.e. both surfaces coated) and double-textured (coating sandwiched between two fabrics) coated fabrics. The definition of coated fabric includes membranes attached by transfer coating, and arguably by several lamination techniques. This is a major revision of an earlier 3-part Standard that has yet to filter down into recognised industry specifications for garments.

A more widely recognised benchmark is EN 343:2003 that 'specifies requirements and test methods applicable to materials and seams of protective clothing against the influence of precipitation (e.g. rain, snowflakes), fog and ground humidity'. This forms part of a set of related Standards and specifications for outdoor protective clothing that includes EN 340:2003 (general requirements), EN 342:2004 (cold weather, $< -5\,°C$) and EN 471:2003 (high visibility). Some relevant tests and fail/pass levels for different classes of fabric are shown in Table 4.2. The highest breathability level of EN 343:2003, Class 3 with an R_{et} pass of $\leq 20\,m^2\,Pa/W$, is rather generous and most modern WBFs with R_{et}s of typically 5–12 $m^2\,Pa/W$ will easily conform. It is highly likely that

Table 4.2 Some performance requirements and recommendations relating to coated and laminated fabrics conforming to EN 343:2003.

Property	Class 1	Class 2	Class 3
Resistance to water penetration (kPa)[#]			
As received	8	—	—
After dry cleaning and/or washing	—	≥ 8	≥ 13
After abrasion	—	≥ 8	≥ 13
After repeated flexing	—	≥ 8	≥ 13
Influence of fuel and oil	—	≥ 8	≥ 13
Water vapour resistance, R_{et} ($m^2\,Pa/W$)	> 40	$20 < R_{et} \leq 40$	≤ 20
Recommendations for wearing time			
at 25 °C (min)	60	105	205
at 20 °C (min)	75	250	*
at 15 °C (min)	100	*	*
at 10 °C (min)	240	*	*
at 05 °C (min)	*	*	*

— indicates no test required, * indicates no limit for wearing time
\# conversion factors: 1 kPa = 10.2 cm H_2O; 1 psi = 70.3 cm H_2O or 6.89 kPa

future revisions of this Standard will tighten up on breathability requirements (cf. the upper level of $\leq 13\,\mathrm{m^2\,Pa/W}$ in EN 342:2004).

There are very few tests for assessing low-temperature performance of fabrics, as distinct from garments. ISO 8096:2005, Annex G assesses the flexibility of samples that have been folded, and exerted to 4 kPa pressure for 48 hours at $-30\,°\mathrm{C}$. The unfolded samples are examined visually for cracks or delamination and subjected to a hydrostatic head test.

4.6 Environmental issues

A typical life cycle for coated and laminated fabrics, applicable to clothing in general, is shown in Fig. 4.6. Eco-friendly labelling schemes administered by government licence or independent test houses are widely available for producers and consumers involved in this sequence (Challa, 2008). For instance, the international Oekotex Standards 100 and 1000 regulate that products contain strictly limited amounts of harmful substances (e.g. certain dyestuffs, pesticides, heavy metals, formaldehyde), and that manufacturing sites are following environmentally sound procedures and practices. Even so, there are negative impacts at all stages, including noise pollution with textile machinery, effluents and energy requirements for heating, lighting, transportation, storage, laundering, etc. The main implications, however, are at both ends of the cycle.

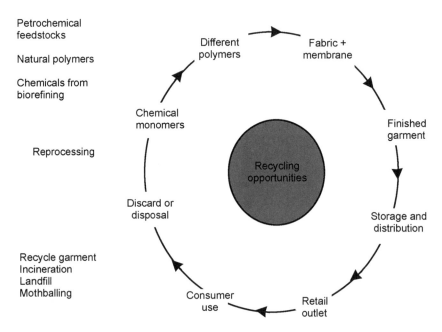

4.6 Typical cradle-to-grave cycle for weatherproof garments based on coated and laminated fabrics.

Most coated and laminated fabrics are fully synthetic, and this leads to depletion of fossil fuels, energy and water use during processing, gas/vapour emissions (including nitrous oxide), and generation of hazardous or non-hazardous waste materials. The interdependences of such manufacturing impacts are extremely complex, and resolving one environmental issue always has repercussions elsewhere (Anon., 2007; Dessler and Parson, 2006). For example, some coating compounds can be made in water instead of solvent. This cuts down on VOC emissions and petrochemical usage, but affects energy consumption because it takes four times as much heat to remove water during processing. Carbon-neutral cotton substrates can be used instead of polyester, but this raises other issues of land usage, pesticide, herbicide and fertiliser contamination, and substantially higher energy costs for processing and laundering.

Replacing natural carbon resources with, e.g., plant biomass gives a different set of chemicals such as ethanol and glucose, which are not primary feedstocks for the existing textile industry (see Fig. 4.1). Major producers such as BASF, du Pont, Cargill and DSM are investing heavily in biorefinery projects for renewable monomers, enzyme technology and new synthetic pathways for plastics, coatings and films. Some progress has already been made in sustainable textile applications. Propan-1,3-diol can be used instead of ethylene glycol to produce modified polyester fibre (Corterra, Sorona) and this raw material is now available from corn starch (Scott, 2006).

Poly(amino-acids) and chitosan, a natural product obtained from crushed mollusc shells have been used in PU coatings and as film laminates (World Pat 2007056348, 2007), and fabrics woven from biodegradable fibres, based on polylactic acid, PLA (Gupta *et al.*, 2007) can be used as substrates. The industry is also moving away from halogen-containing products, such as PVC and brominated flame retardants in membranes, and some fluorocarbon-based water-repellents have been withdrawn because of health concerns over perfluorooctanoic acid (PFOA) used in their manufacture (Betts, 2007).

The usual fate of worn-out or unwanted clothing is also summarised in Fig. 4.6. There are no statistics relating specifically to cold-weather garments or coated and laminated fabrics, but they most likely follow general trends (Wang, 2006). In 2004, for example, the UK generated approximately 2.35 million tonnes of clothing and textile waste, representing only 0.7% of the annual total but still one of the fastest growing sectors (Allwood *et al.*, 2006). A paltry 13% of waste textile material was recycled, 13% was incinerated and the vast majority (74%, 1.8 million tonnes) went for landfill raising questions over degradability.

A great step forward would be to close the loop, by converting unwanted clothing back into monomers for reuse. This is routine for plastic bottles and other packaging materials, but less feasible for coated and laminated fabrics, because (i) it is difficult to separate pure polymer from other additives such as pigments, dyes and other auxiliaries, (ii) many fibres are crystalline or intractable and (iii) most membranes are crosslinked or vulcanised, all of which limits reprocessing and other forms of destructive recycling. As usual, there are notable exceptions that illustrate the way forward.

PET and most polyesters can be broken down into reusable diacid and diol monomers and oligomers by high-temperature hydrolysis or glycolysis reactions. This fact has been promoted in 'green' marketing campaigns for Sympatex garments based entirely on polyester fibres and film (Anon., 2008b). Nylon 6 is readily converted back into starting material, caprolactam, which differentiates it from alternative polyamide fibres such as the more common 6,6 variant. Patagonia, and one of their main fabric suppliers, Toray, have recently announced a 'return and recycle' voucher scheme for clothing based specifically on Nylon 6 (Anon., 2008d). It will start with surf pants and hard-shell alpine jackets sold from Patagonia's 2008/9 collections – when no longer needed these garments can be returned, first to the retail outlet then to the fibre producer. Toray claim that chemical reprocessing of Nylon 6 will afford 70% reduction in energy costs and lower CO_2 emissions compared with normal manufacture from petroleum feedstocks. Finally, reports suggest that *ca.* 20% of clothing remains officially in circulation but not used, i.e., stored in wardrobes, drawers, attics, garages, etc. Weatherproof garments are typical items that are quite literally 'put away and saved for a rainy day'.

4.7 Current applications

The main application is foul-weather clothing for fashion, leisure and workwear activities. A coated or laminated fabric usually forms the outermost layer, protecting a variety of internal clothing components that provide thermal insulation and deal with the primary wet and dry heat flows that regulate comfort. There are several different levels of protection required depending on the activity, anticipated extremes of weather, potential duration, proximity of safe haven or rescue, etc., which means that functionality has to be correctly balanced with style, quality, comfort, weight and price. Leisure pursuits such as cycling, golfing, and angling arguably require a lower level of protection than say, hill-walking, back-packing, caving and mountaineering where survival can be an issue.

High-visibility outer fabrics, often fluorescent yellow or orange woven, textured polyester conforming to EN 471, are widely used for foul-weather and corporate clothing issued to police forces, postal and telecommunications services, emergency services (ambulance, Red Cross, etc.) and workers in the transport, construction, and utilities (gas, electric, water) industries. Conversely, outdoor occupations such as farming, forestry, stalking, hunting and fishing traditionally prefer muted or camouflage colours for their inconspicuousness. Articles of clothing for all applications include anoraks, cagoules, over-trousers, coveralls, salopettes, gloves, hats and integrated hoods, gaiters, sock liners and upper panels for shoes. WBFs using PU or ePTFE layers are a good all-round choice for these items and market sectors. The exceptions are very inexpensive garments where PVC, acrylic and non-breathable PU are more cost effective and certain uses where breathability is sacrificed for the added security of a heavier, impermeable membrane, e.g., of Neoprene, Hypalon or butyl rubber.

Another sector is specialised clothing designed to protect against deliberate or more usually, accidental immersion in cold waters. In this situation it is vital that the wearer keeps dry, because water is 25 times more efficient at conducting heat away from the body than an adhering still air layer. The average survival time decreases rapidly with water temperature, and may only be a few minutes in very cold seas (0–5 °C) or inland lakes. The primary function of the garment is to delay the onset of hypothermia, thus extending the survival time and increasing the chance of escape or rescue. Those potentially at risk include crew members of cargo ships, navy surface craft, submarines and commercial fishing vessels, all aboard passenger ships, workers in offshore oil/gas production and exploration industries, salvage teams, coast guards, first response and search-and-rescue teams, and those undertaking recreational pursuits such as yachting, power boat sailing, windsurfing, jet-skiing and diving.

There are several different types of one-piece 'survival suits' available, depending on the exact safety function, buoyancy aids and level of thermal insulation required. A common feature, however, is that a coated or laminated fabric provides the outer, watertight layer. This is normally Neoprene or PU-coated polyamide fabric often made from high-tenacity Nylon 6,6 yarns, but WBFs, especially flame-retardant ePTFE laminates, are becoming popular for dry-suits and in-flight protection, e.g., for helicopter passengers. Neoprene closed-cell foam (5 mm) laminates are used for stowed immersion suits where long survival times are mandatory. Many of these uses are covered by stringent specifications (e.g. ISO 15027:2002) and other protocols dealing with maritime safety, notably the International Convention for Safety of Life at Sea (SOLAS), first adopted in 1914 after the enquiry into the Titanic disaster (Anon., 2004).

Outer fabrics used in Winter sports clothing, e.g., for downhill skiing, cross-country skiing, skating, tobogganing, snowboarding, etc., tend to be less functional and more fashion oriented, in respect of colour, fit and other aesthetics. These sports are often characterised by periods of intense high activity, especially in speed and endurance events followed by lengthy resting or waiting periods, and large variations in thermal comfort can occur. Functionality is perhaps more important for professionals working in low-temperature climates, and those undertaking, e.g., expeditions in the Arctic, Antarctic and high mountain ranges. Cold, drier climates present different requirements. Absolute waterproofness is less of a concern than minimising wind chill and the ability of the outer layer to shed snow or ice.

Wind-resistant materials such as Ventile and various soft-shell outers are often selected ahead of coated and laminated fabrics. In this case, their low degree of air permeability is an advantage, by helping to dissipate water vapour through forced convection. As sub-zero temperatures decrease, condensation invariably forms inside the clothing and the physics of the environment militates further against Fickian diffusion through the outer layer if membranes are employed. The working temperature of the membrane must be much closer to ambient than body temperature if the insulation layers are working properly. The driving force for diffusion, Δp is extremely small at low temperatures (see Table 4.1), theoretically generating very low WVTRs (Eqn. 4.1). Accordingly, ventilation features and correct use of layering principles become even more important under these conditions. Even so, WBFs are still useful in clothing systems, e.g., for optimum windproofness, when sitting, kneeling or lying on ice or snow for long periods (pressure melting effects) or situations where the wearer may be able to move in and out of a warmer environment, e.g., expedition bases, other refuges or mountain restaurants. This is one reason why Gore-Tex Paclite is still specified by the US Military for their third-generation (2006) Extended Cold Weather Clothing System (ECWCS, pronounced ek-waks), designed for a temperature range of approximately +5 to −40 °C (Anon., 2008c).

4.8 Future trends

Demographically, the coating and laminating industry has changed radically during the last two decades and will continue to do so. For instance, much of WBF production has moved from Europe, US and Japan into Taiwan, Korea, Thailand and increasingly into China, where the sheer scale of technical clothing manufacture can be quite astonishing. It has led to a downward spiral in fabric and finished garment prices, driven mainly by wholesale suppliers. Buyers, specifiers, trading standards and Standards Committees need to be vigilant that technical performance is maintained during such cost-cutting exercises. Western manufacturers will continue with more specialised business where short runs, short lead times, non-standard fabric substrates, novel or gimmick coating effects and higher levels of technical performance are required, and they will of course pursue innovation.

Two areas of immediate interest are weatherproof membranes as carriers (e.g., for phase-change materials, other microencapsulated finishes, delivery systems, biocides, chameleon pigments, biosensors and even integrated circuits) and further development of multi-ply systems, where each polymer layer provides a specific function. Devising a polymer that flows and seals itself after accidental puncture is one such opportunity. These applications cross over smoothly into other, very active textile areas such as intelligent and smart materials, biomimetics, and nanotechnology.

Space exploration and military sectors have been valuable sources of new materials and concepts, partly realised through hefty research budgets. These types of organisations often consider innovation as something that might take a decade or more to become reality, and plan much further ahead than this (Leitch and Tassinari, 2000). The US Army Soldier Systems Centre (Natick, Massachusetts) team working on selectively permeable membranes has already developed laminates that breathe, yet form total occlusive barriers against chemical and biological (CB) agents, including aerosols and vapours (Truong and Wilusz, 2005). They are a significant advance over previous CB clothing based on charcoal adsorption, and will further evolve. The Navy section is looking at SmartSkin, a future material for wetsuits that adjusts the permeability of the fabric's inner layer using a thermally-sensitive polymer hydrogel (Anon., 2008a).

Next generation spacesuits will be for Mars exploration, which draws interesting parallels with extreme cold weather on Earth. Being much further away from the sun, the surface temperatures on Mars vary from $+20$ to $-120\,^\circ\mathrm{C}$, with a mean value of $-63\,^\circ\mathrm{C}$, so that low-temperature flexibility and brittleness of textile and plastic components under these conditions are very important. Clothing materials also need to be impermeable and non-degradable in a predominantly carbon dioxide atmosphere (95.3%), and function correctly at low atmospheric pressure (0.007 bar). Thin membrane technology is at the very forefront of these fascinating challenges (Marcy *et al.*, 2004). Already, solid films are being considered for wicking and WVT layers (PEO-PU and PA copolymers), barriers against internal oxygen and moisture loss (ethylene-vinyl alcohol copolymer), damping coatings for micrometeoroid protection (silicone elastomers), carbon dioxide exclusion (PET-PVDC bilayers) and chemical inertness (the ubiquitous ePTFE). These are just a few examples of membrane developments that might well filter down into future sports clothing and workwear.

4.9 Sources of further information and advice

Recommended textbooks are mentioned in the references. The dedicated periodical is *Journal of Industrial Textiles* (previously *J. Coated Fabrics*), but occasional articles are found in *J. Textile Institute*, *Textile Research J.*, *J. Clothing Science and Technology*, and *Technical Textiles*. Publications aimed at retail and consumers include *WSA* (World Sports Activewear), *Future Materials*, *Textile World* and *Ecotextile News*, and hobbyist magazines (e.g. *Climb*, *Skiing*) offer informative gear reviews. E-mail alerts such as fabricdirector-e or webnews@sportswear.com keep readers abreast of latest industry news, Wikipedia is surprisingly informative on technical textile subjects, and most scientific journals are available online (usually by subscription, or on a fee per article basis). Fabric and garment manufacturers' websites also provide useful, if sometimes biased data and explanations. Information on International Standards is readily available (e.g. at http://www.oeko-tex.com/ and http://www.iso.org/), and Patents can be searched and downloaded free of charge on the European and US Patent Office website (http://ep.espacenet.com/).

Exhibitions and trade fairs offer a different perspective, such as the ability to see, touch and handle the latest materials and talk to key players in the industry. The main fair for technical fabrics is Techtextil (http://techtextil.messe frankfurt.com/), held biennially in Frankfurt and now extended to locations in the US, China, Russia and India. Techtextil, Frankfurt runs in conjunction with the Avantex exhibition, which concentrates on innovation, and two symposia are held. The 12th show in June 2007 attracted more than 1,100 exhibitors and 23,000 visitors. Other fairs such as Winter and Summer ISPO in Munich (http://www.ispo.com/) are orientated towards garments and fashion aspects. Presentations on coated and laminated fabrics are given at a number of conferences on textiles, e.g. the topical theme of the 16th International Conference on Textile Coating and Laminating, held in Barcelona, November 2006 was 'Innovation for a successful future: smart/intelligent coated and laminated fabrics'.

4.10 References

Allwood J M, Laursen S E, de Rodriguez C M and Bocken N M P (2006), *Well dressed? The present and future sustainability of clothing and textiles in the UK*, Cambridge, University of Cambridge Press.

Anon. (2004), *SOLAS* (Consolidated Edition 2004), London, IMO Publishing.

Anon. (2007), *Climate Change 2007*, IPCC fourth assessment report (AR4), http://www.ipcc.ch/index.htm (accessed January 2008).

Anon. (2008a), 'Adaptable 'skin' – hydrogel changes wet suit's water flow for better protection', http://www.natick.army.mil/about/pao/pubs/warrior/02/mayjune/smartskin.htm (accessed January 2008).

Anon. (2008b), 'Germany: Sympatex presents world's first 100% recyclable laminate', http://www.fibre2fashion.com/news_id=46766 (accessed January, 2008).

Anon. (2008c), 'Introducing gen III ECWCS the next generation extended cold weather clothing system for the warfighter', http://www.adstactical.com/about_ads/gen3_ecwcs.htm (accessed January 2008).

Anon. (2008d), 'Toray, Patagonia to jointly work on chemical recycle of Nylon 6', http://www.toray.com/news/fiber/nr071212b.html (accessed January 2008).

Bartels V T and Umbach K H (2002), 'Water vapour transport through protective textiles at low temperatures', *Text Res J*, 72, 899–905.

Betts K (2007), 'PFOS and PFOA in humans: new study links prenatal exposure to lower birth weight', *Environ Health Perspect*, 115, A550.

Braddock S E and O'Mahony M (1999), *Techno Textiles: Revolutionary Fabrics for Fashion and Design*, London, Thames & Hudson.

Challa L (2008), 'Impact of textiles and clothing industry on environment: approach towards eco-friendly textiles', http://www.fibre2fashion.com/ (accessed January 2008).

Dessler A E and Parson E (2006), *The Science and Politics of Global Climate Change: A Guide to the Debate*, Cambridge, Cambridge University Press.

Fung W (2002), *Coated and laminated textiles*, Cambridge, Woodhead.

Fung W (2005), 'Coated and laminated textiles in sportswear' in Shishoo R, *Textiles in sport*, Cambridge, Woodhead, 134–175.

Gretton J C (1999), 'Condensation in clothing systems', *World Sports Activewear*, 5, 38–43.
Gretton J C, Brook D B, Dyson H M and Harlock S C (1998), 'Moisture vapor transport through waterproof breathable fabrics and clothing systems under a temperature gradient', *Text Res J*, 68, 936–941.
Gupta B, Revagarde N and Hilborn J (2007), 'Poly(lactic acid) fiber: an overview', *Prog Polym Sci*, 32, 455–482.
Holmér I (2005), 'Protection against cold' in Shishoo R, *Textiles in sport*, Cambridge, Woodhead, 262–286.
Holmes D A (2000), 'Waterproof breathable fabrics' in Horrocks A R and Anand S C, *Handbook of Technical Textiles*, Cambridge, Woodhead, 282–315.
Leitch P and Tassinari T H (2000), 'Interactive textiles: new materials in the new millennium. Part 1.', *J Ind Tex*, 29, 173–190.
Lomax G R (1985a), 'The design of waterproof, water vapour-permeable fabrics', *J Coated Fabrics*, 15, 40–66.
Lomax G R (1985b), 'Coated fabrics. Part 2. Industrial uses', *J Coated Fabrics*, 14, 127–144.
Lomax G R (1992), 'Coating of fabrics', *Textiles*, 21, 18–23.
Lomax G R (2001), 'Poly(ethylene oxide)-based polyurethanes for breathable, waterproof fabrics and garments', *Proceedings of the 40th International Man-Made Fibres Congress*, Dornbirn.
Lomax G R (2007), 'Breathable polyurethane membranes for textiles and related industries', *J Mater Chem*, 17, 2775–2784.
Lyman D J and Loo B H (1967), 'New synthetic membranes for dialysis. IV. A copolyether–urethane membrane system', *J Biomed Mater Res*, 1, 17–26.
Marcy J L, Shalanski A C, Yarmuch M A R and Patchett B M (2004), 'Material choices for Mars', *J Mater Eng Perf*, 13, 208–217.
McCann J (2005), 'Material requirements for the design of performance sportswear' in Shishoo R, *Textiles in sport*, Cambridge, Woodhead, 44–69.
McCullough E A, Kwon M and Shim H (2003), 'A comparison of standard methods for measuring water vapour permeability of fabrics', *Meas Sci Technol*, 14, 1402–1408.
Oakes J, Wilkins H, Riewe R, Kelker D and Forest T (1995), 'Comparison of traditional and manufactured cold weather ensembles', *Clim Res*, 5, 83–90.
Payne A R (1970), *Poromerics in the shoe industry*, London, Elsevier.
Ruckman J E (2005), 'Water resistance and water vapour transfer' in Shishoo R, *Textiles in sport*, Cambridge, Woodhead, 287–305.
Scott A (2006), 'Down on the plastics farm', *Chemistry World*, (August).
Sen A K (2007), *Coated textiles: principles and applications*, 2nd edition, Lancaster, Technomic.
Tanner J C (1979), 'Breathability, comfort and Gore-Tex laminates', *J Coated Fabrics*, 8, 312–322.
Träubel H (1999), *New Materials Permeable to Water Vapor*, Berlin, Springer-Verlag.
Truong Q and Wilusz E (2005), 'Chemical and biological protection' in Scott R A, *Textiles for protection*, Cambridge, Woodhead.
Umbach K H (1986), 'Evaluation of textile and garment comfort' in Blum C and Wurm J G, *Proceedings of the European textile research symposium Competitiveness Through Innovation*, London, Elsevier.
US Pat 4194041, 1980, W L Gore & Associates.
Wang Y (2006), *Recycling in textiles*, Cambridge, Woodhead.
World Pat 2006075144, 2006, Baxenden Chemicals Ltd.
World Pat 2007056348, 2007, E I du Pont de Nemours & Co.

5
The use of smart materials in cold weather apparel

J. HU and K. MURUGESH BABU, Hong Kong Polytechnic University, Hong Kong

Abstract: This chapter has introduced adaptive protective clothing under cold weather. The chapter has a comprehensive introduction about the design requirements for cold weather clothing firstly. There are several kinds of smart fabrics and textiles which can be used to protect people from cold weather, including breathable fabrics, far-infrared fabrics, and other intelligent textiles. Shape-memory polymers can be used as adaptive textiles under cold conditions as well. Phase-change materials (PCMs), substances with a high latency, are introduced and reviewed in this chapter subsequently. Some novel fabric systems used currently are also discussed. Finally, the future of smart clothing for cold weather conditions will be consolidated by the integration of new materials and new technologies.

Key words: adaptive systems, cold weather clothing, breathable fabrics, far-infrared fabrics, intelligent textiles, shape-memory polymers, phase-change materials.

5.1 Introduction

Protective clothing is needed for working at temperatures below 4°C. The selection of proper clothing should be made according to conditions such as temperature, weather, the kind of activity being performed and its duration. Activity levels influence the amount of heat and perspiration generated while working. Excessive sweating may occur if the work pace is too fast or if the type of clothing is not selected properly. This results in the clothing next to the body becoming wet, decreasing the insulation value of the clothing dramatically. There is an increase in the risk of cold-related injuries in this situation (CCOHS, 2008).

Currently, the use of both shape-memory polymers, with their thermally sensitive response characteristics, and phase-change materials with their quick phase-change responses to environmental changes, is being widely discussed with regard to smart adaptive systems for cold weather clothing. An understanding of the principles and underlying mechanisms involved will help fabric manufacturers and designers produce newer fabrics and design suitable cold-weather clothing incorporating these smart materials to combat the severe cold conditions faced by soldiers, sportsmen, cold-store workers and space research professionals. A description of the various adaptive systems such as shape-memory polymers and phase-change materials for cold weather clothing are presented in this chapter, together with the currently available adaptive systems for cold weather clothing and their possible limitations.

5.2 Design requirements for cold weather clothing

Generally speaking, cold weather clothing refers to garments that are used in extreme climates such as those found in arctic regions and in mountainous areas. Cold weather garments include those for skiwear and mountaineering, and for army personnel posted in hilly areas at high altitudes. Other environments where cold weather clothing is required include garments for those working in cold stores.

The most important requirements for clothing in extremely cold conditions are protection and comfort. Clothing must reduce heat loss from the body and permit the controlled transmission of moisture produced by perspiration (Wang et al., 2006). In order to provide the required level of warmth, clothing should provide the highest degree of insulation with the least amount of bulk. In addition, air permeability and resistance to water penetration are particularly important for outdoor use (http://5pr.kpk.gov.pl/prog_tem_3/m&t/i/topic03.htm). Clothing needs to be appropriate for different environmental conditions and to accommodate individual requirements. Individuals may not require the same layers to sustain them for the same temperature range. Clothing therefore needs to be flexible, allowing users some choice over which layers to wear. Water resistance, the ability of a material's surface to repel moisture, will help maintain the insulation value of the material and thus the comfort level of the garment. Resistance to airflow (wind) through the material is very important as airflow carries heat away from the skin surface via convection. It is also important to take into account a smooth surface to shed snow, minimum restriction of movement, durability, ease of donning and removal of clothes, and overall cost.

Protection from the cold mainly depends on the following factors: (i) metabolic heat, (ii) outside temperature, (iii) wind chill, (iv) thermal insulation, (v) air permeability and (vi) moisture vapour transmission (Morris, 1975). Survival depends on the balance of heat losses from the body to the environment (Ukponmwan, 1993; Mathur et al., 1997; Goldman, 1978; Jian, 2000) and heat output due to metabolic activity (Bajaj and Sengupta, 1992). Clothing comfort is closely related to the microclimate temperature and humidity between clothing and skin (Chung and Cho, 2004). To develop comfortable and functional cold weather clothing, it is necessary to consider the physical properties and wear performance of fabric and clothing during exercise at increasing activity levels.

The insulating value of a garment is mainly dependent on: (i) material thickness; (ii) the ability of the garment to trap air, which itself acts as an insulator; and (iii) how dry the material is. Insulation is generally referred to in Clo units which measure resistance to heat transfer. One Clo unit is the insulation provided by a business suit and the usual undergarments worn with it, and can be defined as: 1 Clo = $0.18\,°C/kcal/m^2/hr$.

Protection and comfort are strongly influenced by moisture produced by the skin through sweating. As the body heats up, liquid sweat evaporates, releasing heat due to the evaporation which increases heat loss from the skin surface. It is important to transfer as much water as possible away from the skin through clothing, either in vapour or liquid form. The permeability of a material is the ability to allow vaporized moisture to pass through it. Condensed (liquid) moisture transfer is also important, allowing liquid to pass along material fibres to the clothing surface where it can evaporate. Thermal insulation may be decreased by as much as 30–50% if the moisture remains within the material, due to an increase in conductive heat loss within the wet clothing layers (Giesbrecht, 2003). Liquid condensing in the layers near the surface may also freeze in cold conditions. Specific design requirements may be summarized as follows (CCOHS, 2008):

- Clothing with multiple layers generally provides better protection than a single thick garment. The air present in these multiple layers provides better insulation than the clothing itself. It is also easy to remove layers if a person is active before the body gets too warm and produces too much sweat, and conversely add layers when the person is less active and the body cools. This enables the clothing to accommodate varying temperature and weather requirements.
- It is important to provide looser outer clothing to prevent the outermost layers compressing the inner garments, resulting in a decrease in the insulation properties of the clothing assembly.
- The inner layer must be able to provide insulation and be able to 'wick' moisture away from the skin to help keep it dry. Thermal underwear made of polyester or polypropylene fabrics is most suitable for such a purpose. Since polypropylene has low density, 'fishnet' underwear made from polypropylene fibres wicks perspiration away from the skin. The moisture evaporates easily and is captured on the next layer, away from the skin. The holes present in the fishnet structure contribute to the insulation properties of the fabric and the second layer helps to cover these holes to maintain proper insulation. Denier gradient fabrics work in much the same way, transferring moisture outwards.
- Additional layers in the clothing should provide adequate insulation during severe weather conditions. Outer jackets must contain methods for closing and opening certain sections such as the hood, neck, front body panel and sleeves to help control how much heat is retained or released. For added ventilation, netted pockets and vents around the trunk and under the armpits (with zippers or Velcro fasteners) can be used.

- For working in wet condition, the outer layer of clothing should be waterproof. An easily removable windbreak garment should be used if it is not possible to shield the work area against wind. Under extremely cold conditions, heated protective clothing may be needed.
- It is extremely important to be able to keep clothing clean since dirt fills air cells in the fibres and destroys their insulating ability.

Until recently, the reduction of heat loss was accomplished by wearing heavy garments and/or by putting on multiple layers of garments. The transmission of moisture has been achieved in various ways, such as wearing vapour-permeable garments, providing ventilation at the cuffs or collars, deliberately providing vent holes in the garments, etc. With the advent of newer technologies and the introduction of smart polymers and materials in the last 15 years, new concepts have emerged for providing enhanced comfort and/or survivability, even in extremely cold and harsh environments.

Clothing systems used to counteract extreme cold environments have traditionally been based on clothing with multilayer fibrous systems. Fibrous materials are effective since they have the capacity to trap huge volumes of air because of their high bulk. For example, a suiting fabric is 25% fibre and 75% air; a blanket is 10% fibre and 90% air; and a fur coat is 5% fibre and 95% air (Bajaj and Sengupta, 1992). The thermal resistance of a fabric depends mainly on the content of trapped air within it. When the ambient air temperature is higher or lower than the body's skin temperature, a temperature gradient is formed between the outer layer of the clothing and the skin (Fig. 5.1). If there is no convection, the larger the volume of air trapped, the thicker the batting and the higher the insulation value. Fibrous insulation systems with fine denier

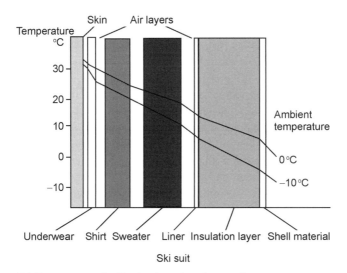

5.1 Temperature distribution in various layers of a garment.

fibres, incorporating air spaces and increasing loft, provide increased insulation.

Using multiple layers increases the amount of air trapped because it is trapped between the layers as well as by the fabric. By varying the amount of trapped air in the layers of fabric, it is possible to alter the thermal protection offered by the protective garments. The different layers also serve other purposes. In the polartech-type fleece, for example, the durable nylon outer layer is wind and abrasion resistant whilst the soft polyester inner layer pulls moisture away from the skin. The fine polyester fibres trap as much air as fur, providing exceptional insulation relative to the weight and thickness of the fabric.

The Extended Cold Weather Clothing System (ECWCS) is designed to maintain adequate environmental protection in temperatures ranging between $+4$ and $-51\,°C$. It is a protective clothing system developed in the 1980s by the United States Army Natick Soldier Research, Development and Engineering Center, Natick, Massachusetts. The ECWCS system uses moisture management principles to transfer perspiration away from the skin so the wearer remains dry and warm. Components include a lightweight undershirt and drawers, midweight shirt and drawers, cold-weather fleece jacket, cold-weather wind jacket, soft-shell jacket and trousers, extreme cold/wet weather jacket and trousers, extreme cold weather parka, and trousers which are used in various combinations to meet the cold weather environmental requirements of the US military (and others). The latest ECWCS is the 3rd-generation system (Gourley, 2009). The seven-layer, 14-component multilayered insulating system allows the soldier to adapt to varying mission requirements and environmental conditions.

In some circumstances, people need to use unitary garments rather than layers. Typically these garments consist of three components: an outer or shell layer for providing ruggedness together with a high permeability to air and moisture vapour; a middle layer with an approximately one-inch thick layer of soft and flexible polyurethane open-cell foam; and an inner layer with a woven or knitted lining fabric. All these three layers are stitched together to form a single garment.

In addition, breathable membranes and coatings, and high-loft battings of special fibres in conjunction with reflective materials have been tried. A textile-based thermal insulator should be capable of trapping the maximum amount of air. The contribution of the fibre itself to thermal insulation comes second. Convection can be reduced by optimizing the inter-fibre space. The heat loss due to radiation is not significant and in clothing for extremely cold conditions, it can be further reduced by incorporating a thin layer of metal-coated film. It is also possible to increase thermal comfort by interactive insulation using smart materials such as shape-memory polymers and phase-change materials.

5.3 Types of smart fibres and fabrics

5.3.1 Natural versus synthetic fibres

Natural fibres are made from plant, animal and mineral sources. According to their origin, natural fibres can be classified as vegetable fibres or animal fibres. Vegetable fibres include cotton, ramie, jute, flax and so on. Animal fibres include wool, silk, mohair, alpaca and so on. The most used natural fibres are cotton and wool. Cotton absorbs moisture, but clings to the skin when wet and should not be used next to the skin in cold environments.

According to the type of material used, synthetic fibres can be divided into mineral fibres and polymer fibres. Mineral fibres include glass fibre, metallic fibre and carbon fibre. The most widely used synthetic fibres are polymer fibres such as polyamide (e.g. nylon), polyester, polyolefin, acrylic and polyethylene fibres, elastomers and polyurethane fibres.

Wool can be used next to the skin and may keep the skin relatively dry. Textiles such as wool and wool blends possess a high absorbing capacity and can handle small amounts of moisture without losing their insulation properties as well. When the fabric becomes saturated, however, moisture control is reduced. The moisture absorbed by clothing layers, in addition to causing potential discomfort, adds to the weight being carried and also gradually reduces the thermal insulation of that particular layer. When activity drops and sweating ceases, the drying of wet clothing layers may deprive the body of more heat than is generated by its metabolic rate. The result is a post-chilling effect that may endanger heat balance and result in hypothermia. For activities producing sweating, and particularly for longer outdoor excursions, the advantages of using light-weight, strong and hydrophobic materials as part of the clothing ensemble should be recognized.

Many synthetic textiles are hydrophobic and moist air is transported from the skin through the fabric to the next layer. When synthetic fibres are considered as insulating materials, a quilted fabric structure appears to offer the best all-round performance. After considerable experimental work a polyester-fibre batting, sandwiched and quilted between a continuous-filament nylon face and backing fabrics, has been developed (Ross *et al.*, 2004). This type of integrated fabric is considered the best in terms of warmth/weight ratio and is highly insulating. Torn garments do not suffer in the same disastrous way as do down-filled garments. A major advantage of these fabrics is that they lend themselves particularly well to the layer principle, as the smooth surface of the outer fabrics enable one liner to be worn on top of another without discomfort or constriction.

Aerogel is among the best solid thermal insulators. Aerogels are a form of insulation based on nanotechnology. They can be anything from 2–8 times more effective than traditional insulation, thereby providing extensive heat and energy savings, and consequently benefiting the environment. The reason they are so effective is that they are made up of very little solid material; approximately 96%

of their volume is air. Most aerogels are made from silicon or carbon. Aerogel is a silica gel formed by supercritical extraction which results in a porous open cell solid insulation with a thermal conductivity as low as 0.013 W/m K. Aerogels have a wide range of uses such as insulation for windows, vehicles, refrigerators/ freezers, etc. (Bardy *et al.*, 2007).

Wool and down fabrics are highly insulating due to the very nature of their fibres. Modern synthetic textiles, such as battings made of hollow polyester fibres or polyolefin micro-fibres, resemble natural materials and provide good insulation per unit thickness. As a spin-off from space technology, reflective materials (mostly aluminized fabrics or fibres) are used in garments and survival kits. The idea is that much of the heat the body loses through radiation will be reflected back to the skin. Such an ensemble will transmit less overall heat and its net insulation is higher than a similar one without a reflective layer. However, practical tests demonstrate that the net effect is small, and in certain conditions negligible (Holmes, 2007). There are several reasons for this. Radiation heat loss is only a minor part of the overall heat loss in the cold, particularly in the presence of wind and/or body movements (10–15%). Reflection of radiation requires spacing of the layers, which is difficult to achieve and maintain. Most reflective fabrics are impermeable and interfere with moisture transfer. Aluminized insoles for shoes are common but provide no additional insulation compared to soles of similar thickness without aluminium. Gloves and socks with aluminium threads in the fabric do not provide higher insulation than those without them.

5.3.2 Breathable fabrics

In bad weather, protection against rain and snow is required. However, waterproof fabrics may interfere with evaporative heat exchange, and activity causes a person to get wet from inside their clothing instead of from outside. Microporous materials help to solve this problem to some extent. The small pores allow water vapour to pass through but prevent liquid water from doing the same. This works reasonably well in temperate and warm climates, but due to the 'cold wall' principle, it becomes less effective in colder climates. In cold conditions vapour turns rapidly to liquid which can start to build up in clothing layers. Exposure to alternating warm and cold conditions (e.g. moving in and out of cold environments) may allow absorbed moisture to escape during warm conditions. Textiles of this kind are often highly windproof, which is beneficial in cold environments. Breathable monolithic fibres like sympatex utilize a monolithic membrane fabric that has no pores. Other waterproof/breathable fabrics include Gore-Tex which is manufactured from polytetrafluoroethylene (PTFE). It is used in a wide variety of applications such as high-performance clothing, medical implants, filter media, insulation for wires and cables, gaskets, and sealants.

5.3.3 Far-infrared fabrics

Any regular fabrics, such as cotton, polyester or acrylics, made with fibres implanted with various metals of ceramic compounds, such as platinum derivatives or alumina and silica, may be referred to as far-infrared (FIR) fabrics (Marji Graf). Fabrics with fibres containing ceramic compounds can absorb, reflect and emit FIR waves. The mean temperature of the human body is 36.5 °C, equal to 9.4 μm of wavelength. When far infrared rays radiate on the human body, their frequency is consistent with that of the cellular movement in the human body, so that it is easy for the human body to absorb the far infrared ray and the heat it generates. Wearers of such a fabric claim to feel a gentle, soothing heat. After reaching the storage capacity of the ceramics, excess heat is allowed to leave the fabric and is vented. This results in the fabrics maintaining a comfortable temperature without overheating.

Breathable FIR fabrics can be used to warm up the body quickly in extremely cold weather and will allow excess heat to escape during warm conditions. This may be achieved by using multiple methods to store and generate heat energy. The fabrics simultaneously make use of the three primary means of heat conservation, i.e., reflection, insulation, and emission. The reflection and insulation characteristics of FIR fabric allow the body to heat and warm itself using its own natural FIR waves to prevent heat loss. The emission property can help fabrics to emit wavelengths on the 5 to 15 micron wavelength (far infrared), as FIR fabrics can normally absorb heat from sources such as light, room temperature and the human body (Marji Graf, http://www.etechnologypad.com/info/Technology/Infrared-Technology.html).

5.3.4 Intelligent textiles

In recent years, several new types of material and fabric containing active components, e.g., phase-change materials (PCM), inflatable tubings and electrical heating, have been put on the market. PCM fabrics react to heating by absorbing heat. They respond to cooling by releasing the absorbed heat from a range of waxes in the fibre or fabric. By choosing a certain temperature for the phase change, the fabric in a garment could assist the wearer's thermoregulatory adjustment to hot and cold environments.

Inflatable fabrics, in principle, should allow the thickness of the ensemble to expand, thereby increasing the effective insulation. Fabrics with a system of thin tubing can be inflated by the mouth. The effect is a thicker layer that should add insulation for that particular garment.

Fabrics incorporating electrically heated elements have been available for many years. The drawbacks, so far, have been the low capacity of portable batteries and the durability of the wiring system. The rapid development of mobile phones and portable computers has resulted in the availability of powerful, long-lasting, portable batteries that can also be used for auxiliary heating. This concept is likely to be most beneficial for the heating of hands and feet. Garments are already available on the market with built-in batteries that can be charged from the mains supply. Szmocha and Sudol (2003) have developed heated clothing for use in cold weather. The assembly consists of an insulated suit with a transparent infrared window to transmit heat; a light, portable, low temperature, infrared heater; a heat collector, and a heat distribution system. This type of clothing can be used to protect people exposed to cold environments such as construction workers, mineral exploration and oil field personnel, and other workers required to work in adverse weather conditions or those exposed to cold surroundings.

The heat generated by the low-temperature heating unit, which is portable and located outside the suit, is transferred to the interior of the suit via a special window by means of infrared radiation. Infrared radiation energy is absorbed by the heat collecting plate and transferred inside the suit to the different parts/regions of the human body by a heat distribution system. The control system module is equipped with temperature sensors to measure the temperature of different parts of the wearer's body, and provides for the control of the stream of heat delivered to particular parts of the body. The temperature of individual parts of the body is kept within preset temperature limits set individually by the suit user.

New clothing systems have been developed based on microencapsulated phase-change materials. These materials are regarded as suitable components for fabric coatings when exceptional heat transfer and storage capabilities are desired. As an example US Pat. No. 5,290,904 (Colvin et al., 1994) describes substrates coated with a binder containing microencapsulated phase-change materials (MicroPCMs) filled with energy absorbers, exhibiting extended or enhanced heat retention or storage properties. These MicroPCMS can also be incorporated into fibres. US Pat. No. 4,756,958 (Bryant and Colvin, 1988), for example, describes a fibre with integral microspheres filled with phase-change materials with enhanced thermal properties at predetermined temperatures (Fig. 5.2). These fibres may be woven into fabrics which can exhibit superior thermal storage properties.

5.4 The use of shape-memory materials

5.4.1 Principles of shape-memory materials

Shape-memory polymers (SMPs), a type of shape-memory material, are defined as polymeric materials with the ability to sense and respond to external stimuli with a predetermined shape (Wei et al., 1998). Polymers such as polynorborene, trans-polyisoprene, styrene-butadiene copolymer, crystalline polyethylene, some block copolymers, ethylene-vinyl acetate copolymer and segmented poly-

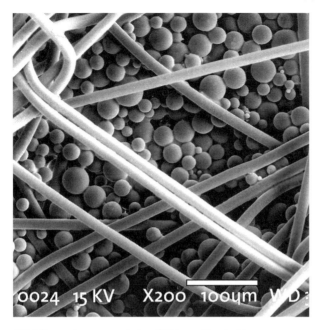

5.2 Microspheres containing PCM used in clothing.

urethane have been discovered to have a shape-memory property (Lendlein and Kelch, 2002). Shape-memory materials can revert from their current shape to a previously held shape, usually due to the action of heat. In contrast to shape-memory alloys, shape-memory polymers are easy to shape, have high shape stability and an adjustable transition temperature. The air gaps between adjacent layers of clothing increase to give better insulation when these shape-memory materials are activated in garments. The incorporation of shape-memory materials into a garment is extremely important when attempting to confer greater versatility in the protection it could provide against extreme cold (Zafar, 2008).

Great changes in the thermomechanical properties of SMP films occur across the glass transition temperature (T_g) of the soft segment crystal melting point temperature (T_{ms}) (Lomax, 1990). In addition to these changes, it has also been found that an SMP has a large change in moisture permeability above and below T_g/T_{ms}. It demonstrates low moisture permeability below T_g/T_{ms}, while in a glassy state, whereas it shows high moisture permeability above T_g/T_{ms}, in its rubbery state. This is normally based on T_g/T_{ms} set at room temperature. For SMP-laminated textiles this behaviour is very useful, because they can provide thermal insulation at cold temperatures and high permeability at room temperature or above (Mondal and Hu, 2006).

Cold weather clothing that can regulate its temperature and moisture content can be produced using membranes combined with suitable SMPs. A shape-

94   Textiles for cold weather apparel

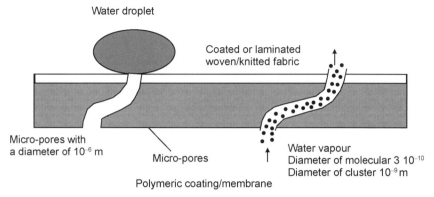

5.3 Principle of water vapour permeability in a textile product with a hydrophobic micro-porous coat of open through-pores.

memory polyurethane polymer, exhibiting permeability to water vapour and the ability to retain or release heat, can be laminated onto a fabric (Fig. 5.3), resulting in a fabric which is waterproof, windproof and breathable (Brzeziński et al., 2005). The polymer changes its structure at a transition temperature, similar to shape-memory alloys. When the garment reaches the glass transition temperature, it turns into a fabric permeable to water vapour and heat, cooling the wearer's body heat, especially after hectic activity or due to an increase in the external temperature. The material returns to a less permeable structure as the internal temperature falls, providing the required warmth to the user (Bowen et al., 2003).

5.4.2 Types of shape-memory materials (SMMs)

A variety of alloys, ceramics, polymers and gels have now been found to exhibit shape-memory effect (SME) behaviour. The fundamental and engineering aspects of SMMs have been investigated extensively and some of them are currently available as commercial materials. Some SMMs can be easily fabricated into thin films, fibres or wires, particles and even porous bulk materials, making it feasible to incorporate them with other materials to form hybrid composites.

Shape-memory polymers (SMPs) are a class of smart materials. After being heated above a characteristic transition temperature (T_{trans}), they can develop large elastic deformation, which can be fixed by subsequently cooling the SMPs to below T_{trans}. The deformation can be recovered by reheating the SMPs up to T_{trans} (Takahashi et al., 1996; Lin and Chen, 1998; Kim et al., 2000; Kim and Lee, 1996; Lendlein and Kelch, 2002). The temperature T_{trans} is therefore called the shape-memory temperature for triggering shape memorization. It is essential for defining their shape-memory properties. The unique thermal-response features of SMPs have attracted increasing attention from the technical

| ← Soft segment → | ← Hard segment →

═══ = Long chain diol   ──── = Short chain diol   ---- = diisocyanate   •= Urethane group

5.4 Segmented polyurethane.

community. They are considered promising materials in many fields involving smart textiles (Ding *et al.*, 2004; Hu *et al.*, 2003; Tobushi *et al.*, 1996), medical materials (Lendlein and Kelch, 2002; Lendlein and Robert, 2002), sensors and actuators (Tobushi *et al.*, 1996; Monkman, 2001) and damping materials (Tobushi *et al.*, 1998).

Shape-memory polyurethane (SMPU) demonstrates a different structure from conventional polyurethanes, which are generally segmented (Fig. 5.4). The polymer has a wide range for the glass transition temperature (T_g) and a soft segment crystal melting point temperature (T_{ms}). Because of this property, it is able to remember its original shape, and return to it under external stimuli from its deformed state. The chemical composition and chain length of the soft segments (blocks) influence the morphology of these polymers. An SMPU is characterized by a micro-phase separated structure as a result of the thermodynamic incompatibility between the hard and soft segments. Hydrogen bonding and crystallization processes can help to bind hard segments, resulting in a solid polymer below their melting temperature. The shape-memory effect is obtained from the reverse phase transformation of the soft segments. The molecular weight of the soft segments, mole ratio between soft and hard segments, and polymerization process may all play a part in controlling the shape-memory effect (Hayashi, 1995). Shape-memory properties can be triggered by various external stimuli (He *et al.*, 2001; Behl and Lendlein, 2007) such as temperature, pH, electric fields, magnetic fields, etc. Of these temperature is most significant for cold weather clothing applications.

Stimuli-sensitive polymers (SSPs) or shape-memory polymers (SMPs) can be used to produce intelligent textiles that exhibit unique environmental responses, such as protection against the cold. The molecular structure of SSPs facilitates phase-change behaviour in response to environmental stimuli and allows SSP textiles to change their structures and properties. SSPs create fabrics whose air permeability, hydrophobic characteristics, heat transfer properties, shape and light reflectance are responsive to environmental stimuli such as temperature, pH, moisture, light and electricity. These properties can be utilized to design garments containing suitable SMPs for protection against extreme cold weather conditions.

Temperature-sensitive shape-memory polyurethanes (TSPUs) are a special class of adaptive materials that can convert thermal energy directly into mechanical work. Their molecular structure is the same as that of SMPUs, but in addition they have a micro-phase separated structure due to the thermodynamic incompatibility between their hard and soft segments. These polymers are generally solid below their melting point due to the binding of hard segments through hydrogen bonding and crystallization. This results in the release of stored heat energy which helps to protect against cold environments.

Generally, TSPUs consist of two phases, i.e., a thermally reversible phase (soft segment) and a fixed phase (hard segment). Hence, the polymer has often been observed to have a block or segmented structure. A phase transition temperature (glass transition temperature) is generally shown by the thermally reversible phase in many cases. This can be used as a switch temperature (T_s). The reversible phase is subject to softening when heated above the T_s or hardening when cooled below T_s. It is well-known that changes in the physical properties of polymeric materials, especially a large increase of free-volume (FV) hole size and an enhanced micro-Brownian motion when heating through the glass transition temperature (T_g), are accompanied by a phase transition. This property can be used in the molecular design of smart membranes with functional gates and controllable gas permeation (Chen et al., 2007).

The water-vapour permeability and waterproof properties of these materials are important for them to have a broad range of applications in the textile industry, especially in designing cold weather clothing. Suitable combinations of ordinary fabrics with TSPU membranes can lead to the development of smart textiles, which can exhibit waterproof behaviour at any temperature and also provide variable breathability in response to the climate temperature (Ding et al., 2006). It is possible to incorporate films of a temperature-sensitive polymer in multilayer garments as the polymer reverts within a wide range of temperatures. This holds great promise for these SMPs in making cold weather clothing with adaptable features. Using a composite film of shape-memory polymer material as an interliner in multilayer garments, in outdoor clothing such as soldier protective garments and in sportswear for cold regions, can provide adaptable thermal insulation and be used to obtain effective protective clothing.

A successful and widely used design approach for cold-weather garments is to use a fairly permeable outer fabric sheath over a layer of fibrous insulation batting. The batting layer provides most of the thermal insulation while the outer sheath protects the batting and minimizes wind-induced airflow through the batting. The latter function is very important because airflow in the batting layer greatly reduces its insulating effectiveness. From this perspective, an impermeable sheath would be best. However, a certain amount of permeability is essential so that moisture produced by perspiration can escape. Unfortunately, permeability of the sheath allows penetration of cold ambient air in windy conditions and this tends to substantially increase heat losses.

Today's modern technologies offer various possibilities for making cold-weather clothing, such as waterproof and breathable constructions using SMPs

(Holmes, 2000). Waterproof, breathable textiles can be produced by coating/laminating breathable nonporous membranes (films) or laminating microporous films of SMPs onto a suitable base fabric. The presence of permanent air-permeable pore structures enables the microporous film laminates to breathe. The micropores, of which there are many, are sufficiently large to allow the penetration of perspiration molecules, but are small enough to prevent water droplets from penetrating the fabric (Mondal and Hu, 2006). In SMPs films, great changes in thermomechanical properties occur across the glass transition temperature (T_g) or soft-segment crystal melting-point temperature (T_{ms}) (Lomax, 1990). In addition, SMPs also have a large change in moisture permeability above and below the T_g/T_{ms}. Based on the T_g/T_{ms} set at room temperature; SMPs have low moisture permeability below the T_g/T_{ms}, during the glassy state, and high moisture permeability above T_g/T_{ms}, during the rubbery state. This behaviour is very useful for SMP-laminated fabrics, which could provide thermal insulation at cold temperatures and high permeability at room temperature or above (Mondal and Hu, 2006).

A fabric that is laminated/coated with a porous film or membrane of SMP is able to prevent the penetration of water droplets and therefore could provide a good overall balance between breathability and waterproofness. The dimensions of the pores (0.1–10 μm) are larger than those of perspiration molecules and much smaller than those of water droplets. Their structure may be symmetric, i.e., the pore diameters do not vary over the membrane cross-section, or asymmetrically structured, i.e., the pore diameters increase from one side to the other by a factor of 10 to 1000 (Fig. 5.5).

Nonporous/dense films or membranes consist of a dense film through which permeates are transported by diffusion under the driving force of a pressure, concentration, temperature, or electrical potential gradient. The membrane struc-

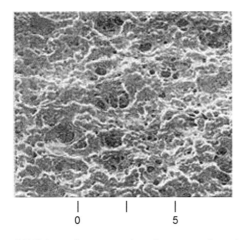

5.5 Schematic presentation of porous polyurethane topcoat.

ture can be in a rubbery state or a glassy state, depending on its glass transition temperature, with the exception of a crystalline dense membrane, which do not exhibit any macroscopic pores. The diffusion mechanism determines the transport behaviour, which means that the components first dissolve in the membrane due to some driving force. Permeation generally occurs due to the diffusivity and/or solubility (Shannon, 2004).

Thermax and Thermolite are two very successful speciality fibres produced by DuPont. These are designed to take advantage of the cross-sectional shapes of the fibres to provide comfort at high and low temperatures. Both Thermax and Thermolite are hollow core fibres. In addition, Thermax includes channels in the fibre surface. These fibres prevent heat loss by trapping a layer of warm air within the hollow core while transporting perspiration to the outer layer of the moisture-absorbing fabric. They offer great softness and wicking ability. They combine the advantages of being a lightweight fibre providing breathability and warmth while moving moisture from the body by combining a thermally efficient polymer and micro-tested hollow core fibres that heat up quickly and retain warmth. They use the reduced thermal conductivity of the air trapped in the fibre for protection from cold weather. The disadvantage of such weather-specific fabrics is the need for two separate sets of apparel designed specifically for cold weather.

5.4.3 Applications of shape-memory polymers

Although there has been a lot of research on shape-memory polyurethane, it is still in its initial stages. Shape-memory fibres usually have outstanding mechanical properties because of their molecular orientation. Using SMPs, it could be possible to design fibres to maximize comfort in all weather conditions. The basic idea is to design a fibre with a cross-section that opens up at higher temperatures to form a channel in the fibre as the 'permanent shape', which closes at lower temperatures to mimic a closed hollow fibre. Selecting polymers with significantly different thermal expansion coefficients and coupling them with polymers with higher thermal expansion coefficients to comprise the inner section will form a fibre system very sensitive to temperature changes.

Temperature-sensitive polyurethane (TSPU) offers tremendous opportunities for developing intelligent, waterproof, breathable fabrics (IWBF) suitable for cold weather clothing because of its unique smart property (Subrata, 2006). When the temperature is lower than the transition temperature of T_m, the water vapour permeability of the TSPU membrane is extremely low, which prevents air and water molecules from passing through it. As the temperature rises, the sharp increase in free volume will trigger a significant increase in water vapour permeability (Hu, 2007). A schematic illustration of this smart property is given in Fig. 5.6. Hence, when the smart membrane is laminated with fabrics, it will

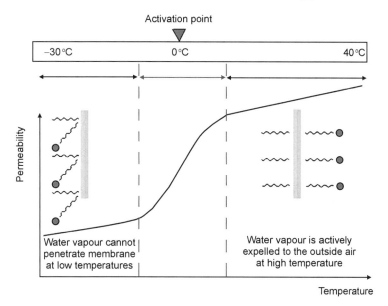

5.6 Schematic of the smart property of a TS-PU.

prevent air and water molecules passing through it at low temperatures and keep the body warm. At the same time, it will have more breathability at high temperatures and keep the body comfortable. The working mechanism of IWBF is shown in Fig. 5.7. A TSPU membrane can be laminated with fabrics by four different methods to produce a suitable thermal insulating fabric: Direct Laminates, Insert Laminates, Lining Laminates, and Three-layer Laminates (Fig. 5.8).

When shape-memory polyurethane is laminated to a fabric, a smart fabric is formed. Mitsubishi Heavy Industries have developed their own range of active sports clothing called 'DiAplex' (Gupta, 2008). This is created by laminating an

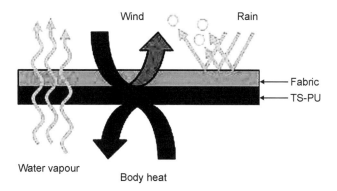

5.7 Working mechanism of IWBF.

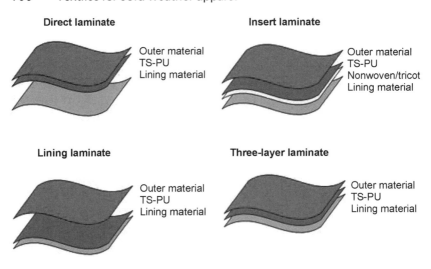

5.8 Different lamination methods.

SMP between two layers of fabric, creating a membrane. It can simultaneously be waterproof, windproof and breathable. The membrane works by applying micro-Brownian motion (Hayashi *et al.*, 1994; Kobayashi and Hayashi, 1992). When the temperature rises above a predetermined value, the DiAplex membrane is stimulated. In addition, micro-pores are created in the polymer membrane due to the Micro-Brownian motion, which allow perspiration and body heat to escape.

In order to form an ideal combination of thermal insulation and vapour permeability for cold weather clothing, it is also possible to coat SMPU onto a fabric such that its permeability changes as the wearer's environment and body temperature changes. The fabric remains less permeable and keeps the body heat in when the body temperature is low, but allows the water vapour to escape into the air when the body is sweating. This happens because its moisture permeability becomes very high with increasing body temperature. The moisture permeability results in the release of heat from the apparel. Since the fabric is waterproof, apparel made with coated or laminated SMPU fabric can be used regardless of the weather.

Veriloft, a USA-based company, has developed cold-protection clothing using a smart system based on the development of thermally responsive fibres. By using smart materials in the form of shape-memory polymers, the fibre structure alters as the temperature changes. This concept may be integrated into the production of co-extruded, bi-component fibres. Fibre layouts are optimized for transition from a flat, two-dimensional configuration to one that is three-dimensional. With the correct design, a network of fibres creates a self-regulating batting that gradually changes from a flat, low-loft, low-thermal resistance configuration to a high-loft, high-thermal resistance configuration, effectively regulating thermal protection in response to environmental and body temperatures.

5.5 The use of phase-change materials

5.5.1 Principles of phase-change materials

Phase-change materials (PCMs) possess the ability to absorb energy when the phase-changes from solid to liquid, and release energy when the phase changes from liquid to solid (Vijayaraaghavan and Gopalakrishnan, 2007). In 1987, the technology for using phase-change materials (PCM) in clothing was developed and patented for the purpose of improving the thermal insulation of textile materials during changes in environmental temperature conditions (Bryant and Colvin, 1992). Many PCMs change phases within a temperature range just above and below human skin temperature, which is used for making protective all-season outfits, and those for abruptly changing environments. Materials such as fibres, fabrics and foams with built-in PCMs store the warmth of the body and then release it back as the body requires it. Because of the dynamic nature of phase changes, these materials are continuously turning from solid to liquid and back according to the physical movement of the body and the outside temperature.

When condensed PCMs are heated to melting point, they absorb heat energy as they move from a solid state to a liquid state and a short-term cooling effect is produced in the clothing layers by this phase-change effect. The heat energy required for this purpose may come from the body or from a warm environment, and once the PCMs have totally melted, the storage of heat stops. When a PCM garment is worn in extremely cold environments, where the temperature is below the PCM's freezing point and the fabric temperature drops below the transition temperature, the micro-encapsulated liquid PCM will convert to a solid state, generating heat energy and producing a momentary warming effect. In other words, when used in sufficient quantities, this heat exchange creates a buffering effect in clothing, minimizing changes in skin temperature and maintaining the thermal comfort of the wearer. The fabric layers in PCM garments must go through the transition temperature range before the PCMs change phase and either produce or absorb heat. Hence, the wearer has to perform some activity for the temperature of the PCM fabric to change. Generally, PCMs exhibit transient phenomena and have no effect in a steady-state thermal environment.

For applications where temperature stabilization is desired, a substrate can be coated with a suitable PCM or a fibre incorporating a phase-change material (Pause, 2000, 2001). Examples of paraffinic hydrocarbon phase-change materials suitable for use in coatings or in fibres are shown in Table 5.1. The number of carbon atoms in such materials is directly related to the respective

Table 5.1 Temperature range of paraffinic phase-change materials

Compound	No. of carbon atoms	Crystallization point	Melting point
n-Eicosane	20	30.6 °C	36.1 °C
n-Octadecane	18	25.4 °C	28.2 °C
n-Heptadecane	17	21.5 °C	22.5 °C
n-Hexadecane	16	16.2 °C	18.5 °C

Source: Pause (2000)

melting and crystallization points. Phase-change materials, such as paraffinic hydrocarbons, can be formed into microspheres and encapsulated in a single or multi-layer shell of gelatin or other material. Although encapsulated microsphere diameters of 1 to 100 microns are possible, diameters from 10 to 60 microns are preferred. It is possible to bind microspheres in a silica matrix of sub-micron diameters.

Phase-change materials can provide thermal protection due to their high thermal inertia (Cabeza *et al.*, 2007). By the appropriate selection of the phase-change temperature, a fabric containing an appropriate quantity of PCM can act as a transient thermal barrier, protecting the wearer from the effects of cold environments. When a PCM fabric is exposed to an environment where the temperature is below its crystallization point, the cold will interfere with the cooling effect of the fabric structure by changing its phase from liquid to solid, and the temperature of the fabric will be kept constant at the crystallization point. The fabric temperature will drop when all the PCM has crystallized, and the PCM will have no further effect on the fabric's thermal performance. Hence, the thermal performance of a PCM depends on the phase-change temperature, the amount of phase-change material that is encapsulated, and the amount of energy it absorbs or releases during a phase change (Ghali *et al.*, 2004).

5.5.2 Types of phase-change materials

For textile applications, the PCM may be enclosed in small polymer spheres of only a few micrometres in diameter (Li and Zhu, 2004; Mukhopadhyay and Vinay Kumar Midha, 2008). Alternatively, fibres containing microPCMs could be spun using melt-spinning technology. The technology for incorporating micro-encapsulated PCM inside textile fibres to improve their thermal performance was developed and patented by scientists in the 1980s (Shim *et al.*, 2001). They reported that heat released by a PCM in a cold environment decreases body heat loss by an average of 6.5 W for a one-layer suit and 13.2 W for a two-layer suit compared with non-PCM counterparts. Ying *et al.* (2004) reported a study on assessing the temperature-regulating performance of textiles incorporating PCM.

The thermal energy or heat can be stored as either sensible or latent heat. Storage of sensible heat depends upon a change in temperature and occurs with all materials and temperatures. Storage of latent heat, however, depends upon a change in phase of a material, such as a solid to a liquid or a liquid to a gas, and occurs at specific temperatures for individual materials. An unusually high quantity of energy needs to be transferred when a phase change occurs in a material to move from one physical state to another. In the case of ice, when solid ice changes to water, there is an absorption of approximately 80 cal/g and when water changes back to ice, it must also give up that much thermal energy. In a latent phase-change material, the energy storage is based upon the heat absorbed or released when a material passes through a reversible phase transition. Generally, latent heat storage systems have a higher storage capacity per unit weight or volume compared to sensible heat storage systems. They have greater system efficiency and require a simpler control system (Colvin and Colvin, 2001).

When selecting a PCM for a specific application, the operating temperature of the heating or cooling cycle must be matched to the phase transition temperature of the material. In this respect, PCMs are less versatile than sensible heat storage materials. However, the PCM may be used to store large quantities of thermal energy or heat if it is found to be operating at the desired temperature. For example, water can store sensible heat up to 1 calorie/g, but is capable of storing latent energy almost 80 times larger. The inadequacy of water is that the solid/liquid phase change only occurs at 0 °C, which limits its application as a PCM to a great extent. Other existing PCMs however, change phase at different temperatures; the phase-change temperature of some pure paraffins ranges from sub-ambient to greater than 60 °C, depending on the length of their carbon chain. The phase-change temperature from a solid state to a liquid state for paraffin such as octadecane is around 26–28 °C, depending upon its purity. In addition, octadecane can store up to 58 calories/g depending upon its purity, and other paraffins and materials exhibit different phase-change temperatures, but can exhibit similar quantities of heat storage (Colvin and Mulligan, 1989).

The total thermal capacity of the PCM in many products is influenced by its specific thermal capacity and quantity. The required quantity may be determined by considering the application conditions, the desired thermal effect and its duration, and the thermal capacity of the specific PCM. The thermal efficiency of the PCM, which has to be measured with respect to the material selection and the product design, is also affected by the structure of the carrier system and the end-use product.

Due to their high heat-storage capacities, PCMs have been applied to textiles in many ways to improve the thermal comfort of end-use products. Coating, lamination, finishing, melt spinning, bi-component synthetic fibre extrusion, injection moulding, and foam manufacturing are some common methods for incorporating PCMs into textile structures (Nihal and Emel, 2007). MicroPCMs may be incorporated into the textile structure in several ways. For example, they can be incorporated into the spinning dope of man-made fibres like acrylics, embedded into the structure of foams, and coated onto fabrics for smart textile applications (Shim *et al.*, 2001). PCM garments exhibit dynamic behaviours and behave like smart garments. This means that the thermal properties of such garments are related to changes in temperature and time (Ying *et al.*, 2004). For example, when the environmental temperature reaches the PCM melting point, the physical state of the PCM in the garment will change from solid to liquid and absorb heat, while the temperature of the PCM in the garment remains constant at its melting point. Hence, the PCM garment can provide a cooling effect caused by heat absorption and a heating effect caused by heat emissions of the PCM to the human body. Wash-fastness tests reveal that the material maintains the PCM effect even after ten washes. Phase-change materials are at the moment being used in textiles designed to protect the extremities; gloves, boots, hats, etc. These PCM materials can be useful down to 16 °C, enough to ensure the comfort of someone wearing a ski boot in the snow, for example. They are increasingly applied in body-core protection, blankets, sleeping bags and mattresses (Doshi, 2006).

Certain products, like clothing, footwear, foam and bedding materials use microencapsulated phase-change materials (or microPCMs) for the regulation of temperature. These so-called microPCMs contain encapsulated paraffin wax, which absorbs and releases heat to maintain a regulated temperature (Fig. 5.9). In special garments, such as ski jackets, the paraffin wax present in the microPCM initially absorbs the skier's body heat and stores it until the body temperature drops due to the outside environment, when warmth is imparted to the skier by the process of releasing heat. Thus, the body temperature is regulated for the comfort of the skier. MicroPCM products may be supplied as a wet filter cake or as a dry powder and can be applied to fabric, or other types of material as a coating. The microPCM products contain small bi-component particles consisting of a core material – the PCM – and an outer shell or capsule wall.

MacroPCMs are characterized by the presence of large 3-mm spherical beads and contain high concentrations of phase-change materials for use in cooling

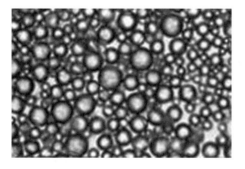

5.9 Micro PCMs.

5.10 Macro PCMs.

vests and garments (Fig. 5.10). The principle of macroPCMs is that the particles normally absorb excess heat and permit the user to function for a longer time at a more comfortable temperature. Microtek vests worn as an undergarment can be loaded with these particles to allow for close skin contact.

Most of the widely used and well-known PCMs are linear chain hydrocarbons known as paraffin waxes (or *n*-alkanes), hydrated salts, polyethylene glycols (PEGs), fatty acids and mixtures or eutectics of organic and non-organic compounds. They have melting points ranging from 0 to 50 °C, which can absorb and release large amounts of heat. These substances can also be used as core materials in microcapsule production (Chen and Eichelberger, 1985; Hartmann, 2004; Zalba *et al.*, 2003).

Phase-change materials, such as paraffin wax, have a high latent heat-storage capacity. They have the ability to store/release heat as the latent heat of fusion when they undergo a liquid/solid phase transition at a constant temperature. While many heating systems are based on sensible heat (the heat of an object that is changing temperature), applications of latent heat are becoming more and more popular because they have a much greater volumetric heat storage capacity. Figure 5.11 presents a comparison of the heat capacity of different common materials. The Δt is 15 k and the unit of heat capacity is KJ/kg.

The exact temperature of the liquid/solid transition is dependent on the number of carbon atoms in the polymer chain backbone for materials like paraffin wax. Octadecane ($C_{18}H_{38}$) has a melting temperature of 27 °C, which is close to the temperature of the human body (37 °C). This means that the PCM can be re-melted just from the heat produced by the body. Two important advantages of using octadecane are its high heat of fusion and its commercial availability at a reasonable cost.

5.5.3 Applications of phase-change materials

Wang *et al.* (2006) reported a special clothing system for thermal protection against extreme cold-weather conditions, consisting of four layers. The first

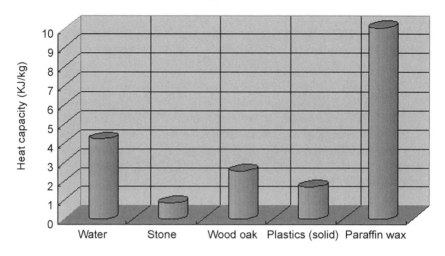

5.11 Comparison of heat capacity of different common materials.

layer is cotton fabric, and the second layer a non-woven polyester fabric treated with PCM enclosed in small polymer spheres with diameters of only a few micrometres. Non-woven polyester fabric makes up the third layer, and the outermost layer consists of a waterproof breathable fabric (Figs 5.12 and 5.13). The second layer keeps the first cotton layer from becoming saturated with perspiration. This fabric assembly was tested on a bionic skin model in a climate chamber where the temperature was controlled at $-15 \pm 5\,°C$. When the PCM layer's temperature increases above the PCM's melting point (28 °C), the PCM melts and becomes liquid. Thermal energy is absorbed and stored during this

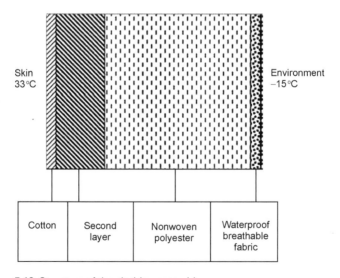

5.12 Structure of the clothing assembly.

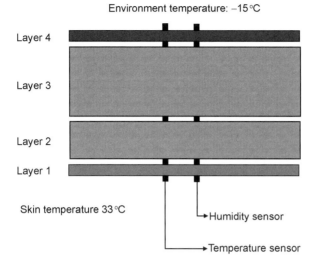

5.13 The location of the sensors.

process. When the temperature of the PCM layer falls below 27 °C, the liquid PCM becomes solid and releases heat energy. The PCM acts as a thermal buffer material by releasing stored heat.

Outlast Technologies' PCMs, incorporated into clothing, interact with the skin temperature to provide a buffer against temperature swings. The microencapsulated PCMs (mPCMs), produced by Ciba and Microtek, are called Thermocules. These are used for finishing fabrics or are infused into fibres during the manufacturing process. The technology employed in designing fabrics for cold environments continuously interacts with the unique microclimate of the human body and the environment to moderate temperature from being too hot or too cold.

PCMs can be incorporated into fibres during spinning for certain applications. Microcapsules can be seen located inside the fibre as can be seen in Figs 5.14–5.16. These fibres are later spun into yarns for the manufacture of socks, underwear or knitwear. All these products are suitable for wearing next to the skin. It is also possible to coat textiles such as non-woven fabrics with PCMs (e.g. sleeping bags). For clothing such as jackets coated linings are used but PCMs can also be applied to other layers, such as mid-layers between the first layer and the lining, leaving the manufacturers free to choose from a range of design options.

5.6   Future trends

New techniques for temperature control are becoming popular in cold weather clothing including silver-based fabrics, adjustable insulation systems and bionic

108   Textiles for cold weather apparel

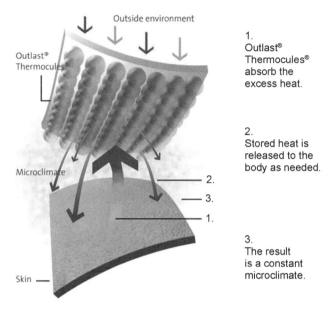

5.14 The smart PCM fabric Outlast Adaptive Comfort inside the Tempex polar coveralls helps to keep the body comfortable all day.

5.15 Outlast's PCM acrylic fibre.

climate membranes, although more work needs to be carried out to improve the temperature control behaviour, expense and wearability of these garments (Sawhney et al., 2008). Currently, there is an acute need to develop all-weather clothing systems that provide optimum comfort and any future research must concentrate its efforts on developing temperature control systems for performance apparel with overall optimum comfort. There is also increasing interest in biomimetic systems that mimic the insulating properties of wool and fur. With

The use of smart materials in cold weather apparel 109

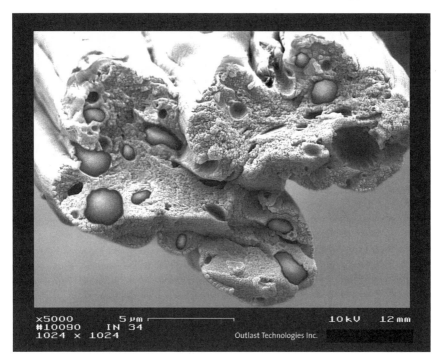

5.16 Outlast's PCM viscose fibre.

such technologies people who perform various outdoor activities can now combat cold temperatures more easily than in the past.

5.7 References and further reading

Bajaj P and Sengupta A K (1992), Protective clothing, *Textile Progress*, 22, 2–4, 1–110.

Bardy E R, Mollendorf J C and Pendergast D R (2007), Thermal conductivity and compressive strain of aerogel insulation blankets under applied hydrostatic pressure, *Journal of Heat Transfer*, 129, 2, 232–236.

Behl M and Lendlein A (2007), Shape memory polymers, *Materials Today*, 10, 4, 20–28.

Bowen W E, Murrey L J and McCarry P M (2003), Membrane permeable to aromatic products, US Patent No. 6902817.

Bryant Y G and Colvin D P (1988), Fiber with reversible enhanced thermal storage properties and fabrics made therefrom, US Patent No. 4756958.

Bryant Y G and Colvin D P (1992), Fibres with enhanced reversible thermal storage properties, *Textile Symposium*, 1–8.

Brzeziński S, Malinowska G, Nowak T, Schmidt H, Marchinkowska D and Kaleta A (2005), Structure and properties of microporous polyurethane membranes designed for textile-polymeric composite systems, *Fibers and Textiles in Eastern Europe*, 13, 6, 53–58.

Cabeza L, Castellón C, Nogués N, Medrano M, Leppers R and Zubillaga O (2007), Use of microencapsulated PCM in concrete walls for energy savings, *Energy and Buildings*, 39, 2, 113–119.

CCOHS (2008), http://www.ccohs.ca/

Chen J C H and Eichelberger J L (1985), Method for preparing encapsulated phase change materials, US Patent No. 4505953.

Chen Y, Liu Y, Fan H, Li H, Shi B, Zhou H and Peng B (2007), The polyurethane membranes with temperature sensitivity for water vapour permeation, *Journal of Membrane Science*, 287, 192–197.

Chung H and Cho G (2004), Thermal properties and physiological responses of avporpermeable water-repellent fabrics treated with microcapsule-containing PCMs, *Textile Research Journal*, 74, 571–575.

Colvin D P and Mulligan J C (1989), Method of using a PCM slurry to enhance heat transfer in liquids, US Patent No. 4911232, filed on Apr. 27, 1989, issued March 27, 1990. Assigned to Triangle Research and Development Corporation.

Colvin D P, Bryant Y G and Mulligan J C (1994), Thermally enhanced heat shields. US Patent No. 5290904.

Colvin V and Colvin D (2001), Microclimate temperature regulating pad and products made therefrom, US Patent No. 6298907, Delta Thermal Systems.

Craig F N (1972), Evaporative cooling of men in wet clothing, *Journal of Applied Physiology*, 33, 331–336.

DiAplex, the Intelligent Material, http://www.diaplex.com/extremehp.html, 8 March 2006.

Ding X M, Hu J L and Tao X M (2004), Effect of crystal melting on water vapor permeability of shape memory polyurethane film. *Textile Research Journal*, 74, 39–43.

Ding X M, Hu J L, Tao X M and Hu C P (2006), Preparation of Temperature-Sensitive Polyurethanes for Smart Textiles, *Textile Research Journal*, 76, 406–413.

Doshi G (2006), PCM in textiles, http://ezinearticles.com/?Pcm-In-Textiles&id=367030.

Ghali K, Ghaddar N, Harathani J and Jones B (2004), Experimental and numerical investigation of the effect of phase change materials on clothing, *Textile Research Journal*, 74, 205–214.

Giesbrecht G (2003), Cold weather clothing, Winter Wilderness Medicine Conference, Snow King Resort, Jackson Hole, Wyoming, 13–18 February.

Goldman R F (1978), Tolerance limits for military operations in hot and/or cold environments. Paper presented in 12th Commonwealth Defence conference on Operational Clothing and Combat Equipment, Ghana, 1978.

Gourley S R (2009), Soldier armed: extended cold weather clothing system. Army. FindArticles.com. 12 March.

Gupta S (2008), All weather clothing, http://www.techexchange.com/thelibrary/allweather.html

Hartmann M H (2004), Stable phase change materials for use in temperature regulating synthetic fibres, fabrics and textiles, US Patent No. 6689466.

Hayashi S (1995), room-temperature-functional shape-memory polymers, *Plastics Engineering*, 51, 2, 29–31.

Hayashi S, Kondo S and Giordano C (1994), Properties and applications of polyurethaneseries shape memory polymer, *Technical Papers of the Annual Technical Conference: Society of Plastic Engineers*, 52, 2, 1998–2001.

He C Q, Dai Y Q, Wang B, Wang S J, Wang G Y and Hu C P (2001), Temperature dependence of free volume in cross-linked polyurethane studied by positrons, *Chinese Physics Letters*, 18, 1, 123–125.

Holmes D A (2000), Performance characteristics of waterproof breathable fabrics, *Journal of Coated Fabrics*, 29, 4, 304–316.

Holmes G T, Marsh P L, Barnett R B and Scott R A (2007), Clothing materials – their required characteristics. In *Handbook of Clothing: Biomedical effects of Military clothing and equipment systems*, 2nd edn. Ralph F. Goldman, Bernhard Kampmann.

Hu J L, Zeng Y M and Yan H J (2003), Influence of processing conditions on the microstructure and properties of shape memory polyurethane membranes, *Textile Research Journal*, 73, 172–178.

Hu J L (2007), *Shape memory polymers and textiles*. Woodhead Publishing Limited, Cambridge, 331–333.

Jian R (2000), Thermal storage and thermal insulating fibres and their manufacture and applications, CN 1,229, 153, 22 Sep 1999, *Chemical Abstracts*, 133, 31770.

Kim B K and Lee S Y (1996), Polyurethanes having shape memory effects, *Polymer*, 37, 5781–5793.

Kim B K, Shin Y J, Cho S M and Jeong H M (2000), Shape-memory behavior of segmented polyurethanes with an amorphous reversible phase: the effect of block length and content, *Journal of Polymer Science B: Polymer Physics*, 38, 2652–2657.

Kobayashi K and Hayashi S (1992), Woven fabric made of shape memory polymer, US Patent No. 5128197, 7 July.

Lendlein A and Kelch S (2002), Shape-memory polymer, *Angewandte Chemie*, International Edition, 41, 2034–2057.

Lendlein A and Robert L (2002), Biodegradable, elastic shape-memory polymers for potential biomedical applications, *Science*, 296, 1673–1676.

Li Y and Zhu Q (2004), A model of heat and moisture transfer in porous textiles with phase change materials, *Textile Research Journal*, 74, 5, 447–457.

Lin J R and Chen L W (1998), Study on shape-memory behavior of polyether-based polyurethanes. I. Influence of the hard-segment content, *Journal of Applied Polymer Science*, 69, 1563–1574.

Lomax G R (1990), Hydrophilic polyurethane coatings, *Journal of Coated Fabrics*, 20, 88–107.

Marji Graf, *Infrared Technology*, http://www.etechnologypad.com/info/Technology/Infrared-Technology.html

Mathur G N, Hansraj and Kasturia N (1997), Protective clothing for extreme cold regions, *Indian Journal of Fibre and Textile Research*, 22, 292–296.

Mondal S and Hu J L (2006), Structural characterization and mass transfer properties of nonporous-segmented polyurethane membrane: influence of the hydrophilic segment content and soft segment melting temperature, *Journal of Membrane Science*, 276, 1–2, 16–22.

Monkman G J (2001), Advances in shape memory polymer actuation, *Mechatronics*, 10, 489–498.

Morris J V (1975), Developments in cold weather clothing, *Annals of Occupational Hygiene*, 17, 279–294.

Mukhopadhyay A and Vinay Kumar Midha (2008), A review on designing the waterproof breathable fabrics. Part 1: Fundamental principles and designing aspects of breathable fabrics, *Journal of Industrial Textiles*, 37, 3, 225–262.

Nihal S and Emel O (2007), The manufacture of microencapsulated phase change materials suitable for the design of thermally enhanced fabrics, *Thermochimica Acta*, 452, 149–160.

Pause B (2000), Interactive thermal insulating system having a layer treated with a coating of energy absorbing phase change material adjacent to a layer of fibres containing energy absorbing phase change material, US Patent, 20 June.

Pause B (2001), Air-conditioning of automotive interiors with phase change materials, *11th Techtextil Symposium*, No. 504, April, Frankfurt.

Ross G, Zhang H and Valentine C (2004), Nylon monofilaments and process for preparing nylon monofilaments, US Patent No. 10/860,229.

Ross J H (2004), Heat transfer characteristics of flight jacket materials, *Fire and Materials*, 4, 3, 144–148.

Sawhney A P S, Condon B, Singh K V, Pang S S, Li G and Hui D (2008), Modern applications of nanotechnology in textiles, *Textile Research Journal*, 78, 731.

Shannon D (2004), *Silica Derived Membranes for Liquid Separation*, Ph.D. Thesis, Department of Chemical Engineering, The University of Queensland, 26 May, 8.

Shim H, McCullough E A and Jones B W (2001), Using phase change materials in clothing, *Textile Research Journal*, 71, 6, 495–502.

Subrata M (2006), *Studies of structure and water vapour transport properties of shape memory segmented polyurethanes for breathable textiles*, Ph.D. Thesis, HK Polytechnic University, Hong Kong.

Szymocha K and Sudol T (2003), Heated clothing for use in cold weather and cold climate regions. US Patent No. 6550471.

Tailorable Insulation Materials (http://www.dodsbir.net/sitis/archives).

Takahashi T, Hayashi N and Hayashi S (1996), Structure and properties of shape-memory polyurethane block copolymers, *Journal of Applied Polymer Scence*, 60, 1061–1069.

Tobushi H, Hara H, Yamada E and Hayashi S (1996), Thermomechanical properties in a thin film of shape memory polymer of polyurethane series. *Smart Material and Structure*, 5, 483–491

Tobushi H, Hashimoto T and Ito N (1998), Shape fixity and shape recovery in a film of shape memory polymer of polyurethane series, *Journal of Intelligent Material Systems and Structures*, 9, 127–136.

Ukponmwan J O (1993), The thermal insulation properties of fabrics, *Textile Progress*, 24, 4, 1–54.

Vijayaraaghavan N N and Gopalakrishnan D (2007), Application of PCM to textiles, *Indian Textile Journal*, January, 95–99.

Wang S X, Li Y, Hu J Y, Hiromi Tokura and Song Q W (2006), Effect of phase-change material on energy consumption of intelligent thermal-protective clothing, *Polymer Testing*, 25, 580–587.

Wei Z G, Sandstro R and Miyazaki S (1998), Shape-memory materials and hybrid composites for smart systems, *Journal of Materials Science*, 33, 3743–3762.

Ying Bo-an, Kwok Y L, Li Y, Yeung C Y and Song Q W (2004), Thermal regulating functional performance of PCM garments, *International Journal of Clothing Science and Technology*, 16, 1/2, 84–96.

Zafar J (2008), Recent and future developments in health care smart garments, *Pakistan Textile Journal*, Jan.

Zalba B, Marin J, Cabeza C F and Mehling H (2003), Review on thermal energy storage with phase change: materials, heat transfer analysis and applications, *Applied Thermal Engineering*, 23, 251–283.

6
Biomimetics and the design of outdoor clothing

V. KAPSALI, University of the Arts London, UK

Abstract: Biomimetics is the transfer of technology from biology into the man-made world; this chapter focuses on applications specific to outdoor clothing. An introduction to the discipline along with some key developments is followed by an outline of the requirements of clothing performance specifically designed for protection in cold outdoor conditions and examples of biomimetic technology that offers such functionality.

Key words: biomimetics, protective clothing, adaptive textiles, physiological comfort.

6.1   Introduction

During the 1930s, Otto Schmitt coined the term *biomimetics* in his doctoral thesis, to describe an electronic feedback circuit he designed to function in a similar way to neural networks, this invention later became known as the *Schmitt Trigger*. Over the coming years several synonyms such as *bionics*, *biomimesis*, *biomimicry*, *biognosis* cropped up in various parts of the world to describe developments inspired by the functional aspects of biological structures. Biomimetics is a compound word of Greek origin: bio- meaning life and -mimesis meaning to copy: the outcome is the interpolation of natural mechanisms and structures into engineering design. The cross-disciplinary nature of the field has established a platform for technology transfer that transcends subject specific 'cultural' barriers such as technical language thus functioning as a vehicle for ideas from biology to find useful applications in other fields.

Several historical examples are believed to be inspired by natural mechanisms, such as the design of the Eiffel Tower and the glass roof of the Crystal Palace that housed London's Great Exhibition of the 1850s. The first textile innovation linked to Biomimetics was the invention of the dry adhesive tape known as Velcro that was inspired by the hook mechanism found on the surface of burrs which enables them to attach onto animal fur in a way that is difficult to remove. However, such examples are probably serendipitous; biomimetic innovations today are the product of systematic study, with wide-reaching applications.

This chapter focuses on the role of biomimetics in outdoor clothing applications. An introduction to the discipline along with some key developments is followed by an outline of the basic requirements of clothing functionality specifically designed for protection in cold outdoor conditions. Opportunities for biomimetics are illustrated using some current developments that highlight the potential for innovation.

6.2 Inspiration from nature

Biology has always been a rich source of visual and aesthetic inspiration for the design of clothing, common to every culture and era. There are countless examples of motifs such as flowers, insects and various animals, incorporated into the design of textiles either through structural patterning (e.g. jacquard weave), print or embroidery. Elaborate floral motifs expressed in print, embroidery and embellished with precious stones is a key trademark of London Fashion Week designer Mathew Williamson. The replication of animal markings such as the 'leopard print' has become a trademark for Italian fashion house Dolce & Gabbana.

Man has sourced materials for clothing since prehistoric times and gradually developed sophisticated technology that enabled survival in the most extreme conditions. Ancient Inuit hunters for example, used the protective functionalities of seal and bird skins in their clothing systems to create the highly insulating and water resistant clothing necessary for survival in the freezing conditions of their natural habitat (Ammitzboll, Bencard *et al.* 1991).

The desire to transfer various properties from biological materials to the textile sector is not entirely novel. In fact attempts to imitate the functionality of silk have led to great turning points in the history of textile technology. The strength and lustre of the silk fibre was the object of man's obsession for centuries. Efforts to synthesize a material that imitates these properties date as far back as 3000 BC in China. It was not until the early twentieth century that these efforts were successful and the first man-made fibre, Rayon, was mass produced. Although rayon imitated the lustre of silk, it lacked its strength (Cook 1984). It was not until a few years later that the industry was revolutionized with the mass production of synthetic fibres.

The first synthetic fibre was commercially produced in 1939 by E.I. Du Pont de Nemours and Company. Following an extensive research programme, the company synthesized a polyamide fibre they branded Nylon. Nylon fibres were long, smooth and offered a silk-like handle to textiles but with much superior tensile strength. (Handley 1999).

By the 1950s more synthetic fibres were commercially produced such as polyester and acrylic. Unlike natural and regenerated fibres, they absorb only nominal quantities of moisture (Cook 1984) creating quick drying textiles that require no ironing. Crisis struck the synthetic fibre industry in the 1970s as consumers rejected products made from these fibres and sales plummeted. The fibres caused a range of new discomfort sensations such as cling, clammy, static and various skin irritations (Kemp 1971). The hydrophobic nature of synthetic materials that created a revolution in the 1950s was the cause of their demise twenty years later as consumers began to favour the properties of natural fibres over their synthetic counterparts.

The 1970s was a very important time for the textile industry. During this decade great losses were made in the man-made fibre sector which drove researchers to investigate the causes of comfort/discomfort and technologists to find ways of manipulating the performance of synthetic materials to imitate the properties of natural fibres. By the end of the 20th century synthetic fibres had made a total recovery in the clothing sector and in some cases, synthetic textiles could command higher prices than those made of natural fibres (Handley 1999).

The application of biomimetic technologies in the clothing sector has resulted in the introduction of new functionalities to garments such as performance enhancement. Speedo pioneered a range of swimsuits branded FastSkin that use a textile system with small ridges designed into the surface texture of the textile, similar to the surface morphology of shark's skin. Sharks can swim remarkably fast for their size and shape, the simple mechanism in the texture of the animal's skin is believed to reduce drag thus increasing speed. Speedo claims that the FastSkin product can offer this functionality to a swimmer. Although there is no scientific evidence proving the textile system's functionality, the products have become widely accepted by athletes internationally.

Biomimetic technologies also offer ideas for new methods of implementing existing processes such as stain/soil resistance. The Lotus effect, for instance, was inspired by the water-repellent and self-cleaning properties of the lotus leaf. The functionality was found to be due to a layer of epicuticular wax crystals that covered the sculptured surface of the leaf (Barthlott and Neinhuis 1997). Originally this mechanism was interpreted into technology adopted by the paint industry to produce a paint that would self-clean every time it rained. Recently, this has found application in the textile sector as a finish that delivers water, stain and dirt resistant properties to clothing without affecting the appearance or handle of the cloth. Although there are conventional finishing processes that achieve this effect (e.g. Teflon coating), they use highly toxic chemicals, whereas the methods (e.g. plasma treatment) used to create the Lotus effect are low energy thus offering an environmentally sound alternative (Slater 2003).

6.2.1 Biomimetic principles and methods

Our knowledge of evolution depicts the natural environment as a testing ground for design and development in nature. Selective pressures are exerted onto the organisms of an ecosystem, for example, through limited reserve of nutrients vital to sustain life. In order to survive, plants and animals evolve mechanisms and structures that enable them to make optimal use of minimal resources

(Beukers and Hinte 1998; Vincent, Bogatyreva et al. 2006), thus successful 'design' survives and bad 'design' disappears.

The link between energy/resource in nature and cost in engineering is believed to be a common agenda between design in biology and engineering. The optimization of limited resources in biological materials and structures was originally interpreted as opportunities to develop clever yet cheap materials and structures (Beukers and Hinte 1998). This notion of energy = money evolved to encompass the greater cost to the environment and the consumption of natural resources in the construction of man-made products (Benyus 1997). Biomimetic scientists today believe nature is a rich source of clever and sustainable design that can offer new properties as well as clean methods for existing processes.

Functionality through design

Engineers rely on material properties to deliver desired functions such as stiffness, strength or elasticity to structures. Functionality in biological materials is incorporated into the structure through the design and distribution of basic building blocks (Benyus 1997; Beukers and Hinte 1998). Whenever a new property is required in the man-made world, a new material is usually synthesized; as a result there are over 300 man-made polymers currently available. There are only two polymers in the natural world – protein and polysaccharide – whose structural variations offer a vast range of properties superior to their man-made counterparts (Vincent, Bogatyreva et al. 2006). Insect cuticle, for instance is made from chitin and protein and can demonstrate a host of mechanical properties, it can be stiff or flexible, opaque or translucent, depending on variations in the assembly of the polymer (Vincent 1982).

Conditions of manufacture

The production of conventional man-made materials and structures is generally a 'costly' procedure in terms of energy consumption and resource waste. Extreme temperatures, pressures and toxic chemicals are often required during production. Man-made fibres are a prime example of 'high-energy', 'high-waste' production processes. Natural materials require low energy conditions for their 'production' and normal temperatures and pressures no different to those necessary for life. There is also no need for harmful chemicals; usually water is adequate for the creation and growth of structures (Benyus 1997).

Multifunctional/adaptive structures

Biological materials and structures are designed to perform multiple functions as a way of maximizing the use of resources. Key properties are introduced into a material or structure through clever design and application of available resources. The texture of a surface, for instance, is engineered to provide self-cleaning properties to a plant or animal. Functional surfaces often occur on the surface of leaves, to protect the plant from contamination; the Lotus mentioned earlier is a plant well known for this property. Several species of insect also employ the same principle to render their wings hydrophobic. A similar mechanism is found on the shell of the dung beetle providing anti-adhesion and anti-wear properties (Nagaraja and Yao 2007).

Multifunctional textiles are currently created by bonding layers of materials with different properties together to form a composite textile. Membranes such as Goretex and Sympatex are laminated with knitted or woven textiles to create multifunctional systems whose properties amount to the sum of the individual properties of each component. Often the outcome is a textile system predominantly used in outer shell clothing that offers breathability, wind and water resistance.

Adaptive

Biological materials are created by the organism and are in constant flux with the conditions in the environment, whereas their man-made counterparts are fabricated by external efforts (Hollington 2007). The design brief for man-made structures is predetermined and aimed to satisfy a specific set of requirements that remains unaltered during the useful life of a product. The structure of biological materials is defined partly by DNA and partly by the environment – 'Nature and Nurture'. The survival of plants and animals depends on the ability of their structures to adapt to the changing demands of the environment; some key properties are self-assembly, reproduction, self-repair and redistribution of vital resources (Beukers and Hinte 1998).

6.2.2 Development models

Biomimetics is a relatively new field with a short history; as a result there is currently no standard methodological approach to the transfer of technology. A popular model is one adopted by the Biomimetic Guild that illustrates a linear progression of ideas from biology to engineering (Gester 2007). According to this method Biomimetic developments can follow one of two directions: bottom up or top down (Fig. 6.1).

The bottom-up approach (Fig. 6.1(a)) denotes a development or innovation that has been instigated by a single biologist or a team. The biologist(s) identify an interesting mechanism in nature they believe would potentially have a beneficial application in industry. The property is studied to create an understanding of the operational aspects of the mechanism. The principles are abstracted into a model that is taken up by a team of engineers who identify methods of interpreting the technology into useful man-made products.

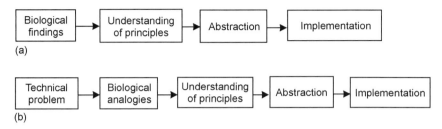

6.1 Biomimetic development model: (a) bottom up, (b) top down (source: Gester, 2007).

The top-down process (Fig. 6.1(b)) is initiated by industry need or a gap identified in the market. This need or gap is defined in terms of a technical problem for which analogies are sought in biology. Once suitable paradigms are identified a process similar to the bottom-up approach is followed where a team of biologists study the mechanism(s), identify how they work and pass on the information to engineers who interpret the ideas into solutions to the technical problem.

This model succeeds in creating a simple illustration that is reflective of the technology transfer process among many biomimetic teams today. However, it is limited by the fact that both bottom-up and top-down directions rely on a serendipitous and non-systematic approach to problem solving (Vincent and Mann 2002). An alternative model, currently under development at the University of Bath's Centre for Biomimetic and Natural Technologies, has adopted TRIZ (Russian acronym for Theory of Inventive Problem Solving) framework, a verified methodology used by engineers for decades that offers a systematic approach to the definition and solution of problems. Researchers at Bath University are currently developing this tool (Vincent, Bogatyreva *et al.* 2006).

6.3 Biological paradigms for outdoor clothing

There are a number of opportunities for biomimetics with specific applications in outdoor clothing and the factors affecting the physiological comfort of the wearer. An overview of garment requirements is followed by developments in the biomimetic sector that can offer innovation to the design and performance of clothing for the outdoors.

6.3.1 Overview of clothing system requirements

A clothing system can be made of one or more layers (base, mid, external shells) extending from the surface of the skin to the face of the outer garment creating a portable environment (Watkins 1995) of fibrous material and air. The role of the system is to satisfy the physiological and psychological needs necessary for the individual to function within the physical and social environment. The dynamic

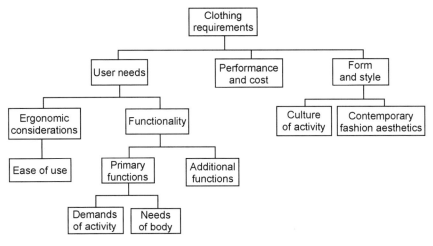

6.2 Clothing requirements (source: Black, Kapsali *et al.*, 2005).

micro-climate created within the system controls the physiological comfort of the wearer and is influenced by external factors (climate, activity of wearer, etc.) and internal factors (fibre properties, textile structure, design of garment, etc.) (Black, Kapsali *et al.* 2005).

The end use of the system dictates whether emphasis during the design and development is placed on the physiological or psychological functionality of the clothing, however all garments must satisfy some basic requirements (Fig. 6.2). The form and style of each item within a clothing system must suit the culture of the activity and meet basic contemporary design aesthetics (Black, Kapsali *et al.*, 2005). Clothing must also satisfy basic ergonomic considerations to avoid inhibiting general activities and functions. It is vital that each item of clothing is easy to use (adding and removing garments) and does not restrict movement. The product also needs to balance performance with cost; successful design convinces the consumer that its price is suitable to the performance of the garment.

Additional functional requirements represent possible future demands from clothing enabled by new and emerging technologies. The advancing fields of bio-, nano-, electro- textiles are introducing new properties to apparel that could supplement the functionality of conventional clothing to meet changing needs of the consumer's lifestyle. Remote connectivity, for instance, enabled by innovations in wearable electronics offers clothing able to take on additional roles currently performed by devices such as mobile phones, PDAs and satellite tracking devices.

In the context of clothing engineered for protection in outdoor activities, the most important factor in the determination of the functional profile of the system is the external conditions that affect the physiology of the user, i.e., environmental temperature, moisture concentration in the atmosphere, weather conditions (rain, snow, sun, wind). Protective functionalities associated with psychological hazards and other potential hazards such as microbes, chemicals, physical impact, etc., are not exclusive to cold weather clothing and will not be discussed in this chapter for purposes of simplicity.

From the perspective of the wearer, the preservation and protection of physiological comfort is paramount and relies on the flexibility of the clothing system to accommodate changes in the system's microclimate as well as protection from external factors (wind, rain, snow). Garments engineered for protection in cold weather must retain enough heat within the system to ensure the wearer's comfort while at the same time manage the penetration of water and cold air from the external environment and moisture produced by the individual.

The design and selection of materials composing the clothing system are the key factors in the management of microclimate conditions. The clothing system's permeability to heat, moisture and air can be controlled to a certain extent by the properties inherent in the materials used for the composition of the garment and various design features.

Design features

Collars, cuffs and belts are some structural features that enable the management of a garment's insulation properties by trapping volumes of air. Layering is another technique used to vary the insulation properties of a system; more layers equate to more trapped air thus more insulation. This technique is the most flexible for accommodating any changes in external conditions or wearer activity. The individual assesses the level of insulation required to maintain his/her levels of physiological comfort and adds or removes layers accordingly.

It is well known that saturated air trapped in the microclimate is the key factor causing physiological discomfort. This is often the problem with clothing that offers high insulation. Design features such as zips and openings enable the saturated air to be replenished, a method known as periodic ventilation (Ruckman, Murray *et al.* 1999). A prime example is the design of traditional Inuit clothing that was made from highly insulating furs and feather pelts that offered protection from the extreme conditions of their natural habitat and ensured their survival. Although the materials used to construct the garments offered the necessary insulation properties, it was clever design of the upper garment that enabled extremely efficient ventilation to accommodate changes in the individual's activity (Ammitzboll, Bencard *et al.* 1991; Humphries 1996). The traditional Inuit hood was closely fitted around the face with no front opening and air was trapped in the system at the chin and waist. The sleeves on the parka were long enough to cover the hands and fitted to prevent cold wind from penetrating the system. During periods of activity when the microclimate was threatened by saturation, ventilation was achieved by pulling the garment forward at the front of the throat, pushing the hood back or loosening the closure at the waist (Humphries 1996).

Textile properties

There are several methods used to manipulate the permeability (heat, air, moisture) properties of a textile structure; the shape or cross-section of a man-made fibre, for instance, can be engineered to trap air (i.e. hollow fibres) or to introduce a crimp along the length of the fibre (imitating the morphology of wool fibres) to increase the volume of air trapped when applied to a textile. A fibre can also be designed to maximize the rate of moisture evaporation by increasing the surface area. A popular example is the Coolmax fibre whose cross-section is often said to represent Micky Mouse ears. This particular configuration is calculated to increase surface area by 30% compared to a standard circular cross-section of the same diameter. Cellular and double knit/weave structures can maximize the insulation properties of the textile while the tightness or openness of the structure affects air permeability while wind and water resistance can be managed by the incorporation of specialist membranes into the textile system. These are a few examples indicative of the sector and by no means exhaustive.

6.3.2 Opportunities for biomimetics in the design of cold weather apparel

Conventional clothing systems require the wearer to initiate any adaptation to heat, air and moisture permeability, this method poses several limitations such as the ease involved in the removal and storage of necessary layers of clothing during wear and the individual's ability to sense the onset of physiological discomfort. It is often the case that the cue for adaptation is the discomfort sensation itself; this can be prevented if the system's properties are adjusted in response to initial changes in the microclimate before the temperature and moisture concentration reaches a point where discomfort occurs. A clothing system with the ability to sense changes in the microclimate and alter its properties without the intervention of the wearer would be ideal. Materials and structures in biology rely on adaptive mechanisms that respond to changes in environmental moisture and temperature for survival.

Adaptive insulation

During the 1990s a team of British biomimetic investigators at the University of Reading became fascinated with penguins and their ability to survive in extreme conditions. Penguins must withstand extreme cold for up to 120 days without food and then be able to dive up to 50 m into freezing waters in order to feed. The team found that the secret behind their survival was the structure of their coat (feather and skin combination) and its ability to switch from an insulating barrier to a waterproof skin.

When necessary the penguin coat provides highly efficient insulation that minimizes heat loss through radiation and convection, with structural properties that function as an excellent wind barrier eliminating heat loss though convection. Yet when the animal needs to dive for food, the coat transforms into a smooth and waterproof skin eliminating any trapped air. This switch in functionality is achieved by a muscle attached to the shaft of the feather, when the muscle is locked down the coat becomes a water-tight barrier and when released the coat transforms itself into a thick air filled windproof coat (Dawson, Vincent *et al.* 1999).

The feathers in a penguin's coat are packed evenly over the animal's body averaging between 30 and 40 per cm^2. Dawson, Vincent *et al.* (1999) identified that the mechanism responsible for the remarkable insulation properties are found in the afterfeather (Fig. 6.3). The afterfeather consists of approximately 47 barbs averaging 24 mm in length. Each barb is covered with around 1250 barbules that are about 335 μm in length. This structure creates air spaces of

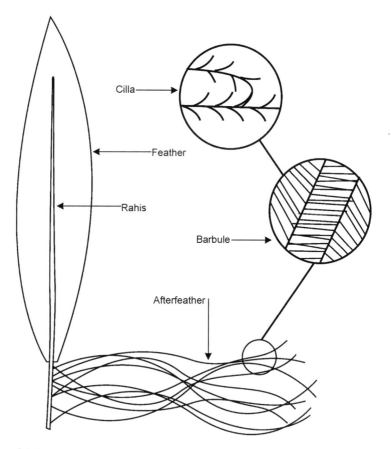

6.3 Penguin feather (source: Dawson, Vincent *et al.*, 1999).

around 50 μm in diameter, which provide an enormous surface for trapping air thus creating a structure capable of providing such high levels of insulation.

A key factor in the success of the adaptive mechanism is the ability to recreate a uniform division of air space every time the coat's functionality alters from waterproof to high insulation. The mechanism that enables this is found on the surface of the barbules; Dawson, Vincent *et al.* (1999) noticed that tiny hairs known as cilia (Fig. 6.3) covered the barbules that function as a stick slip mechanism to keep the barbules entangled and maintain the movement in directions relative to one another to ensure uniformity in creation of air pockets during the coat's function change.

The insulation properties of the penguin coat adapt by varying the volume of air trapped within the system by drawing the feather towards the skin when the waterproof functionality is required and releasing it when the penguin needs to be kept warm. Attempts to interpret this functionality into garments has led to the creation of an experimental textile system referred to as *variable geometry*; some development has been carried out by N. & M.A. Saville Associates (Figs 6.4 and 6.5).

The structure consists of two layers of fabric, which are joined together by strips of textile at an angle to the plane of the two fabrics. By skewing the two parallel layers the volume of air between them reduces, this results in the reduction of thermal resistance. The idea was used in the design of military uniform systems that can be adapted to function in both extreme cold and hot conditions.

Adaptive insulation has recently been commercialised by Gore & Associates who have created an ePTFE membrane and polyester structure (76%PE

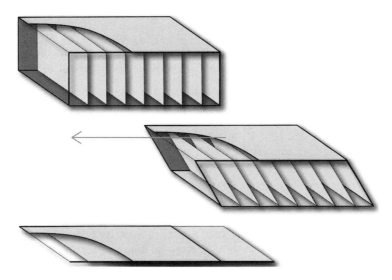

6.4 Variable geometry textile.

124　Textiles for cold weather apparel

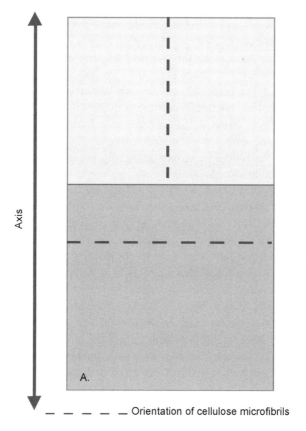

6.5 Bilayer configuration.

24%PTFE) used as a garment insert under the brand name Airvantage. This product allows the user to inflate/deflate the jacket thus controlling the necessary amount of air for the provision of adequate thermal resistance.

Solar radiation for advanced thermal protection

The brilliant white coat of the polar bear provides effective camouflage in his arctic habitat and conceals him from his prey. The pelt is an extremely effective mechanism that supports the animal's thermal regulation by insulating the animal for extreme cold conditions.

Grojean, Sousa *et al.* (1980) studied the mechanism and found that the pelt itself consists of thick hairs approximately 100–150 μm in diameter and 6–7 cm in length, a dense layer of fur about 1 cm long with fine fibres (25–75 μm), these are attached to a thin layer of black skin approximately 1 mm thick. There has been little study on the structure of the polar bear fur but it is believed to trap air and prevent heat loss in a similar way to the pelts of other animals; however, the absorption and conversion of solar energy is attributed to the morphology of the longer, thicker hairs.

The hairs are hollow in structure along their length and taper to a solid edge at their tip. Although they have a smooth external surface, the core is very rough while the hairs themselves contain no pigmentation. The air pockets created in the hair's core offer additional insulation while Grojean found that the light energy from the sun is drawn into the core of the hairs and the anomalous surface scatters the energy downwards toward the skin. This 'solar lumination' system absorbs UV light from the sun to support the animal's temperature regulation. As a result polar bears require 12–25% less effort to maintain a comfortable temperature (Grojean, Sousa *et al.* 1980).

Hollow fibres are extensively used to engineer highly insulating garments, sleeping bags and other products for cold weather protection. The concept of using textile systems to reflect heat to or from the body has recently been implemented into textiles using thin aluminium films or textiles impregnated with ceramic particles. Additional heat can be introduced into a clothing system via heating elements in electronic textile configurations and paraffin filled phase-change microcapsules (PCMs by Outlast); both products are limited to the provision of a finite amount of heat energy dictated either by the power source (electronic textiles) or capacity to store heat energy (PCMs). Although there are no existing developments using polar bear hair as a model for technology transfer, advances in optical fibres could enable the interpretation of solar illumination technology into clothing and other systems to provide additional 'free' heat in extreme cold environments.

Smart microclimate ventilation

The replenishment of saturated air is a key factor in the maintenance of physiological comfort. Open textile structures such as loose weaves and knits allow the movement of air between the microclimate and the environment. Although these structures provide effective ventilation, heat is lost to the environment creating a system that provides poor insulation. In the case of clothing for cold weather protection, it is essential that textiles prevent cold air from penetrating the system. Currently, ventilation of cold weather clothing can only be achieved manually with the aid of design features such as those discussed earlier.

Membranes such as Diaplex by Mitsubishi Industries and C-change by Schoeller are made from a type of polyurethane that is claimed to alter its porosity to moisture at different temperatures. These products are ideally situated on external layer garments as they can respond to changes in external temperatures; cold environments require less porosity from the clothing while an individual would benefit from a more porous garment in warmer conditions. Temperature changes in the clothing microclimate are not as representative of comfort sensation as moisture concentration (Li 2005) especially during higher levels of activity. These products are therefore ideal for low-level outdoor activity.

Several plants use environmental moisture conditions to trigger seed dispersal. These hygroscopic mechanisms hold some ideas for creating textile structures that can alter their permeability to air in response to changes in humidity. Work conducted by Dawson *et al.* in 1997 as part of the Defense Clothing and Textiles Agency (DCTA) studied the opening and closing mechanism of the pinecone that is triggered by changes in environmental moisture concentration.

Dawson, Vincent *et al.* (1997) found that the bract is composed of two types of wood one type (active tissue) demonstrated great dimensional swelling when exposed to moisture while the other remained unaffected. Although both types of wood were constructed from cellulose, the microfibrils in the cell wall of the active tissue type are positioned at 90° to the axis of the bract (Fig. 6.5) while the microfibrils in the other type are orientated more or less parallel to the central axis (Harlow, Coté *et al.* 1964). The coefficient of hygroscopic expansion was found to be three times greater in the active cells than that of the non-swelling tissue (Dawson, Vincent *et al.* 1997) which explains the greater longitudinal swelling shown by the active tissue. Dawson *et al.* noticed that the mechanism operated in an analogous method to a bimetallic strip where the switch was ambient moisture content instead of changes in electrical current (Dawson, Vincent *et al.* 1997).

A textile system was developed using a light-weight synthetic woven structure laminated onto a non-porous membrane such as those marketed under the Sympatex brand. Small u-shaped perforations were cut into the surface of the composite textile demonstrated in Fig. 6.6(a). An increase in relative humidity caused the loose sections of fabric created by the incisions to curl back (Fig. 6.6(b)) thus increasing the system's permeability to air. When the microclimate becomes damp the textile alters its porosity to allow the renewal of saturated air, the structure resumes its original properties when conditions near the skin are dry and ventilation is no longer required. Nike recently implemented a similar concept into a clothing system; the technology was incorporated into a tennis dress worn by Maria Sharapova at the 2006 US Open. The garment featured a fish scale pattern on the back panel that opened up as the athlete perspired to increase local ventilation and maintain the wearers comfort.[1]

Researchers at the University of Bath have developed a prototype textile based on the work of Dawson that applies the principle to a yarn able to increase the porosity of a textile structure in damp conditions and reduce permeability when dry. This technology is currently being developed for applications in the commercial sector.

---

1. US Patent Application 20050208860.

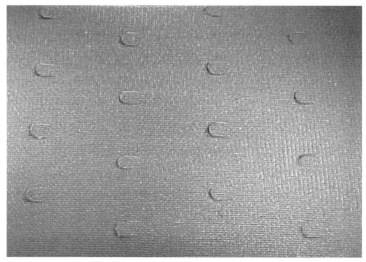

(a) Dry conditions

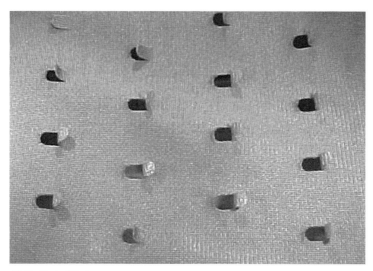

(b) Damp conditions

6.6 Adaptive ventilation textile: (a) dry conditions, (b) damp conditions.

Functional surfaces

Nature uses surface texture as a tool to introduce important functionalities that protect the organism from contamination, impact, etc. The self-cleaning properties of the lotus leaf, discussed earlier, have been interpreted into paints for building exteriors, finishes for textiles and more recently a coating for window glass panes. Superhydrophobic surfaces are used in textiles to protect garments from dirt and liquid contamination as well as penetration from water. This functionality can be extended to benefit cold weather apparel, sleeping bags, tents, etc., especially in rain or snow; droplets of water simply roll off the surface of the article preventing the structure from moisture penetration while removing dirt or other contaminants in their path.

6.4 Future trends

The boundary between wearer and clothing is undergoing a reform fuelled by advances in material science and technologies that could have been extracted from science fiction. Shape-memory alloys are used to alter the shape of a garment in response to heat or electrical currents. Microcapsules filled with various substances such as essential oils and synthetic wax can be incorporated into clothing through foams and fibres. Washable electronic circuitry is integrated into clothing through textiles that enable the system to operate as a mobile phone, MP3 player, GPRS etc. Some futurologists even predict that garments of the future will imitate the behaviour of living organisms able to adapt, self-heal and even reproduce (Tastuya and Glyn 1997).

For new innovations to integrate successfully into the clothing sector, technology push needs to be met by consumer pull. There is little value in introducing functions to clothing that the consumer is not ready for or indeed do not need. It is possible that consumer expectation from apparel will evolve; individuals will require their clothing systems to sense and respond to changes whether they are physiological or psychological.

Traditional clothing for protection in cold weather requires the wearer to manage the functional profile of the system either by adding/removing layers or by using openings designed into the garment, this is often not practical or initiated once discomfort sensations are well established, which is too late. This chapter has examined a range of biological paradigms that would improve the existing 'state of the art' in the cold-weather sector, it should be noted that the technology discussed is not limited to garments and can be adapted to accessories (shoes, gloves, etc.) and other apparatus (i.e. shelter, portable storage systems). Although some of the technologies discussed are at prototype stage and require significant development before they can enter the commercial sector, they offer a glimpse of potential properties of cold-weather clothing in the future and highlight the significance of adopting ideas from nature.

6.5 Sources of further information and advice

- Centre for Biomimetic and Natural Technologies at Bath University
  http://www.bath.ac.uk/mech-eng/Biomimetics/
- Biomimetics at Reading University
  http://www.rdg.ac.uk/Biomim/

- Online resource from the US-based Biomimicry Institute
  http://www.biomimicry.net/indexbiomimicryexp.htm
- German Biomimetic network
  http://www.biokon.net/index.shtml
- Biomaterials Network
  http://www.biomat.net/

6.6   References

Ammitzboll, T., M. Bencard, *et al.* (1991). *Clothing*. London, British Museum Publications.

Barthlott, W. and C. Neinhuis (1997). 'Characterisation and distribution of water-repellent, self-cleaning plant surfaces'. *Annals of Botany* **79**: 667–677.

Benyus, J. M. (1997). *Biomimicry: Innovation Inspired by Nature*. New York, William Morrow & Company.

Beukers, A. and E. V. Hinte (1998). 'Smart by Nature'. *Lightness; the Inevitable Renaissance of Minimum Energy Structures*. Rotterdam, 010 Publishers.

Black, S., V. Kapsali, J. Bougourd and F. Geesin (2005). 'Fashion and function: factors affecting the design and use of proective clothing'. *Textiles for Protection*. R. A. Scott. Cambridge, Woodhead Publishing Limited.

Cook, G. J. (1984). *Handbook of Textile Fibres: Natural Fibres*. Wiltshire, Merrow Publishing Co.

Dawson, C., J. F. V. Vincent, *et al.* (1997). 'How pine cones open'. *Nature* **390**: 668.

Dawson, C., J. F. V. Vincent, *et al.* (1999). 'Heat transfer through penguin feathers'. *Journal of Theoretical Biology* **199**(3): 291–295.

Gester, M. (2007). 'Integrating biomimetics into product development'. *Biomimetics: Strategies for Product Design Inspired by Nature*, DTI Global Watch Mission Report: 38–41.

Grojean, R. E., J. A. Sousa, *et al.* (1980). 'Utilization of solar radiation by polar animals: an optical model for pelts'. *Applied Optics* **19**(3): 339–346.

Handley, S. (1999). *Nylon, the Man-made Fashion Revolution*, London, Bloomsbury.

Harlow, W. M. C., W. A. J. Coté and A. C. Day (1964). 'The opening mechanism of pine cone scales'. *Journal of Forestry* **62**, 538–540.

Hollington, G. (2007). 'Biomimetics and product design'. *Biomimetics: Strategies for Product Design Inspired by Nature*, DTI Global Watch Mission Report.

Humphries, M. (1996). *Fabric Reference*. London, Prentice-Hall International.

Kemp, S. (1971). 'The consumer's requirements for comfort', in *Textiles for Comfort*. Manchester, Shirley Institute, New Century Hall.

Li, Y. (2005). 'Perceptions of temperature, moisture and comfort in clothing during environmental transients'. *Ergonomics* **48**(3): 234–248.

Nagaraja, P. and D. Yao (2007). 'Rapid pattern transfer of biomimetic surface structures ontho thermoplastic polymers'. *Materials Science and Engineering: C* **27**(4): 794–797.

Ruckman, J. E., R. Murray, *et al.* (1999). 'Engineering of clothing systems for improved thermophyhysiological comfort: the effect of openings'. *International Journal of Clothing Science and Technology* **11**(1): 37–52.

Slater, K. (2003). *Environmental Impact of Textiles*, Cambridge, Woodhead Publishing Limited.

Tastuya, H. and P. Glyn (1997). *New Fibres*. Cambridge, Woodhead Publishing Limited.
Vincent, J. F. V. (1982). *Structural Biomaterials*. Princeton, NJ, Princeton University Press.
Vincent, J. F. V. and D. L. Mann (2002). 'Systematic technology transfer from biology to engineering'. *Philosophical Transactions of the Royal Society* **360**(1791): 159–173.
Vincent, J. F. V., O. A. Bogatyreva, *et al.* (2006). 'Biomimetics: its practice and theory'. *Journal of the Royal Society Interface* **3**: 471–482.
Watkins, S. M. (1995). *Clothing: The Portable Environment*. Iowa, Iowa State University Press.

7
Designing for ventilation in cold weather apparel

N. GHADDAR and K. GHALI, American University of Beirut, Lebanon

Abstract: The fundamental criterion for clothing comfort of cold weather garments is dictated by insulation and permeability (breathability) to maintain a warm dry skin for the active wearer. This chapter reviews water vapour and moisture transport from the skin through clothing to the environment in cold weather and its dependence on clothing thermal and evaporative resistances and layering in cold clothing designs. The chapter then describes the mechanisms and physical model of microclimate (skin-adjacent air layer) ventilation in cold weather apparel designed for active people. It emphasizes factors affecting enhanced ventilation through controllable clothing apertures, clothing physical properties, size of microclimate air layer, and human motion including frequency and swing of limbs. Finally, the chapter forwards recommendations for design of versatile/ adjustable clothing for highly active people and future research trends to improve ensemble design and develop convenient tools for assessing clothing ventilation and performance.

Key words: clothing ventilation in cold weather, modelling of microclimate ventilation; versatile clothing design for active people; low temperature water vapour transport.

7.1 Introduction: importance and function of ventilation in cold weather apparel

The two main properties of cold weather apparel are insulation and breathability. In cold weather the clothing should provide the required insulation for protection against the cold and it should allow the transmission of perspired sweat from the skin to the environment. The insulative property of the cold clothing system is accomplished by the relatively thick material constituting the cold fabric and its ability to trap air. While the breathability is accomplished by the vapour permeability property of the fabric which is related to the ability of the clothing system to allow vaporized moisture to pass through the material. Those two properties are highly relevant factors in determining human thermal comfort in cold weather. The fundamental criterion for clothing comfort of cold weather garments is dictated by insulation and permeability that determines both the amount of heat and moisture that could be exchanged between the human body and the environment to maintain a warm dry skin of the wearer. The heat and moisture transfers can be obtained by performing a thermal energy balance between the wearer at a given activity level and the environmental conditions specified by temperature and wind speed.

For the cold weather apparel to be considered comfortable it should provide cold protection at low activity level by reducing as much as possible heat and moisture loss to the environment and at higher activity level it should allow better heat and moisture dissipation. However, the insulation and permeability properties are not adjustable with the varying weather conditions and with the changes of the wearer's activity involving different levels of body heat production. If one is performing work in cold weather clothing while wearing heavy insulative clothing, heat will accumulate in the body causing the skin temperature to rise due to the resistance of the clothing to heat and moisture transport between the skin and the environment. If the skin temperature is too high, sweating starts to function as an evaporative cooling system. As the moisture cannot be transferred to the outside, because it is blocked by the heavy clothing, it will condense in the clothing layers on its way to the clothing surface and ambient air. The condensed moisture within the clothing material can reduce the insulation by as much as 30–50% (Fan *et al.*, 2003).

Sweating increases heat losses through evaporation and reduces the insulation due to the wetting of insulation layers. Condensation can lead to an increase in the thermal conductivity of the insulating material, as the thermal conductivity of water is approximately 24 times that of the conductivity of the air. Moreover, moist clothing causes a strong sensation of chilling discomfort and usually results in the decomposition and deterioration of the quality of the insulating material. Therefore, the insulating material would lose its role in providing warmth in cold weather as it gains water. In addition, the heavy insulative clothing which would have been comfortable for a person performing light work or at rest will cause cooling of the skin and a sensation of cold during light work phases after becoming wet.

The main problem of foul-weather clothing is that its thermal and moisture transport properties are not adaptable with the wearer's changes in activity levels for the different tasks that the person is performing in the cold weather and they are not adaptable to changes in the environmental conditions. There is no all-purpose garment that is versatile for cold clothing, suitable for all weather conditions, and suitable for all tasks and activities performed by the wearer which could vary along the working day. Traditionally, most of the cold clothing is geared towards high insulation. It provides warmth and is capable of transporting the perspired sweat to the environment during light work but it causes excessive sweating, wetting of the clothing and discomfort during heavy work. To design a versatile garment that is adjustable to the activity of the wearer at mild and very cold conditions for the purpose of ensuring comfort during its wearing, it is important to transfer as much moisture as possible from the skin through the clothing, either in liquid or vapour form to the external surface of the clothing. The design of an all-purpose foul-weather garment could be accomplished by the following.

1. Physical modelling of the condensation phenomena of perspired sweat; this would explain the dependence of water vapour transmission through the clothing ensemble on both the thermal resistance of the clothing as well as its vapour transport properties.
2. Designing the cold weather apparel in multiple layers rather than in one thick layer; this would allow the wearer to fine tune the thermal insulation value of the clothing in accordance to work loads and environmental conditions at which the wearer is performing the load.
3. Facilitating the transmission of moisture by wearing vapour-permeable garments, providing ventilation at the cuffs or collars and deliberately providing vent holes in the garment; the vapour-permeable garments will increase vapour permeation through the clothing fabric and the vent holes will decrease vapour resistance of the garment by allowing air exchange between the air layer adjacent to the skin and the environment. The air exchange occurs without having the air pass through the thick layered clothing system. In addition, ventilation will increase convective heat loss from the skin causing lower skin temperature and therefore leading to a lower rate of respired sweat from the skin.

7.2 Water vapour transport through cold weather textiles at low temperatures

In cold weather conditions two gradients are present along the thickness of the cold weather garment: temperature and moisture. For a cool environment at low moisture content and a warm skin with higher moisture content, moisture content and temperature of air will decrease along the path from the skin to the environment. Also the local water vapour pressure and the saturation vapour pressure determined by the temperature of the air will decrease along the same path. Condensation in the cold weather garment could occur at any location when the local air vapour pressure rises above the saturation vapour pressure at that location temperature.

Location of condensation can be predicted by utilizing the saturation vapour line and water vapour pressure line (Keighley, 1985, Ruckman, 1997). Thus by plotting the water vapour pressure distribution in the cold weather apparel against its temperature variation, as shown in curve A of Fig. 7.1, and if the saturation vapour pressure corresponding to the garment temperature distribution is plotted in the same figure, as shown in curve B of Fig. 7.1, then the location of condensation can be predicted. The saturation line curve B shows the water vapour pressure corresponding to a 100% relative humidity at a specific temperature, and if the water vapour pressure at that temperature exceeds the saturation temperature, condensation will occur at that location.

134   Textiles for cold weather apparel

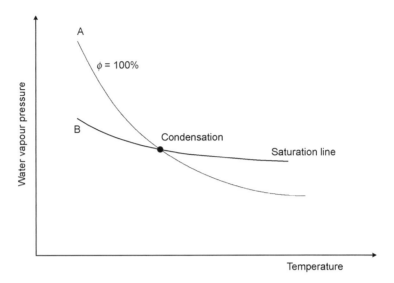

7.1 Schematic of the water vapour pressure distribution in a fibrous medium against its temperature variation (curve A) and the corresponding saturation vapour pressure distribution (curve B).

In mild cold conditions and low workloads, evaporated perspiration from human skin will escape through clothing without the occurrence of condensation since the rate of perspiration is low. At a higher activity level, the perspiration will increase to a certain level that would cause condensation to occur within the clothing system. The occurrence of condensed sweat in a clothing system is generally affected by the vapour permeability of the different fabric layers constituting the clothing ensemble, the skin vapour concentration, which is mainly related to activity, and the outside environmental temperature. If condensation occurs, liquid moisture should reach and pass through the outer layer before it accumulates and freezes to maintain the insulation value of the clothing garment. However, the outer layer that covers the cold weather apparel is usually a waterproof breathable layer that repels water to protect the garment from rain but contains micro-pores for water vapour transmission. As the presence of this protective layer will prevent the condensed sweat from reaching the clothing surface where it can evaporate, the moisture will accumulate and eventually it might freeze (Havenith *et al.*, 2004), therefore, adding weight, decreasing resistance to heat transfer and compromising the safety of the clothing system in the outdoors.

The outer protective layer is quite important in the overall performance of the cold weather apparel for providing protection against wind, rain and snow. But it is also important to permit condensed sweat to reach the surface of the clothing. This could be accomplished if the garment is furnished with lower heat resistance paths during heavy workloads. This would increase the air temperature through the garment and would increase the saturation vapour pressure along the path from the skin to the environment. The lower insulation value causes a rise in the temperature of the garment that helps keep perspired sweat in vapour form as it passes through the garment. However, there is no known garment that is capable of self-adjusting its insulation value in accordance to the needs and personal preferences of the wearer. It is worth mentioning that there is a general trend of decreasing the thermal insulation and vapour resistance in the wind and during walking.

McCullough and Hong (1992) showed that the walking effect correlated strongly with the level of static insulation and developed a regression equation to predict the insulation values under dynamic conditions. However, the decrease in the insulation value during dynamic conditions may not be sufficient to prevent the water vapour pressure in the cold weather clothing from reaching the saturation pressure. In addition, in many instances heavy workloads might not involve walking at high speeds or a high level of extremities motion. The transient changes in the transport properties of the foul weather clothing caused by changes in the human activity level might be small in the absence of vents, wind, proper permeability of the fabric and sufficient gap width between the clothing layers. The issue of ventilation will be discussed in more detail in the coming sections. Since a clothing system is incapable of adjusting itself, its transport properties with changes in weather and wearer's workloads, different methods should be pursued to prevent sweat condensation.

7.3 Layering cold weather clothing

The accepted cold weather clothing is a layered system consisting of different fabrics that are worn one over the other. Wearing clothing in multiple layers rather than one large layer allows the wearer to adjust the insulation of his clothing ensemble to changes in environment or workload. The multi-layered clothing system will offer better insulation for an individual performing a light workload or at rest because it can trap more air between the layered clothing system. During heavy workloads, the individual can remove some of the clothing layers to assist the transport of moisture away from his skin also to allow for more heat to dissipate to the environment (Wu and Fan, 2008; Fohr *et al.*, 2002). The multi-layered clothing system is usually comprised of three main layers: an inner layer adjacent to the skin, a middle insulating layer and an outer protective layer (Geisbrecht, 2003).

The middle layer can also comprise several insulating layers. The inner layer is preferably made of a highly permeable synthetic fibre that is capable of transporting liquid and vapour moisture away from the skin but at the same it can be dried quickly at body heat temperatures. The middle layer(s) is a permeable insulation layer that is characterized by a high resistance to heat flow. The outer layer is usually a Gortex layer, laminated polytetrafluroethylene, which provides protection against wind, rain and snow. The outer layer offers impermeability to the rain but breathability for the respired sweat. Therefore a multi-clothing system permits the wearer to remove or add layers to adjust the thermal properties of the clothing system in accordance to the wearer's needs and preferences. In addition, if condensation and freezing of moisture occurs, the multi-layered clothing increases the chances that the freezing occurs between the clothing layers where it can be easily removed instead of being trapped within the material as the case in one thick-layered clothing system.

7.4 Mechanism of ventilation in cold weather

The multi-clothing system offers a solution to the problem of fixed thermal properties of clothing which prevent the clothing system from being a self-adaptable clothing system to adjust to warmer or colder conditions in accordance with the wearer's needs. However, the multi-clothing system still requires the interference of the wearer in the fine tuning or adjustment of the thermal insulation of the clothing. Also in certain circumstances or workloads, it might not be convenient for the wearer to remove or add clothing layers. For this reason, it is important to consider a cold clothing garment design that is capable of ventilating itself with little interference from the wearer. Such small interferences could be to open a sleeve or unbutton a collar. The skin accumulated sweat will be able to evaporate if clothing allows ventilation. Ventilation increases sensible heat loss from the skin, causing lower skin temperature which triggers thermoregulatory control to halt sweating. Proper clothing design is constructed of material that allows the passage of water vapour and allows the wearer to unzip and open the clothing periodically to increase ventilation. Thus the performance and applicability of clothing to different activity levels and weather conditions will increase.

For a given ensemble, the amount of ventilation depends on wind, wearer displacement due to motion causing a wind effect, and relative motion of clothed body parts (limbs) with respect to their clothing cover. Both arms and legs swing during walking while only legs move during cycling. Different body parts are subject to different mechanisms of wind, swinging motion, and combined wind and swinging motion that stimulate ventilation at different rates. In general, motion has an effect on all trapped and surrounding air layers while wind mainly affects the surrounding air layer and the layer directly underneath the garment (Havenith, 1999). Few studies have examined the clothing micro-climate air layer ventilation and even fewer investigations dealt with the mechanism of microclimate ventilation by wind and motion. The properties and thickness of the air layers (microclimate) between and on the outside of the material layers and the skin influence clothing insulation, particularly when movement and wind are present.

Using static thermal insulation data in environmental assessment of cold stress may lead to substantial errors (Havenith *et al.*, 2000; Bouskilla *et al.*, 2002). Measurements on thermal Manikins were reported by Holmér and Nilsson (1995) showing that body movements and wind cause reductions of up to 50% in thermal insulation and by Havenith *et al.* (1990) showing up to 88% in evaporative resistance mainly due to increased ventilation of the clothing microenvironment in such conditions. Havenith *et al.* (1990) found that inside the clothing ensemble, motion is a stronger factor in increasing the ventilation rate than the adjacent air layer. Lotens (1993) has empirically derived a correlation for clothing ensemble ventilation as a function of effective wind speed and air permeability for air-permeable fabrics.

Researchers have also addressed the induced ventilation through a clothed body segment due to a moving limb that changes the microclimate air layer size, and ventilation due to open clothing apertures. Previous work of Ghali *et al.* (2006) and Ghaddar *et al.* (2005) addressed the effect of the changing gap width induced by an oscillating body part (cylinder) within a fixed single clothing cover at uniform external environment pressure with close and open clothing apertures. The ventilation model of Ghaddar *et al.* (2005) of the fabric-air-layer annulus was extended by Jaroudi *et al.* (2006) for the case where the gap width varies in both the angular and axial direction due to the rotation of the limb within a clothing cylinder. The Jaroudi *et al.* (2006) model assumed uniform external pressure distribution without wind and was restricted to small swinging amplitude (maximum swing angle is less than 5°) of the clothed limb.

The external air flow due to wind or displaced by body motion impinges on the clothed body, and some air flows around the body, while some penetrates through the clothing. The flow around the body alters the external pressure distribution over the clothing outer surface. The flow through the fabric depends on the pressure difference between the external pressure on the clothing and the internal air layer pressure on the fabric skin side. The microclimate flow resistance depends on the microclimate layer time-space-varying size caused by the motion of the body relative to the fabric, the occurrence of local skin-clothing contact regions during motion, the presence of openings, as well as interfaces and closures between clothing systems components. Ghaddar *et al.* (2008) derived a model from conservation laws to estimate the ventilation rates for any limb motion configuration taking into consideration the periodic motion of the limbs and clothing, their geometric interaction at skin-fabric contact or no contact for open or closed apertures, and in presence of wind. Their model was validated by experimentation using tracer gas method.

The model of Ghaddar *et al.* (2008) is important in assessing factors affecting clothing ventilation since it provides a method derived from first principles for calculating ventilation rates of a moving clothed body element as a function of microclimate air layer thickness (loose or tight clothing), clothing permeability, walking speed represented by the swinging frequency of the limbs (no. of steps per minute), swing angle, external cross wind velocity, and presence of open or closed apertures. These factors influence clothing design for better ventilation in both stressed hot and cold environments because the magnitude of potential heat and moisture transfers is a direct result of the microclimate air renewal (ventilation). The Ghaddar et al. (2008) ventilation model of the microclimate will be discussed in some detail since it is comprehensive and is derived from conservation principles. The model is then used to assess factors that enhance ventilation.

7.4.1 Mathematical model of microclimate ventilation

Figure 7.2(a)–(c) depicts the schematic of the physical domain of a limb-fabric moving system where an enclosed microclimate air layer annulus of thickness $Y(x,\theta,t)$ and length L. The outer cylindrical porous fabric boundary radius is R_f and the human limb skin of radius is R_s. The domain at $x = 0$ is open to the atmosphere (loose clothing openings) and at $x = L$ is closed (tight joint). The mean air layer thickness is Y_m $(= R_f - R_s)$. Motion of a clothed swinging limb during walking is described in two phases. Starting from a concentric position of the clothed limb ($\phi = 0$), it swings at angular velocity $\dot{\phi}$ within the cylindrical fabric boundary changing the microclimate air thickness in time and space until the limb touches the fabric at $x = 0$ with the contact angle $\phi_{contact}$ $(= \tan^{-1}(Y_m/L))$ and both the inner limb and the clothing boundary move together to the maximum swing angle ϕ_{max} and return together to detach again $\phi = \phi_{contact}$ as the limb swings in the other direction.

The no touch phase I of the motion could be of small duration compared to the local touch phase II duration depending on the air layer mean thickness, frequency and swing amplitude. During phase II the air layer thickness does not change in time and its variation in space remains the same as that reached at the end of the non-contact interval. The modelling of the ventilation rates in the air layer depended on the representation of the air layer thickness as a function of space and time during the different phases of the motion. A sinusoidal angular position ϕ can be assumed for the swinging motion of the limb at rotational speed ω related to the frequency of motion. The angular position ϕ and angular velocity $\dot{\phi}$ are given by

$$\phi = \phi_{max} \sin(\omega t) \qquad 7.1a$$
$$\dot{\phi} = \phi_{max}\omega \cos(\omega t) \qquad 7.1b$$

During the non-contact interval (Phase I) of the motion, the air layer thickness Y at any spatial position is given by

$$Y(x,\theta,t) = Y_m - (L-x)\tan\phi\cos\theta \qquad 7.2$$

where Y_m is the mean air layer spacing. The non-contact interval duration is

Designing for ventilation in cold weather apparel

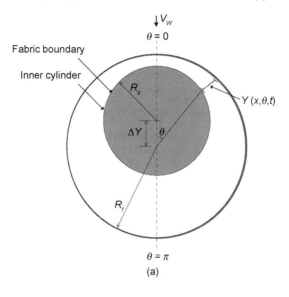

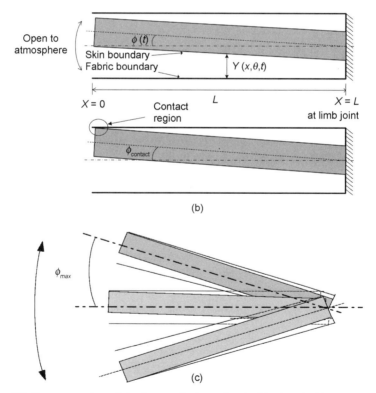

7.2 Representation of the human limb motion inside the clothing cylinder showing (a) front view of geometry, (b) side view of the geometry, and (c) swinging motion.

constrained by the condition that

$$-\phi_{contact} \leq \phi \leq +\phi_{contact} \quad \text{or} \quad 0 \leq Y(x=0, \theta = 0 \text{ or } \pi) \leq 2Y_m \quad 7.3$$

A dimensionless amplitude parameter ζ can be defined at the opening $x = 0$ as

$$\zeta = \frac{\tan \phi_{max}}{\tan \phi_{contact}} \quad 7.4$$

If the amplitude parameter is smaller than unity ($\zeta < 1$), then no skin-fabric contact is present and air layer thickness is calculated from equation 7.2. When $\zeta \geq 1$, the fabric-skin contact is present locally at $x = 0$ and $\theta = 0$ or π. The air layer thickness $Y_{contact}$ during contact interval does not vary with time and is given by

$$Y_{contact}(x, \theta) = Y_m - (L - x) \tan \phi_{contact} \cos \theta \quad \zeta \geq 1 \quad 7.5$$

Air renewal in phase I is induced by changing gap width that forces skin adjacent air in and out of the microclimate. Air renewal in phase II is similar to the situation of a clothed cylinder subject to external wind. The relative cross wind V_∞ around the clothed limb is given by

$$V_\infty = V_w \cos \phi \pm \dot{\phi}(L - x) \quad 7.6$$

where V_w is the free stream wind. Radial flow through the fabric is governed by the pressure difference between the external fabric adjacent air layer pressure and the internal microclimate pressure. Ghaddar et al. (2008) used the external pressure distribution around a clothed cylinder to determine air renewal rates in the microclimate separating the skin and the fabric. They assumed the external angular pressure distribution around the clothed cylinder is not influenced by the presence of the clothing cover as can be deduced from study on flow around cylinders sheathed by clothing by Sobera et al. (2003) and that the internal air flow within the clothing microclimate is laminar. Gibson (1999) and Sobera et al. (2006) reported CFD simulation results that show for a given external air velocity, the amount of air which flows around the body, and the amount of air which penetrates through the clothing layer is determined by the air permeability of the clothing layer and that the flow inside the microclimate is laminar for low and medium permeability fabrics.

Air layer mass balance in phase I of limb motion

The microclimate air layer is formulated as an incompressible lumped layer in the radial direction. During the non-contact time interval of the motion, the general air layer mass balance of Ghaddar et al. (2005) is given by

$$\frac{\partial(\rho_a Y)}{\partial t} = -\dot{m}_{aY} - \frac{\partial(Y\dot{m}_{ax})}{\partial x} - \frac{\partial(Y\dot{m}_{a\theta})}{R_f \partial \theta} \quad 7.7a$$

where $\dot{m}_{ax}$ is the mass flux in the axial direction in kg/m²s, $\dot{m}_{a\theta}$ is the mass flux in the angular direction, and $\dot{m}_{aY}$ is the radial air flow rate through the fabric given by

$$\dot{m}_{aY} = \frac{\alpha \rho_a}{\Delta P_m}(P_a - P_s) \qquad 7.7b$$

where α is the fabric air permeability in m³/m²s, ΔP_m is obtained from standard tests on fabrics' air permeability [ASTM D73775, 1983], P_a is the air pressure in the microclimate trapped air layer (kPa), and P_s is the external adjacent air layer pressure (kPa). The external air pressure is assumed uniform at the environment conditions at P_∞ when no external wind is present and the radial air flow is only induced by limb motion with respect to the fabric boundary.

The boundary conditions for the air flow are

$$\dot{m}_{ax}(x=0,\theta) = C_D \left[\frac{2\rho_a}{|P_L - P_\infty|}\right]^{1/2}[P_\infty - P_L] \text{ and } \dot{m}_{ax}(x=L,\theta) = 0 \qquad 7.8a$$

$$\dot{m}_{a\theta}(x,\theta=0) = 0 \quad \text{and} \quad \dot{m}_{a\theta}(x,\theta=\pi) = 0 \qquad 7.8b$$

where equation 7.8b is derived from the pressure drop at the opening by applying Bernoulli's equation from P_∞ in the far environment ($x \to -\infty$) to the opening at $x = 0$, and C_D is the discharge loss coefficient at the aperture of the domain dependent on discharge area ratio of the aperture to the air layer thickness Y. In case wind is present, the dynamic pressure term is added to P_∞ to give the environment total pressure. The 3-D cylinder model of Ghaddar et al. (2005) was used to estimate the ventilation rates where the mass balance equation on the air layer is transformed to a scalar pressure equation that can be solved numerically to calculate the pressure at any spatial location within the air layer as a function of time. The method of solving for the pressure distribution for the case when the fabric is not in contact with the skin during the periodic motion of the inner cylinder can be found in detail in the work of Ghaddar et al. (2005).

Air layer mass conservation in phase II of limb-clothing motion

In this phase both the limb and fabric move together without any change in the air layer thickness $Y_{contact}$. However, ventilation flow rates in the air layer are not zero due to pressure differences driven by the clothed limb motion relative to the environment air and wind if present. Assuming a Poiseuille flow model in the axial and angular directions (neglecting the fluid inertia associated with flow modulation), the mass balance yields the following pressure equation:

$$0 = -\frac{\alpha(P_s(t) - P_a)}{\Delta P_m} - \frac{\partial}{\partial x}\left(\frac{Y^3}{12\mu}\frac{\partial p}{\partial x}\right) - \frac{\partial}{R_s^2 \partial \theta}\left(\frac{Y^3}{12\mu}\frac{\partial p}{\partial \theta}\right) \qquad 7.9$$

where P_s is the external adjacent air layer pressure at the fabric surface (kPa). The same boundary conditions in equations 7.8a and 7.8b apply to the mass balance.

The external fabric pressure that drives the radial flow is assumed to be the same pressure distribution that results from a flow around circular cylinder at flow condition V_∞. To solve the pressure equation, the time-dependent external pressure distribution around the clothed cylinder must be known as a function of angular and axial position. A curve fit is derived from the data of Fransson et al. (2004) on the pressure coefficient distribution around a circular cylinder with suction or blowing at radial velocity ratio of $\Gamma = V_Y/V_\infty$ smaller than 6%.

The solution for the pressure distribution in the microclimate layer would provide the basis for estimating the ventilation rates for any general movement of the clothed body element and wind conditions. Note that the external pressure distribution P_s in phase I is constant and the time dependence of the microclimate pressure stems from the air layer thickness change due to the limb motion while in phase II the microclimate pressure time dependence is due to external pressure P_s change in time.

Definition of microclimate ventilation through fabric and through apertures

The total ventilation rate is calculated as the positive flow of air into the microclimate integrated over the oscillation period (τ) per unit area of the clothed surface as follows:

$$\dot{m}_a = \frac{1}{2\pi RL\tau} \int_0^\tau \int_0^L \int_0^{2\pi} \max(0, \dot{m}_{aY}) R d\theta\, dx\, dt \quad (\text{kg/s.m}^2) \quad 7.10a$$

In the above integral, $\dot{m}_{aY}$ is equated to zero every time it is negative (flowing out of the microclimate). The net flow of air over one period is zero. The ventilation rate inflow to the microclimate air layer through the open aperture per unit area of clothed surface during the period of motion is calculated as

$$\dot{m}_o = \frac{1}{2\pi RL\tau} \int_0^\tau \int_0^{2\pi} \max(0, \dot{m}_{ax(x=0)}) Y(\theta, x=0, t) R d\theta\, dt \quad (\text{kg/s.m}^2)$$

7.10b

where $\dot{m}_o$ is the net flow rate through the open aperture.

The steady periodic ventilation rates can be used in the water vapour mass and energy balances of the microclimate air layer during contact and non-contact intervals to predict the associated heat loss from the limb (Ghaddar et al., 2006).

7.5 Factors affecting ventilation

During walking, ventilation mechanism is dependent on the specific motion of the clothed body segment. Microclimate clothing ventilation in stagnant environment for open or closed apertures could be determined from uniform relative wind (walking speed) around a fixed clothed cylinder (trunk) and swinging motion of a clothed cylinder at small and large swinging angles (limbs). If walking or activity takes place in the presence of wind, then the limbs will be subjected to a combined effect of wind and swinging motion.

Table 7.1 presents some results on the percentage of the ventilation of each period for closed aperture at wind speeds of 1, 3, and 5 m/s during one cycle at $R_s = 0.038$ m, $R_f = 0.0575$ m, $\alpha = 0.05$ m/s, $\phi_{max} = 20°$ and $L = 0.48$ m.

During the no touch period the variation in the air gap thickness creates a high pressure difference between the air gap layer and the atmosphere so the majority of the ventilation rate is found during the no touch period because of the high pressure difference across the fabric. As the wind speed increases the fraction of the touch period increases because a higher pressure difference is created between outer clothing surface and the microclimate surface causing higher ventilation rate during the touch interval. At low frequencies, it is obvious that the wind speed has a larger contribution to the total ventilation rate due to the lower ventilation induced by a smaller dY/dt term. With the understanding of the mechanism of ventilation, the effect of swinging motion and wind, the air layer thickness, clothing apertures, fabric permeability, and the swing amplitude will be further elaborated.

Table 7.1 The percentage of the ventilation of each period for closed aperture at wind speeds of 1 m/s and 3 m/s during one cycle at $R_s = 0.038$ m, $R_f = 0.0575$ m, $\alpha = 0.05$ m/s, $\phi_{max} = 20°$ and $L = 0.48$ m

f (rpm)	Fraction of $\dot{m}_a$ No-touch interval (%)	Fraction of $\dot{m}_a$ Touch interval (%)	$\dot{m}_a$ (kg/s·m²) During one period
Wind at $V_w = 1$ m/s			
10	60.59	39.41	1.72×10^{-04}
25	91.01	8.99	7.78×10^{-04}
40	97.20	2.80	2.75×10^{-03}
80	99.16	0.84	1.22×10^{-02}
Wind at $V_w = 3$ m/s			
10	15.75	84.25	1.15×10^{-03}
25	52.88	47.12	2.97×10^{-04}
40	81.25	18.75	1.19×10^{-04}
80	95.13	4.87	6.50×10^{-05}
Wind at $V_w = 5$ m/s			
10	8.86	91.14	1.85E-03
25	29.49	70.51	2.39E-03
40	61.12	38.88	4.39E-03
80	87.79	12.21	1.38E-02

7.5.1 Open vs. closed aperture design for ventilation

Ghaddar et al. (2008) reported the variation of ventilation rates over a swinging clothed cylinder by cotton fabric of medium intrinsic permeability α of 6.7×10^{-10} m^2. Figure 7.3 shows the variation of ventilation rates as a function of swinging frequency at various cross-wind speeds for (a) closed aperture and (b) open end aperture of a swinging clothed cylinder. The ventilation rate increases with swinging frequency. The effect of the wind speed is important at low swinging frequencies (low walking speed). The enhancement of renewal rate by introducing an open aperture to the swinging arm is more effective at higher cross-wind speeds and lower walking frequencies as can be seen in Fig. 7.4 where the ratio of the total ventilation rate of open aperture to closed

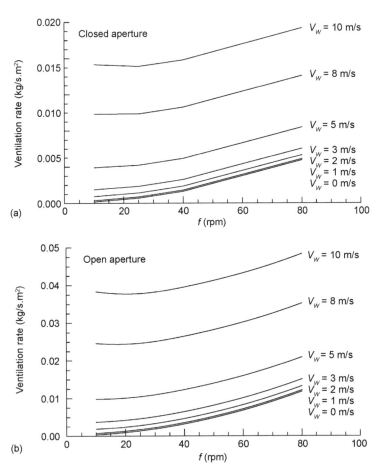

7.3 A plot of ventilation rate as a function of swinging frequency f in rpm at different wind speeds for (a) closed aperture and (b) open aperture at $R_s = 0.038$ m, $R_f = 0.0575$ m, $\alpha = 0.05$ m/s, $\phi_{max} = 20°$ and $L = 0.48$ m.

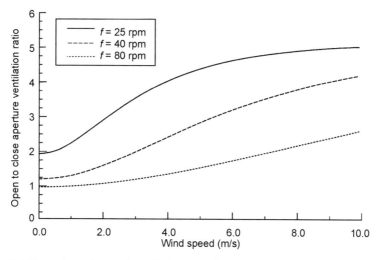

7.4 The ratio of the total ventilation rate of open aperture to closed aperture clothed swinging cylinder as a function of cross-wind velocity at three swinging frequencies of a clothed swinging cylinder of $R_s = 0.038\,\text{m}$, $R_f = 0.0575\,\text{m}$, $\alpha = 0.05\,\text{m/s}$, $L = 0.48\,\text{m}$, $f = 25, 40$, and 80 rpm.

aperture clothed swinging cylinder is plotted against cross-wind velocity at three swinging frequencies.

Adding sleeve openings at lower walking frequencies more than doubles ventilation rate from the limb. Body trunk does not swing during walking and can be represented by cross-flow around a fixed clothed cylinder. The presence of a neck collar opening or possibility of connective underarm microclimate opening to the limbs would help circulate moisture due to sweating to the environment through direct paths different from the clothing. However, this effect cannot be directly assessed from the model of Ghaddar *et al.* (2008) unless the various body parts ventilation models are integrated.

7.5.2 Microclimate air layer thickness

Under static conditions, the vapour resistance of the air layer increases proportionately with air layer thickness for small thickness layer, but are reported to deviate for clothing systems with thick air layers as reported by McCullough *et al.* (1989) who developed an asymptotic relationship for calculating evaporative resistance of the air layer. Havenith *et al.* (1990) reported that tight clothing fits have lower insulation values than loose fits and that stronger reductions in insulation are experienced due to posture and movement in ensembles with low insulation values. The total ventilation rate increases with increased microclimate size or air volume in the adjacent skin layer due to reduced flow resistance to air flow in the axial and angular directions around the body part.

146 Textiles for cold weather apparel

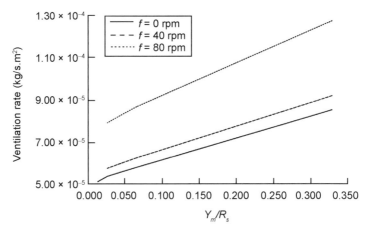

7.5 The effect of the ratio of microclimate layer thickness to inner limb radius Y_m/R_s on ventilation rate for closed aperture is shown at $V_w = 1$ m/s for $R_s = 0.038$ m, $R_f = 0.0575$ m, $\alpha = 0.05$ m/s, $L = 0.48$ m, $f = 0$, 40, and 80 rpm.

An example of this effect based on the model of Ghaddar et al. (2008) can be seen in Fig. 7.5 that shows variation of the ventilation rate of a clothed swinging cylinder with the ratio of microclimate layer thickness to body part radius Y_m/R_s at fixed clothing permeability for closed aperture at $V_w = 1$ m/s for $R_s = 0.038$ m, $R_f = 0.0575$ m, $\alpha = 0.05$ m/s for $f = 0$, 40, and 80 rpm and $L = 0.48$ m. Enhanced microclimate ventilation is necessary in local regions of the body characterized by high sweat rates and one possibility is to increase the microclimate size in those regions. When clothing fits tightly, less ventilation occurs even while moving and moisture diffusion and condensation due to sweating is more likely to take place in high sweat regions.

7.5.3 Clothing permeability

Ventilation through the fabric induced by wind and motion is strongly dependent on air permeability of a clothing ensemble. Using the Ghaddar et al. (2008) model, the ventilation rate is shown to increase with increased fabric permeability. Clothing permeability does not provide a wide margin for clothing design for cold weather conditions since high thermal insulative value is needed. Figure 7.6 shows an example of the ventilation rate variation with fabric permeability for a closed aperture clothed swinging cylinder at $V_w = 1$ m/s, $R_s = 0.038$ m, $R_f = 0.0575$ m, $\phi_{max} = 20°$ and $L = 0.48$ m. The ventilation rate increases significantly with increased permeability for values below 0.6 m/s, but little improvement in ventilation occurs for increased permeability beyond 0.6 m/s in this specific example.

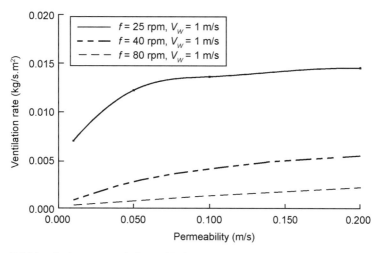

7.6 Ventilation rate variation with fabric permeability for a closed aperture clothed swinging cylinder at $V_w = 1$ m/s, $R_s = 0.038$ m, $R_f = 0.0575$ m, $\phi_{max} = 20°$ and $L = 0.48$ m.

7.5.4 Amplitude of swinging motion

The amplitude of swinging motion is dependent on human activity and is not a clothing design parameter. However, it does impact on ventilation of clothed limbs in a subtle way at high frequencies. The effect of the swing amplitude for closed aperture of clothed cylinder on ventilation rate is shown in Fig. 7.7, where $\dot{m}_a$ is plotted against ϕ_{max} for different ventilation frequencies at $V_w = 1$ m/s, $R_s = 0.038$ m, $R_f = 0.0575$ m, $\alpha = 0.05$ m/s, $\phi_{max} = 20°$ and $L = 0.48$ m. At high frequency, the ventilation rate decreases with ϕ_{max} until $\phi_{max} = 18°$ and then increases for the particular example given in Fig. 7.7. This behaviour is connected to the percentage of time the limb is in touch with the fabric with a fixed gap width. At low amplitude swing, the ventilation rate is governed mainly by the rate of change of the microclimate size (dY/dt) which induces higher ventilation than during the touch interval. At high angles of swing, the amplitude of motion is high giving higher relative wind and the touch period governs the ventilation mechanism where the renewal is due to external flow around a clothed cylinder with fixed gap $(dY/dt = 0)$. These two opposing trends cause a minimum ventilation to occur close to $\phi_{max} = 18°$.

7.6 Recommendations and advice on clothing design for ventilation

Ventilation is an important design parameter for the cold weather clothing of active people to reduce chilling discomfort caused by sweating and buildup of condensate within insulative clothing worn in cold climates. Based on research

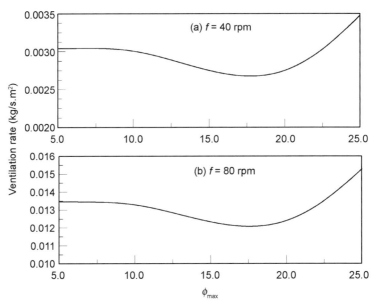

7.7 A plot of $\dot{m}_a$ vs. ϕ_{max} at $V_w = 1$ m/s for a closed aperture clothed swinging cylinder at $V_w = 1$ m/s, $R_s = 0.038$ m, $R_f = 0.0575$ m, $\alpha = 0.05$ m/s, and $L = 0.48$ m for (a) $f = 40$ rpm and (b) $f = 80$ rpm.

work done for modelling and predicting clothing ventilation, several design recommendations can be deduced to enhance ventilation and the comfort of people performing high metabolic heat production activities during their work or exercise in cold weather. These recommendations are summarized below.

1. Clothing ensembles with several layers provide advantages for adjusting thermal insulation and ventilation by the wearer. This may not be a convenient option in all situations since a jogger would prefer to keep a jacket on rather than taking it off and carrying it.
2. Loose clothing fit would perform better than a tight fit for ventilation in reducing moisture buildup in the microclimate and clothing layers. Loose clothing reduces resistance of airflow parallel to clothing surfaces in the microclimates between clothing layers. This facilitates exchange of microclimate air with environment air through apertures.
3. Design strategies that observe flexible operation and control of clothing ensemble apertures can provide versatile clothing that can be adjusted based on activity to maintain comfort and insulation in cold weather. Aperture control can be provided with zippers, pit zips for the trunk, arms, knees, adjustable cuffs, and collars that could be unbuttoned. More research on optimizing opening location and functionality needs to be performed.
4. Enhancement of microclimate exchanges between body and different body segments is desirable in garment design. Walking ventilates moving arms and

legs, but the trunk ventilation is not as significant as the moving limb ventilation. Clothed trunk ventilation can be enhanced with larger microclimate air exchange permitted between body segments to channel the airflow from the relatively high-ventilated limbs microclimate to the less ventilated trunk microclimate.

7.7 Future trends

Research on protective clothing for active people at mild and cold conditions is addressing issues of smart clothing designs that can be adapted to the activity level and to changes in climate conditions with minimum interference to the wearer for the purpose of ensuring comfort. Ventilation of the microclimate air layer is an attractive solution for design of versatile clothing for active people that could result in enhancing the effectiveness of moisture removal away from the skin while providing the necessary warmth. Ventilation rates can be altered where needed by the proper choice of fabric permeability, arrangement of clothing layers, reducing internal microclimate flow resistance for internal air exchanges between connected clothed body segments. Ventilation rates can also be altered by aperture positions, size and operational flexibility of controlling the extent of microclimate direct connection with the environment air through the aperture. Extensive research is needed to understand the relationship between segmental ventilation and local discomfort during walking and when subject to wind for optimizing protective active wear designs.

7.8 References

American Society for Testing and Materials (1983), ASTM D73775, Standard Test Method for Air Permeability of Textile Fabrics, (IBR) approved 1983.

Bouskilla L M, Havenith G, Kuklaneb K, Parsons K C and Witheyc W R (2002), 'Relationship between clothing ventilation and thermal insulation', *J Occup. Envir. Hygiene* 63, 262–268.

Fan J, Cheng X Y and Chen Y S (2003), 'An experimental investigation of moisture adsorption and condensation in fibrous insulations under low temperatures', *Exper. Thermal Fluid Sciences* 27, 723–729.

Fohr J P, Couton D and Treguier G (2002), 'Dynamic heat and water transfer through layered fabrics', *Textile Research Journal* 72(1), 1–12.

Fransson J H M, Konieczny P and Alfredsson, P H (2004), 'Flow around a porous cylinder subject to continuous suction or blowing', *Journal of Fluids and Structures* 19, 1031–1048.

Geisbrecht G G (2003), 'Cold Weather Clothing', *Winter Wilderness Medicine Conference*, Jackson Hole, Wyoming, 13–18 Feb.

Ghaddar N, Ghali K, Harathani J and Jaroudi E (2005), 'Ventilation rates of microclimate air annulus of the clothing-skin system under periodic motion', *Int. J Heat Mass Transfer* 48(15), 3151–3166.

Ghaddar N, Ghali K and Jones B (2006), 'Convection and ventilation in fabric layers',

Chapter 8, *Thermal and Moisture Transport in Fibrous Materials*, edited by N Pan and P Gibson, Woodhead Publishing, CRC Press, pp. 271–307.

Ghaddar N, Ghali K and Jreije B (2008), 'Ventilation of wind-permeable clothed cylinder subject to periodic swinging motion', *ASME J of Heat Transfer*, in press.

Ghali K, Ghaddar N and Jaroudi E (2006), 'Heat and moisture transport through the microclimate air annulus of the clothing-skin system under periodic motion,' *ASME J of Heat Transfer* 128(9), 908–918.

Gibson P W (1999), *Review of numerical modeling of convection, diffusion, and phase change in textiles*. American Society of Mechanical Engineers, Pressure Vessels and Piping Division PVP-397, 117–126.

Havenith G (1999), 'Heat balance when wearing protective clothing', *Annals of Occupational Hygiene* 43(5), 289–296.

Havenith G, Heus R and Lotens W A (1990), 'Clothing ventilation, vapour resistance and permeability index: changes due to posture, movement, and wind', *Ergonomics* 33(8), 989–1005.

Havenith G, Holmér I, Parsons, K C, Den Hartog E and Malchaire J (2000), 'Calculation of dynamic heat and vapour resistance', *Environmental Ergonomics* 10, 125–128.

Havenith G, Hartog E D and Heus R (2004), 'Moisture accumulation in sleeping bags at $-7\,°C$ and $-20\,°C$ in relation to cover material and method of use', *Ergonomics* 47(13), 1424–1431.

Holmér I and Nilsson H (1995), 'Heated manikins as a tool for evaluating clothing', *Annals of Occupational Hygiene*, 39(6), 809–818.

Jaroudi E, Ghaddar N and Ghali K (2006), 'Heat and moisture transport from a swinging limb of a clothed walking human', *Proceedings of the 13th International Heat Transfer Conference*, Australia, 17–22 August.

Keighley J H (1985), 'Breathable fabrics and comfort in clothing', *J. Coated Fabrics* 15(10), 89–104.

Lotens W (1993), *Heat transfer from humans wearing clothing*, Doctoral Thesis, TNO Institute for Perception, Soesterberg, The Netherlands.

McCullough E A and Hong S A (1992), 'Database for determining the effect of walking on clothing insulation', *Proceedings of the Fifth International Conference on Environmental Ergonomics*, 68–69.

McCullough E A, Jones B W and Tamura T (1989), 'A data base for determining the evaporative resistance of clothing', *ASHRAE Trans* 95(2), 316–328.

Ruckman J E (1997), 'An analysis of simultaneous heat and water vapour transfer through waterproof breathable fabrics', *J. Coated Fabrics* 26(4), 293–307.

Sobera M P, Kleijn C R, Brasser P and Van den Akker H E A (2003), 'Convective heat and mass transfer to a cylinder sheathed by a porous layer', *AIChE Journal* 49, 3018–3028.

Sobera M P, Kleijn C R, Brasser P and Van den Akker H E A (2006), 'Subcritical flow past a circular cylinder surrounded by a porous layer', *Physics of Fluids* 18, 106–108.

Wu H and Fan J (2008), 'Study of heat and moisture transfer within multi-layer clothing assemblies consisting of different types of battings', *International Journal of Thermal Sciences* 47(5), 641–647.

7.9 Nomenclature

f	frequency of oscillation of the inner cylinder in revolutions per minute (rpm)
C_D	opening discharge coefficient
L	fabric length in x direction (m)
$\dot{m}_{aY}$	mass flow rate of air in y-direction (kg/m²·s)
$\dot{m}_{ax}$	mass flow rate of air in x-direction (kg/m²·s)
$\dot{m}_{a\theta}$	mass flow rate of air in θ-direction (kg/m²·s)
P_a	air microclimate pressure (kPa)
P_s	external pressure on the surface of the clothed cylinder (kPa)
R_f	fabric cylinder radius (m)
R_s	skin cylinder radius (m)
t	time (s)
Y	instantaneous air layer thickness (m)
Y_m	mean air layer thickness (m)
$Y_{contact}$	air layer thickness during interval of touch (m)
V	velocity (m/sec)

Greek symbols

α	fabric air permeability (m³/m²·s)
ϕ	angular position of the inner cylinder with respect to the axial direction
ϕ_{max}	maximum angular position
$\phi_{contact}$	contact angle
Γ	radial velocity ratio of $\Gamma = V_Y/V_\infty$
ν	kinematic viscosity (m²/s)
ω	angular frequency (rad/s)
ρ	density (kg/m³)
τ	period of oscillatory motion (s)
θ	angular coordinate
μ	air viscosity (N·s/m)

Subscripts

a	conditions of air in the spacing between skin and fabric
o	opening
w	external wind
∞	environment condition

8
Factors affecting the design of cold weather performance clothing

J. BOUGOURD, University of the Arts London, UK and
J. McCANN, University of Wales, Newport, UK

Abstract: This chapter looks at the user-centred design and development process. It discusses factors in creating a microclimate within layered clothing systems to cope with cold for survival and performance during recreation and occupational activities. The process involves gathering requirements, designing and evaluation. It addresses the issue of sustainability in the context of an historically wasteful clothing industry and takes account of current technologies and initiatives driving the cold weather apparel market forward. Case studies look at cold weather clothing for motorcycling and mountaineering and the chapter draws on examples from other types of cold weather activity.

Key words: cold weather clothing, user design, layering system.

8.1 Introduction

The cold weather clothing market, as part of the wider ready-to-wear (RTW) apparel industry, is responding to the global environmental challenge. Major retailers have launched ethical policies, universities have established departments and programmes of study to address sustainability issues, and companies such as Patagonia are leading successful eco-recycling initiatives for cold weather clothing. At the heart of this challenge is a re-assessment of innovation that looks across the life cycle of textile and clothing products (Innov-Ex, 2008) and for the design and development function within that life cycle challenges are being seen as greater than at any other time (Parsons and Rose, 2008).

In this chapter we will review the traditional design and development process, consider the benefits of user-centred design approaches and discuss key factors affecting the design of generic cold weather clothing within the context of a user strategy, gathering and specifying requirements, designing and evaluating cold weather products. It concludes with case studies that highlight differences between design requirements of users for two specified cold weather activities, and presents a profile of an individual's design requirements for cold weather clothing.

8.2 Traditional design development process

This traditional design process model is iterative, using a number of cyclic key stages: designing (problem exploration, idea generation, prototype realisation and evaluation); market analysis to delivery of a product to the consumer; and, in some cases, the setting of this process within a corporate strategy. Such models are generally applicable to fashion and apparel design practices although, as Black *et al.* (2005) observed, few models have specifically been created for use in the fashion and apparel field. Execution of the stages in this process often depends on the size of a company: a small enterprise (prevalent in the fashion and apparel industry) where the individual designer will have total responsibility for the process; while in a large, perhaps international organisation a multi-disciplinary team is likely to be assigned roles. These models take account of target users, but information about the user normally relies on market research following the launch of products – consumer feedback – rather than the case of designers engaging with consumers at the onset of the product development process. Such traditional models, widely used in the RTW fashion industry, embody weaknesses where products fail to meet the needs of users (such as poor fit, inappropriate style or colour). A consequent failure to meet the needs of the user raises questions about sustainability, questions that industry faces today. For example, 900,000 tons of clothing and shoes were thrown away in the UK in the year 2006 (Anderson, 2008).

User-oriented approaches offer alternatives to traditional practices, helping us to address the issues of user satisfaction and subsequent sustainability. They are not new. An early ergonomic systems model was successfully used for such functional clothing design applications as fishing and nursing (Parsons, 2003). Later models of the design and development process involve the user in the process and are sometimes referred to as 'user-centred', 'participatory design' and 'co-design'; each is outlined below.

8.2.1 User-centred design strategies

Central to the concept of user-centred design is a commitment to the belief that the best designed products result from understanding the needs of the users (Black, 2006). Based in ethnographic methods, a designer will immerse him or herself in the user's context, seeking visual and verbal evidence of practical tasks and their social and emotional significance. It is an iterative process where input from users is sought throughout the stages but, as with the traditional process, users are not normally invited to make design decisions.

8.2.2 Participatory design

Here the emphasis is on active co-operation between user and designer. It differs from the concept discussed above in that end-users are directly involved at every stage. They participate in initial problem exploration and definition, help with design ideas, and evaluate solutions. It is an approach that helps to ensure that the designed products meet identified needs and are useable. It is characterised by the involvement of a population of end users brought into the research and design environment – rather than designers and researchers entering the world of the user. It has been used as a design research strategy for clothing with embedded technologies (Uotila *et al.*, 2006).

8.2.3 Co-design and open innovation

The main characteristic of co-design and open innovation models is that they not only acknowledge the involvement of users but also contributions from non-users in the wider community. Originating in mass customised and personalised practices, these models are driven, in part, by sustainability issues, such as:

- the cost of producing prototype design samples;
- material waste;
- garment returns;
- textile landfills.

These issues, considered together with flexible CAD and 3D technologies are recognised as new business opportunities that could support a sustainable future (Allwood *et al.*, 2006).

Co-design models operate in 'bricks and mortar', 'online' and 'click-and-brick' environments. Users are integrated into the value creation by defining, configuring, matching and modifying individual design solutions from a range of options where apparel and footwear are the most common customised products (Piller *et al.*, 2005).

Open innovation is the involvement of users and non-users in the innovation process. This 'crowd sourcing' is an emerging design model, but there are already successful apparel and footwear examples in the market place[1] – other recently launched examples set out to create a 'people's brand'.[2]

Coming full circle, and considering the way in which traditional product design is undertaken, there is Redesign Me, an online company combining open innovation with user co-design to rethink 'bad' products into 'better' ones and turning good ideas into those that are great. Outputs from this model are not only fed back to the companies responsible for the original ('bad') products, but these companies are also being encouraged to work more closely with their end users (Tseng and Piller, 2003), which echoes the earlier view that the best designed products result from understanding the needs of people who will wear them. It is

---

1. E.g. 'Threadless', set up in 2000 by skinnyCorp, in the USA.
2. E.g. NVOHK (pronounced 'invoke').

Factors affecting the design of cold weather performance clothing   155

this more sustainable approach that will be used as a framework to discuss cold weather clothing in this chapter.

8.3 Stages in the process

The ways in which the user becomes involved will vary, as can be seen in the foregoing, but all broadly relate to the four essential activities outlined in ISO 13407: gathering requirements; specifying requirements; design; and evaluation. Each of the four is addressed in turn below.

8.3.1 Stage one: gathering requirements

Clothing designers and ergonomists have long recognised the need to balance form and function. Early publications on functional clothing design proposed that, as the need for specialist clothing grew, so there was a need for a new kind of designer (DeJonge, 1984): one proficient in creative solutions, as well as being equipped to meet a variety of challenges of environments and occupations. More recently, Sperling (2005), in her discussion on ergonomics in user-oriented design, suggested that human factors are in constant development. In the past ergonomics has been mainly concerned with health and safety but is now in what Sperling calls the comfort level of human well being and proposes that ergonomics and aesthetics could meet in user-centred design. This echoes McCann's discussion of sportswear, where design is embedded in function, and vice versa (McCann, 2005).

Notwithstanding these useful developments, the essential criterion for the user is the effectiveness of the clothing design in protecting them from the cold. In gathering requirements for cold weather clothing, background information on climatic conditions, users, their activities and associated apparel performance standards are needed.

Climatic conditions, users and their activity

Cold is a hazard to human health that may adversely affect our physiological functions, work performance and even our lives (Holmér, 2005). A cold environment is characterised by a combination of humidity and wind at a temperature below $-5\,°C$ (BS EN342: 2004), although, for military purposes, further categories are also relevant to some recreational activities that have been identified: 'wet cold' as $\pm10\,°C$; 'very cold' as -10 to $-30\,°C$; 'extremely cold' as -30 to $-60\,°C$ (Thwaites, 2008). These extreme atmospheric conditions can of course have major effects on the body's thermal regulation.

The core temperature of the body needs to be maintained at around $37\,°C$, whilst peripheral parts may have variable temperatures, without ill effects (SHAPCC, 1998). To maintain heat, the body reacts by either conserving or generating it. Parsons suggests that: 'on average the heat transfer into the body and heat generation within the body must be balanced by heat outputs from the body. If heat generation and inputs were greater than heat outputs, the body temp would rise and if heat outputs were greater the body temp would fall'.

Heat loss

Thermal exchanges with the environment are represented in Fig. 8.1. The body can lose heat by convection, radiation, conduction and evaporation.

Convection, one of the major modes of heat transfer, is created by the transfer of heat by movement of the thin layer of insulating air next to the skin. There is a greater cooling effect as the speed of air movement around the body increases, described as 'wind chill' in BS EN ISO 11079 (2008). The effect of wind on air temperature and on the skin is set out in Table 8.1 (it is, however, recognised that heat transfer with clothed surfaces becomes more complex – Holmér, 2005).

Radiation is heat transferred through space to objects that are cooler, and which are not in direct contact with each other, whilst the transfer of heat between objects that are in contact with one another is a form of *conduction*. Air is a poor conductor, whereas solids conduct well and may adversely affect an individual who is seated, standing (with poorly insulated footwear) or in water (where heat is lost at a rate 20 to 30 times faster than in air). *Evaporation* is the transfer of moisture into the air and is a means of reducing body heat by sweating or (as the temperature falls) respiratory means. This is particularly

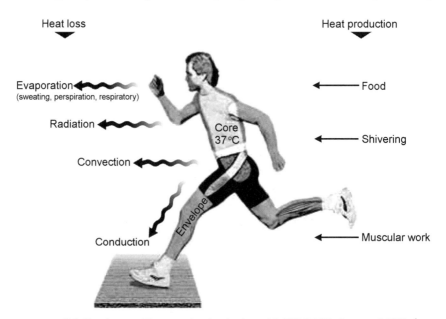

8.1 Heat loss and heat production in the cold. (SHAPCC, January 1998).

Factors affecting the design of cold weather performance clothing 157

Table 8.1 Wind chill temperature (t_{WC}) and freezing time of exposed skin (BS EN ISO 11079, 2008)

Classification of risk	t_{WC} °C	Effect
1	−10 to −24	Uncomfortably cold
2	−25 to −34	Very cold, risk of skin freezing
3	−35 to −59	Bitterly cold, exposed skin may freeze in ten minutes
4	−60 and colder	Extremely cold, exposed skin may freeze within two minutes

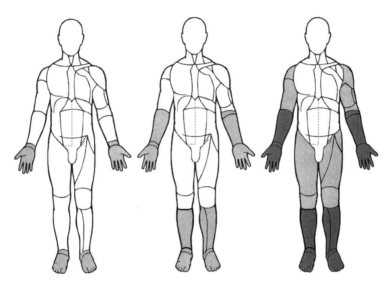

8.2 Progressive local cooling of extremities, as they are sacrificed to protect the core (Lehmuskallio, 2001).

important where there are high levels of heat that need to be transported as water vapour through clothing and air layers adjacent to skin, or by convection through openings in clothing (Holmér, 2005; see Fig. 8.7).

From this we can see that exposure to cold conditions can fall below safe limits when the body loses heat faster that it can be produced. These cooling effects (types of cold stress), are described by Holmér and Rintamäki (2004) as 'general cooling of the body and local cooling of extremities and face' (see Figs 8.2 and 8.7). This progressive cooling of the extremities represents a sacrifice to protect the body core, and can cause discomfort, loss of manual dexterity, forgetfulness and pain, which can, in turn lead to problems such as hypothermia, frostbite, snow blindness and trauma from slips and falls (CWA, 2005).

Heat production

Heat is produced by metabolising food, by involuntary muscular contractions (shivering) and by voluntary muscular exertion and motion. Food provides fuel to be burned and warm drinks provide direct heating, which helps to prevent dehydration, while shivering increases muscle cell activity which, in turn, creates heat although the onset of shivering will depend on fat thickness as subcutaneous fat provides insulation against conductive heat loss. The skin detects cold and triggers a physiological defence against the cold: a restriction of blood flow to the skin that reduces heat loss to the environment (Nakamura and Morrison, 2008). It is important to note here that the requirements for protection are conditioned by the levels of activity involved. Physical work is related to metabolic energy production and equivalent estimation of activities can be found in ISO 8996. Proposed classifications of metabolic rates range between resting and exhaustive (BS EN ISO 11079, 2008; Holmér, 2005). Examples from these sets are shown in Table 8.2.

Table 8.2 Classifications of metabolic rate for kinds of activity extracted and modified from ISO 8996 (2004)

Class[a]	m $W.m^{-2}$	Examples
Resting	65	Sleeping, sitting
Very high metabolic rate	290	Very intense activity at fast to maximum pace; climbing, running; walking in deep, loose snow and cross-country skiing
Very, very high metabolic rate (1 h to 2 h)	400	Very intense activity sustained without breaks; emergency and rescue work and long-distance skiing events

[a]Indicated metabolic rate refers to the average of 60 min over shifts continuous work and different kinds of ski activity

It is important to note that responses to cold stress become impaired with age. There is a reduced perception of cold, body temperature decreases and the intensity of vasoconstriction and shivering are reduced with a consequent effect on metabolic heat production (Frank *et al.*, 2000), although the degree of a user's fitness can affect the extent of impairment.

Standard performance requirements

Requirements for protective clothing ensembles are set within general ergonomic requirements (material selection, design, comfort consistent with activity, size designation and aftercare – BS EN340, 2004). Those requirements, essential against wind, wet and cold, include the following:

- wind may considerably increase convective heat loss, so what is needed is an air permeability (AP) measure of how easily air can penetrate the outer garment material (on a scale one to three);
- resistance to water penetration (WP) has two or three classes that indicate resistance to the passage of water (BS EN343, 2003). For example, for extreme activities, such as skiing. W. L. Gore recommends that a waterproof/breathable membrane should withstand a water entry pressure of 7.1 lbs/square inch;
- water vapour resistance (R_{et}) is a measure of resistance to evaporative heat loss, i.e., how easy it is for moisture to move from the body to the environment. The lower the R_{et} value the less resistance is offered to moisture transfer and, therefore, the higher is the breathability of the clothing. This Hohenstein method was extended to a R_{et} Comfort Rating System – see Table 8.3.
- The most important requirement for cold weather clothing is thermal insulation. There are two common thermal values: one TOG = 0.1 M² kW, 1 CLO = 0.155 M² kW⁻¹. Thermal resistance (R_{ct}) is the temperature difference between two faces of material, divided by a heat flux made up of one or more of conduction, convection and radiation. Effective thermal insulation (I_{cle}) is the insulation from skin to outer clothing surface under defined conditions, measured with a stationary mannequin. Resultant effective thermal insulation (I_{cler}), the insulation from skin to outer clothing surface under defined conditions measured with or calculated for a moving mannequin (BS EN342, 2004; see Table 8.4). Required clothing insulation (I_{REQ}): this standard sets out criteria for the determination and interpretation of cold stress when using I_{REQ} and local cooling effects (BS EN ISO 11079, 2007).

Table 8.3 R_{et} comfort rating system

Rating	R_{et} value	Description
Very good	0–6	Extremely breathable and comfortable at a higher level of activity
Good	7–13	Very breathable and comfortable at a moderate rate of activity
Satisfactory	14–20	Breathable, but uncomfortable at a higher rate of activity
Unsatisfactory	21–30	Slight breathable, giving moderate comfort at low rate of activity
Very unsatisfactory	31 +	Not breathable and uncomfortable, with a short tolerance time

Table 8.4 Resultant effective thermal insulation of clothing I_{cler} and ambient temperature conditions for heat balance at different activity levels and durations of exposure (BS EN 342, 2004)

Insulation I_{cler} m².K/W	Wearer moving activity			
	Light 115 W/m²		Medium 170 W/m²	
	8 hours	1 hour	8 hours	1 hour
0.310	−1	−15	−19	−32
0.390	−8	−25	−28	−45
0.470	−15	−35	−38	−58
0.540	−22	−44	−49	−70
0.620	−29	−54	−60	−83

These temperature values are only valid with even distribution of the insulation on the body and with adequate hand-, foot- and headwear and an air velocity between 0.3 m/s and 0.5 m/s. Higher wind speeds will influence the temperatures because of wind chill effects (see BS EN ISO 11079, 2008).

Cold effects are seen by Holmér and Rintamäki (2004) as being the result of tissue cooling, poor clothing design and difficulties associated with the work environment; they suggest that these effects should be identified, evaluated and controlled as part of a risk management programme. The strategy has three stages: systematic observation of a number of defined risk factors such as air temperature and velocity, clothing, etc; determination of cold stress, using (I_{REQ}); and the assessment of outcomes by qualified experts.

General cooling requires measurement of thermal parameters, activity levels and metabolic rate, thermal insulation of clothing in use, calculation of I_{REQ} (both prediction of insulation required and evaluation of existing clothing), comparison of I_{REQ} with clothing in use, evaluation of conditions for thermal balance and recommended maximum exposure time. Local cooling of the head, hands and feet may produce deterioration of performance and cold injury. Cold is usually determined by the weather and protective measures (adjustment of clothing and control of exposure). It is argued in this standard that there is insufficient knowledge of responses to local cooling for a single, evaluative measure to be used. All types of local cooling may occur including convective, conductive (see ISO 13732-3, 2005), extremity cooling (fingertips, see BS EN511, 2006) and cooling of inspired air (airway).

The principles set out in the I_{REQ} standard can be used either to predict or to evaluate required clothing insulation. Of particular interest to the user and the design team is the classification of metabolic rates for various activities and the basic insulation values for selected garment assemblies, from which a converted value can be obtained for comparison with I_{REQ} (BS EN ISO 9920, 2007). Similar examples are shown in Table 8.5.

Table 8.5 Insulation values of various clothing ensembles – Cold protective clothing – With dynamic insulation values for walking (no wind), body surface area covered (BSAC) and mass of clothing without shoes and belts (BS EN ISO 9920, 2007)

No.	Ensemble description	BSAC (%)	Mass (g)	f_{cl}	I_{cl} (clo)		I_T (clo)	
					Static	Dynamic	Static	Dynamic
601	Extreme cold weather expedition suit with hood (down-filled, one-piece suit), thermal long underwear top and bottoms, mittens with fleece, liners, thick socks, insulated waterproof boots	98.9	2,804	1.5	3.67	3.21	4.12	3.54
604	Ski jacket with detachable fibrefill liner, thermal long underwear bottoms, knitted turtleneck sweater, fibrefill ski pants, knitted hat, goggles, mitten shell with fleece glove inserts, thin knee-length ski socks, insulated waterproof boots	98.5	2,943	1.34	2.3	1.75	2.81	2.12
605	Extreme cold weather down-filled parka with hood, shell pants, fibrefill pants liner, thermal long underwear top and bottoms, sweatshirt mitten shell with inner fleece gloves, thick socks, insulated waterproof boots	98.8	2,916	1.47	3.28	2.53	3.74	2.86

162   Textiles for cold weather apparel

These calculations refer to an average person and serve only as a broad set of guidelines. Physiological capacities, clothing behaviour and individual demands all vary; the final choice and adjustment of clothing to a specific environment will be made according to the experience, needs and preferences of the individual. It is perhaps these variables that need to be addressed if we are to help to create a more sustainable industry. However, when we look at the balance between heat generation and heat exchange, what needs to be achieved is not a constant temperature but rather dynamic equilibrium; as external conditions change, so the body's homeostatic system responds to bring the internal body temperature back to a norm. It is the individual user's homeostasis that is enhanced through the selection of an appropriate layered system.

8.3.2   Stage two: specifying user requirements

In traditional clothing product development for RTW cold weather clothing the process would commence with an exploration of the existing commercial market. The aim would be to identify a gap in the market for new products, as an opportunity for extending the product range or to assess competitive products, created within regular seasonal cycles. Outcomes help to profile the consumer's characteristics, the product and the pricing strategy, which can contribute to the design brief. However, the strategy for a user-oriented approach would be to identify the recreational and commercial occupations that take place in cold weather and then to liaise with users who participate in these activities.

Many kinds of requirement need to be addressed before apparel can be designed and developed. User requirement trees were introduced by one of the authors (McCann, 1999), and have been well documented for functional sportswear. A development of this tree was used for functional clothing (Black et al., 2005). Key issues are function and style/form. Central to both is the need to establish a user profile – gender, age, ethnicity, socio-economic and life style factors – from which size, shape, style and clothing preferences can be determined.

Size and shape

The advent of 3D scanning technologies has enabled recent national anthropometric databases to be created which can be mined for selected user market segment size and shapes, providing opportunity for companies and design teams to identify more carefully user groups to produce better fitting clothes (Williams, 2004; see Fig. 8.3). However, whilst having access to accurate two-dimensional body data relevant to RTW size range compilation and designation labelling requirements for cold weather clothing (BS EN340, 2004), it is the three-dimensional data providing body *shape* for individuals and groups that are beginning to have a greater impact. It is now possible to identify user groups shape ranges to:

Factors affecting the design of cold weather performance clothing 163

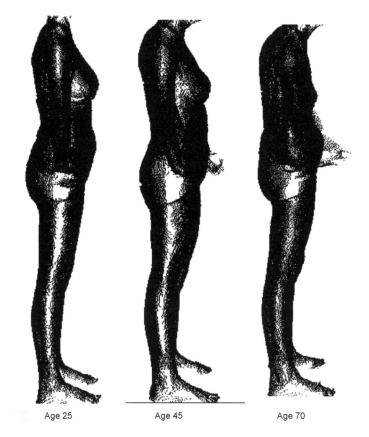

Age 25   Age 45   Age 70

8.3 Images of 3D scans showing changes of posture with age (Sizemic, 2008).

- make 3D mannequins (either for individual companies such as Rohan, whose target market is 35-year-old customers and above, or to reflect national shape characteristics, such as SizeUK and SizeUSA);
- create avatars (TC2, 2008), either for an individual or for user group body shapes, to upload into a virtual environment, so that users can either evaluate garment design and fit,[3] or contribute to the creation of a garment design directly onto a 3D avatar[4] (Optitex, 2008; TPC, 2008).

Some of these systems include fabric libraries, from which appropriate materials can be selected, imported and visualised, as parts of a layered garment system design. In some cases, these are tailored specifically for insulation,

---

3. Traditional garment pattern sections are scanned and scaled so as automatically to wrap around a three-dimensional virtual image of the user's shape.
4. The design is created directly onto the 3D shape of the user, and the pattern sections are unwrapped from the image, flattened and exported for physical cutting and assembly.

evaporative resistance and permeability (Yermakova *et al.*, 2005) as well as a CAD system being used to predict the thermal performance of the human body in a given scenario during simulated wearing periods (Aihua *et al.*, 2008).

These are exciting developments which promise to improve fit and promote user participation in the design and development process but, at present, the majority of mannequins and avatars can only represent static anthropometric stances. Dynamic anthropometry, relying solely on 3D scans to provide anatomic movement and virtual comfort fit analysis, is as yet at an early stage of development (Ruto, 2009). Combinations of scans with software to simulate movement have been made available by Cyberware, and new systems (such as LightStage) allow the capture of three-dimensional, dynamic, digital forms, opening up new methods for documenting and analysing human physical activity.

Whatever the method used, it will be the visual documentation and analysis of a physical activity (including the duration and time of day at which it is undertaken) that will help to establish metabolic rates for specified clothing use, and provide valuable information about the user's body stance (e.g. a ski posture), range of body movements (reaching and bending) and motion (walking and climbing) required to ensure that all garment layers incorporate design lines to accommodate posture and movements (as indicated in BS EN340, 2004), enabling the body to remain covered, and unencumbered during activity. These principles are key to the optimisation of clothing for performance in a cold climate, although a comprehensive study, analysing the biomechanical and locomotion effects on clothing and footwear for cold climates, undertaken by the US military (O'Hearn *et al.*, 2005) suggested that there is a tradeoff between protection and mobility, and concluded that bulky clothing constrains movement, affects resting posture and alters gait. This lends support to the argument put forward by Havenith (2002) that the weight of garments impedes performance, which subsequently increases metabolic rate and, therefore, the volume of moisture to be transported away from the skin.

Little research has been reported on user style preferences and the ways in which the aesthetic requirements of users have been incorporated into design for protective clothing. There has, however, been expert user participation in the development of cold weather clothing for companies, for example, the partnership between the climber Chris Bonnington and Berghaus. There is also the development by experts of their own branded products (e.g. Nick Gill of Gill Marine and Rab Carrington of RAB). Reported user style preferences have been confined mainly to online design (e.g. shoes), where limited options for personal style features have been offered (e.g. colour). When later asked to select a preferred design, users chose their own designs rather than those created by a professional team. This might be thought to support Black's emphasis on the need to consult the user, discussed in Section 8.2.3, and, perhaps, to offer another route to sustainable product design, but whether user design is good design is the subject of current debate.

McCann's requirements for good design include fitness for purpose, appropriateness to task; durability; acceptability to the user (with respect to culture, lifestyle and cost); achieving or exceeding required standards, adding value, provision of innovative solutions and satisfaction of aesthetic criteria – which may or may not have fashion currency (Black *et al.*, 2005). The extent to which fashion considerations impact on the use of protective clothing depends on the activity considered, but it appears to be growing in importance. A leading manufacturer of clothing for ocean sailing reported that cold weather ocean clothing is being influenced by fashion-oriented mainstream sportswear (author interview, 2009). Colour was identified as a key fashion component, affecting sales. It should, however, be noted that the colour of some components is dictated by standards and conventions associated with safety provisions. For example, the use of yellow reflective sailing hoods (Fig. 8.4).

The design of each component must nevertheless be co-ordinated with the overall design of the clothing system in accordance with user style preferences, design trends and existing standards. Health and safety, ease of movement, adjustability of attachments, manipulation of garment components (such as a zip), and ease of donning and doffing must all be considered as a point of departure to inform the design and development of the clothing to be worn in cold climates.

8.4 Styled garments from catalogues of sailing, motorcycle and expedition clothing suppliers (Gill Marine, 2009; Held, 2009; RAB, 2009).

8.3.3 Stage three: design

Cold weather clothing is associated with a wide range of occupational and recreational activities, but we look first at some generic considerations for the design of cold weather protective clothing. A powerful form of thermoregulation is behavioural, as the user puts on and takes off layers of clothing, changes posture, moves or takes shelter, etc. (Parsons, 2003). The user may not be conscious of making such adjustments. Comfort may be considered neutral, in the sense that the user's conscious concern is with feelings of discomfort. There is also what we may call the 'feel-good factor', where the focus is not only on cut, shape, and fit, etc., but also on the wearer's aesthetic and cultural comfort.

An effective clothing system is essential to survival in a cold environment. Systems are characterised by having layers and parts, connected in an organised and complex way, together forming an efficient, interrelated and behavioural whole needed to maintain the efficiency of the clothing microclimate and to provide appropriate levels of protection.

Component layers for cold weather ensembles identified in BS EN342 (2004) include: outer shell; liner (insert with watertight property); thermal lining (non-watertight layer providing thermal insulation); thermal liner (a layer with a watertight property providing additional thermal insulation); lining (innermost material without watertight properties). These components may be combined in a variety of ways (BS EN343, 2003) and, in some cases, increased to meet occupational and recreational needs.

Critical to efficiency of all layers is the way in which various components of clothing are designed to interface with one another at various parts of the body, most particularly at the outer layer:

- upper torso garments/lower torso garments;
- upper torso garments/neck/head/face/eyes;
- upper torso garments/wrist/hand;
- lower torso garment/ankle/foot.

There must be sufficient overlap between components to ensure that the body core and extremities are covered. This particularly applies to neck, wrists and ankles (to prevent warm air escaping and the penetration of water, wind or snow), even during vigorous activity.

Key to the design and development of these garments for the protection of the body core and its extremities is the selection of materials for each of the layers in the system. Shaw (2005) describes a useful approach:

- the assessment of hazards;
- the identification of relevant standards, specifications and guidelines;
- performance testing;
- suitability in performing a job (comfort, cost, durability, use and cultural factors).

Factors affecting the design of cold weather performance clothing   167

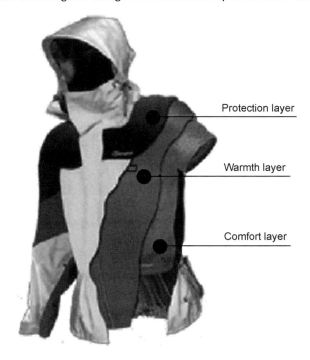

8.5 The Berghaus layering system (Berghaus, 2009a).

The approach does not, however, include reference to the need to minimise environmental impact set out in the standard BS EN340 (2004). Some of these criteria are discussed below among recommendations for generic layered systems, their garments and component parts.

A three-layer system comprises: a base-layer for moisture management, worn close to the skin; a mid-layer for thermal insulation; and an outer layer or shell, to protect the microclimate within the system. In addition, a new category has come into usage, promoted as the soft shell: a hybrid of outerwear protection with the comfort of a mid layer. See Fig. 8.5.

Base layer

When selecting materials to be worn next to the skin, it is essential to select fabrics that do not retain moisture. An absorbent base layer would be dangerous in retaining sweat and would conduct body heat away. If the wearer were to be stationary in a cold environment the moisture might freeze within the clothing. The base layer should therefore be primarily composed of fibres such as polypropylene and polyester that are hydrophobic (moisture repellent). Normally, base layer garments will have texture or hairs on the interior surface to attract moisture from the surface of the body, so as to transport moisture through the fabric and away from the body – propelled by body heat. This process may be enhanced by a hydrophilic application on the inside of polyester fabrics as, typically, polyester would not normally attract moisture. Once attracted, the moisture passes through the capillary structure and evaporates on the exterior surface of the base layer (with the proviso that the mid and outer-layer, garments are designed to be breathable). A denier gradient base-layer structure may be introduced; an inner yarn will have a larger filament and an outside yarn smaller filaments. Capillary action causes moisture to move away from the larger filament to the smaller filament. The base layer may use a weft knit, with garment pieces either cut and sewn or engineered in flat knit construction, in other cases they are seam-free, whole body knits. Stitch structures and yarn properties may be adopted for body mapping, with zones designed to enhance temperature control, breathability, provide support, incorporate antimicrobial finishes and enhanced conductive properties.

Base layer top garments and pants are normally close fitting. Tops for cold weather have crew, polo or turtle neck finishes, long sleeves and an extended back length. The majority of these knit structures allow easy movements; others, more highly styled, may be in elastomeric fabric constructions or have stretch panels inserted to provide additional articulation for the arms. Specific winter sports activities (e.g. snowboarding) may require longer body lengths for upper torso garments. Edge finishes at wrist and ankle are usually fashioned into the seam-free structure or cover seamed in 'cut and sew' assemblies. Where necessary, zip closures are used at the front of smock or jacket-type tops for ease of donning and doffing. Garments are machine washable and quick drying.

Mid layer

The design of garments for the mid layer is, again, fibre and textile driven. Mid layer garments provide an opportunity to use fabrics that increase or decrease levels of insulation. The trapping of still air within the clothing system provides insulation; this layer may be made up of interchangeable combinations of garments using a variety of lofty materials and constructions. The clothing system must be kept dry through the application of hydrophobic fibres in constructions that include fibre piles. Quilted garments may be introduced in varying designs, weights and constructions, using man-made fibres or, for predictably dry conditions, natural goose or duck down. Mid-layer garments are normally protected by an outer waterproof and breathable shell garment to retain the warmth, although some synthetic or down garments have water repellent fabrics that may be worn for a limited duration in damp conditions, while tape seamed wind and rain proof developments offer greater protection but with additional weight and more limited breathability.

Mid-layer insulation includes jackets, vests, smocks and pants. They can be

Factors affecting the design of cold weather performance clothing 169

8.6 Technical base layer and mid-layer fleece (Musto, 2008).

warm and wind or waterproofed, usually cut with a relatively close fit that moves easily over the base layer.

Tops have soft, lined collars and, in some cases, main body sections that cover the lumbar region, and may have additional layers to protect the kidney area and, if necessary, additional layers on the shoulder or seat area, where the insulation may be depressed by equipment being worn or during prolonged seating. Articulated arm lift (enhanced by stretch) is needed to match that of the base and outer layers, all of which can have compatible design elements. See Fig. 8.6. Waist and ankles can have cuffs, be elasticated, have Lycra binding or be closed with Velcro. Waistlines can be controlled with elastic or draw cord features.

Outer layer

High-performance outer layers are the last layer of protection. They are normally unlined, but are the most complex in structure. The most common requirement for an outer shell garment is to be super lightweight with minimum bulk for easy storage, avoiding issues reported on the adverse consequences of bulk. The fabric selected for the shell or protective layer should provide a balance between proofing and breathability for the specified range of activities. The wind chill factor can reduce ambient temperature and threaten the clothing microclimate. Outer fabrics are normally of lightweight nylon (which is stronger than polyester), woven construction with coatings or laminates in two- or three-layer fabric assemblies. Two-layer fabrics have an exposed coating or laminate on the inside, normally protected by a loose mesh lining, and three-layer fabrics have a sandwich construction, with a fine single jersey backing or mesh to protect the laminate.

The outer shell design must incorporate appropriate ventilation as few textile assemblies cope with the moisture produced from extreme workload. Softshell is a new generation of outer layer garment that has emerged as an all-rounder for winter and summer. These are often made of water-repellent, stretch-woven outer-face fabrics and inner pile constructions. A windproof laminate may be introduced into a sandwich type assembly, for example, the three-layer Gore-Tex jacket, which has a fleece lining. As with the base layer, new 'Comfort Mapping Technology' is being introduced, promising not only insulating warmth for the body's core, but also for various temperature requirements for different parts of the body. These new structures offer warmth, breathability, comfort and fit, in particular for snowsport activities.

Typical outer layer garment types include jackets, smocks, trousers and salopettes. One-piece coveralls are a requirement for some applications (such as wind protection or snow camouflage) but, like salopettes, have the advantage of reducing bulkiness at the waist and providing better protection for the body core (Fig. 8.7). All need to be designed and cut to fit comfortably, without friction, over base and mid-layer garments and incorporate the articulation identified through the activity analysis without restriction to major body girths or extremities. Ease of donning and doffing of outer layer garments can be enhanced by material characteristics, degree of fit and tolerance and the position and length of opening. The development of three-layer material assemblies incorporating membranes, has eliminated the need for linings and has reduced friction between outer and mid layers. If linings are required they should be constructed from materials that slide easily over mid layers, in particular if linings incorporate removable mid insulation layers.

Upper torso garments

These have seamless shoulders to prevent penetration of moisture and to ensure comfort when wearing a harness or carrying equipment. Access to pockets need to be considered if any of the harness tapes fit around the waist. Garment length is an important consideration and should meet existing standards (e.g. European Standard, 2004). Overall lengths need to cover the body when seated or engaged in reaching. Sleeve lengths should cover the cuffs of other layers to prevent water seeping up from exposure to snow or rain and may have storm cuffs of varying heights. Jackets may have fixed or removable self or lightweight fabric snow skirt or gaiter, designed to prevent the ingress of snow, with options for a trouser interface and the use of branded trim at the lower, elasticated edge.

Collars are cut very high to protect the neck and parts of the head and face. They have soft linings; with an internal draw cord to reduce the top edge of the

Factors affecting the design of cold weather performance clothing 171

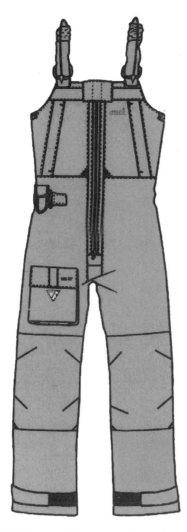

8.7 Musto HPX trousers. Chest fleece, two-way zip and waterproof gusset, thigh pocket with water deflecting inner lip, high chest fit, unlined for easy access and speed of drying. Incorporating Gore-Tex (Musto, 2009).

collar to fit the neck and integral centre zip closure and additional security tabs. Collars act as a platform for a variety of head and face protections (see Head, face and eyes on pages 174–6).

Lower torso garments

These include traditional trousers and salopettes. The latter – all-in-one garments – may include either adjustable elastic straps or, when cut to cover the whole body core, have stretch shoulder yokes to meet bending and reaching requirements. Upper and lower torso garments may be zip linked at the waist, or have fixed or independent gaiters with footwear connections. Further comfort features can include elasticated side sections and side-zipped, drop-down seat for women's wear. The length, width and articulation of trouser legs need to be sufficient to enable bending, squatting and kneeling without limb exposure. Additional insulation or anti-abrasion reinforcements at the shoulder, elbow, knee and seat offer opportunities for design development. Wrists and ankles may be secured with adjustable strap fastenings and can be cut with storm cuffs of varying materials and heights (e.g. waterproof latex neck and wrist seals, for sailing).

Closure systems

These are major protective features on outer garments. In addition to external protection from the wind, wet and cold, closures facilitate donning and doffing, aid body movement and allow control of the microclimate via strategically designed body core and extremity cross-flow venting. Vents make it possible to remove water vapour from inside the ensemble and allow convective, conductive evaporative (and radiant) heat loss from the skin surface. Traditional closure systems facilitate venting at the centre front, neck, wrist and ankle, with additional control gained with mesh-lined zip apertures at back, underarm, outside arm and thigh levels. It has been suggested that these adjustable openings are a more efficient way to control heat and moisture than by passive diffusion through garment layers, and that they can be matched to the temperature at different points on the skin. For example, the skin temperature at the chest has the highest reading during exercise, and that an opening near the chest may be the most effective way of creating ventilation (Ruckman *et al.*, 1999). The lengths of openings need to be designed to allow easy removal of footwear and, if at the neck, the opportunity to pull over the head.

A range of water-repellent zips (YKK and Riri) with rubber or plastic-coated tapes and teeth is readily available. They can be moulded or enclosed with a one- or two-way storm flap that helps to prevent water entry and preserve the microclimate. Two-way zips have the advantage of providing options for venting and increased motion; for example, they can be unzipped at the hem to provide extra room for body movement when climbing. Where appropriate, they are cut with drainage channels, secured with strip hook and loop stays and ends protected with folded sections or moulded rubber pouch. Zip pulls need to be suitable to be manipulated with covered hands, and easy to locate by feel or sight (see Figs 8.4 and 8.8). The placement and number of zipped pockets is determined on the basis of user requirements. They can be watertight, secure (to hold mountain lift pass), designed to allow water to drain or air to pass through, and hold electronic components. They must be large enough to prevent contents spilling out and give adequate access for gloved or mitt-covered hands.

Factors affecting the design of cold weather performance clothing 173

8.8 Zip pullers, enlarged and extended, can assist manipulation when wearing gloves in cold conditions (IndiaMart, 2009).

Hand warmer or mesh linings provide further comfort and in some cases, single-handed access to internal cord pulls for adjusting hem and waistline. Pocket flaps must be designed to meet BS EN342 (2004), i.e., 20 mm wider than pocket mouth. Activity or safety-specific loop fixtures may be included to support a camera, lifejacket, key clip or, in some instances, thumb holders (to keep sleeve extended in extreme conditions). Branded straight or pre-moulded reflective strips may be attached to the head, shoulders and front and back of garments. It is important that all garment components be strong enough to allow the range of body movements to deal with the tasks in hand (BS EN340, 2004).

Seams

The type and placement of seams in an outer layer garment is crucial to the overall performance and integrity of a garment. Several seam junctions at one point need to be avoided as this can degrade the function of the seam and the garment. Designs should also avoid shoulder seams and set in sleeve constructions that risk ingress of moisture. Waterproof and water resistant seaming is continually being developed. Sealed seams with impermeable or micro-porous tapes are widely accepted although as widths of tapes have decreased and use of staple and stretch materials increased seam bulk can be reduced and comfort enhanced, in particular when seams are used in areas such as the underarm. Stretch tapes for use on non-stretch garments give a low profile non-chafing seam. Stitching, welding, radio frequency and ultrasonic techniques can all be used with a variety of seam structures.

Stitch-less seams are usually lighter and more flexible with laser cut edges introduced where textiles have a minimum of 60% synthetic content. Bonded construction can be a 'welded mock-felled seam' with two layers of glue that stretch and are waterproof. Other constructions can be described as 'lap seaming', with adhesive tape sandwiched in seams that can stretch with the wearer, or 'butt seaming' that requires initial ultrasonic joining prior to opening the seam flat and applying tape to strengthen the join on the inner surface. Pockets and zips may be attached by bonded overlap seaming, constructed with a single layer of adhesive (Zweld, 2009). Exterior seam tapes may contribute to the design aesthetic. Bonded methods can be applied to textile assemblies with mesh linings. Developments in surface treatments are intended to enable higher bond strengths in the future that will use fewer adhesives, in saving materials and energy.

Head, face and eyes

One of the body's most effective comfort sensors is the head. Figure 8.9 shows that severe conditions can make it necessary to extend protection to the face, ears and neck. During vigorous activity the head can lose up to 50% of body heat. Garment systems for the head follow the layering principle. These can comprise:

- single garments – close fitting head bands, hard and soft hats, face shields, ear and neck warmers and balaclavas;
- integrating garments with multi-featured components – garments that include hard hat liners with adjustable face and neck shields, chin and nose warmers, or multi-purpose garments that can be used to cover the head or when adjusting body temperature as a neck warmer.

Combinations of these garments (e.g. skull cap, neck gaiter and fleece) can provide a layered system. But some rely on the height of the collar of the torso garment to provide protection for the neck and face. Other types of head, face and neck coverings for severe weather include:

- an independent hood system or hood pod that is layered, water and windproof, with a skirt cut to fit over the shoulders and fasten to the body or torso garment;
- hoods developed as an integral part of a torso garment. These can be cut to form the continuation of a high collar, permanently attached and rolled away within the collar. Hoods are often cut to fit over a helmet or other close fitting head cover (e.g. hat and balaclava) and are normally articulated, with single, internal, hand-operated draw cords that close the outer rim of the hood to the face, often creating a fold from which water can drain (Fig. 8.10). A second

Factors affecting the design of cold weather performance clothing 175

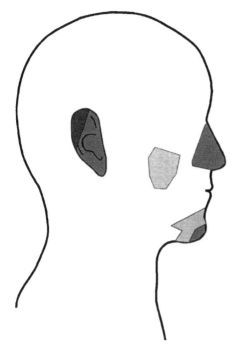

8.9 Local cooling of facial extremities (Lehmuskallio, 2001).

8.10 Cinch with toggle (source: the authors).

176   Textiles for cold weather apparel

8.11 Left: Ice Armor headgear by Clam Corporation, USA (Ice Armor, 2009). Right: Berghaus extreme facial protection (Berghaus, 2009b).

draw cord can be used to cinch the hood to the contoured shape of the head, so as to improve the user's vision.

Additional features include nose and face flaps, fur ruffs and peaks (with wire or reinforced support), secured with Velcro fastenings. Garments can be designed with additional high neck components that include soft chin-support linings, moulded nose covers and perforated front collars for respiratory mouth protection (Fig. 8.11). Covering the nose and mouth also limits drying of the mucus membranes in the nose and throat, caused by climates with low humidity (Castelani *et al.*, 2001). Hoods can also serve as communication or entertainment channels, housing speakers and microphone.

Essential to the design of all head and face wear is the need to ensure an effective interface with any thermal protective eyewear. These need to be designed with either a deeply wrapped polarised lens or side shields that block out ultraviolet radiation and protect the eyes from sun reflection, snow glare and high winds. But, as in the case of motorcycling, eyewear is prone to fogging; experienced users recommend the use of a facemask to deflect moist air from the mouth and nostrils. As a precaution, the wearer can also be equipped with a chamois lens wipe.

Hands and feet

There are many individual requirements for hand and foot protection in the cold. As with general clothing, information – on the environment, activities to be undertaken and on the user – need to be established prior to commencement of the design process. The thermal balance of the body is paramount in preventing falls in temperature and, as can be seen in Fig. 8.2, above, body extremities are extremely vulnerable in the cold. Fingers and toes have a high surface area in comparison with their volume. This means that they lose heat easily, while being poor at generating it.

Hand wear

The hand is the chief means of manipulating the environment and each of its component parts (palm, dorsum or back, digits, thumb, wrist and forearm) need to be considered in the design process. Hand covers need to protect against convective and conductive cold, as well as water penetration. These are linked either to climatic conditions or to industrial and recreational activities (BS EN511, 2006). In understanding design requirements for hand coverings, Abeysekera (1992) suggested that, after protection from the cold, different activities have a different hierarchy of needs, such as wearability needs, requiring good fit, flexibility of use, etc., reinforcing the need to engage with users throughout the design process.

In protecting the hand from the cold, covering systems closely follow the layering principle described above, using similar material structures that wick, insulate and are breathable, wind and water proof (outer shell, moisture barrier, insulation, inner liner, insert and, perhaps, an inner glove). A key difference is those materials used to form part of a mid or outer layer, where individual component parts of the hand require different levels of protection to aid the execution of activities (Fig. 8.12).

The configuration of hand covers can range between two and five layers, with often a different number of layers on the palm to the dorsal and digits. For example, a glove with (i) a palm outer of leather and an inner of acrylic pile and (ii) a dorsal section with an outer layer of Gore-Tex, a mid-layer of polyester batting and a pile fabric (Kasturiya *et al.*, 1999). Systems can comprise mitten, gloves or a hybrid. Mittens allow maximum warmth through the collective, radiated heat of the fingers, but inhibit dexterity. Gloves can optimise dexterity, but have a higher risk of individual finger cooling. In particular, when hands are in contact with cold surfaces, although the degree to which this occurs can be conditioned by acclimatisation to the cold, for example, people who live and work in the arctic region. A hybrid system can have the advantage of individual glove liners or a glove that converts to a mitten (where a mitten is designed with a flap to go over glove fingers), which can in turn aid venting and performance.

Base layers can be a soft single material or a liner bonded with an insert, for example, recent developments for snow wear by Gore-Tex (2009). Insulating layers may be removable, have a fleece layer for palm or a dorsal area lined with loft or sealed foam padding. These insulations can be extended to protect the

178 Textiles for cold weather apparel

8.12 Men's mittens. Lightweight, waterproof, breathable ripstop Ventia shell and fully seam taped. Anatomical boxed construction and 3-panel thumb for dexterity. Removable insulated liner features 10 oz PrimaLoft on the back of the hand and thumb, and Moonlite Pile fleece on the palm. Outdoor Research Alti Mitten (Back Country, 2009).

knuckles and backs of fingers, although insulation that is too thick will reduce dexterity. Shells, in addition to fulfilling waterproof and thermal properties, can be seam sealed, lightweight or have tear-resistant, tough outer properties. Shell linings may (as with jackets) be smooth materials to aid donning and doffing. Palms and thumbs can be seamless, comprise a fleece surface or be reinforced with rubber, leather, neoprene or have additional gripping surfaces. Dorsal areas may be stretch or moulded into a curved shape with pre-curved, seamless fingers (though these must remain flexible to maximise performance). Fingertip grip or thumbs with wipe patches can be included.

Notwithstanding the design challenges of these complex layering and material structures, a key feature of these systems is the way in which the hand covering interfaces with upper torso garments that helps to maintain the thermal balance of the body, which has, in turn, been found to affect the finger temperature by more than 10° (Anttonen *et al.*, 2005). These essential interfaces can be designed as long wrist cuffs or gauntlets with elasticated, adjustable cuff fastenings or one-handed tightening shock cords that cinch to the wrist and upper arm (Fig. 8.10). In addition to maintenance of the microclimate, these interfaces prevent the ingression of cold, wet and wind. Some double, sealable cuff systems (e.g. for ocean sailing) can be designed as an integrated and waterproof garment system.

All hand covering systems need to be designed and cut to fit comfortably with all other layers and to prevent layers from bunching or sliding and, if necessary, secured with Velcro stays. To ensure against loss in windy conditions, they should be worn with a harness or attached with loops to the lower edge of a sleeve. All systems should be easily donned and doffed with cold hands.

Footwear

Feet, as with other extremities, are susceptible to rapid cooling but, unlike the hands and face, they can be in constant contact with cold surfaces and, without protection, exposed to high conductive heat loss. In addition, feet are prone to produce a great deal of sweat, which can double in volume during exercise, requiring an efficient means of moisture transportation to keep feet dry as well as to protect from external conditions. Coverings for feet, ankles and legs comprise traditional layers of socks, boots and gaiters. All need to be designed to aid mobility, support weight and enhance individual activity requirements.

Socks, as with other garment base and mid layers, rely on a variety of knit fibres and structures that are durable and resilient, with anti-microbial and anti-odour benefits and engineered with ventilated panels for improved wicking. The design of each component part (as with hands) needs to be considered according to the requirements of the user. The toe, top, sole, heel, ankle, shin and calf may all need different design to accommodate degrees of cushioning, padded channels or linings to prevent bangs, absorb shock, shear or impact. For example, snowboarding socks may need lining to prevent toe bang, need thicker protection on the ball and heel and thinner cushioning for the shin, instep and arch, with thick pads to protect the calf from boot pressure (Thorlo, 2009; Fig. 8.13). Elasticated sections can help to brace ankles and prevent socks from moving during activity.

Boots

The human foot combines mechanical complexity and structural strength and, with the ankle, provides support, balance and mobility for the body. These physiological requirements, together with climatic conditions and user activity needs, form a basis for the design of cold weather footwear. User requirements can range from traditional uniquely decorated fur snow boots, such as those by Mukluks, to high-tech ski boots that can be created from 3D whole foot scanning, 3D maps of the sole of the foot and designed in a virtual 3D

8.13 Left: boots with built-in gaiters; quick release buckles; Velcro® Straps; Cordura® Nylon Upper; ankle stay support; 1.5" VÆTREX™ foam liner; moisture trap; thermal insole. Arctic boot. (Northern Outfitters, 2009). Right: SNB snowboarding socks – thick cushion, with exclusive wool/THOR·LON® Blend (Thorlo, 2009).

environment. The anatomy of footwear usually falls into two – upper and lower sections – each with their own design challenges. Components are required to help meet both traditional cold, wet and thermal layering requirements, as well as those needed to support and cushion the foot and grip on slippery surfaces. Key factors affecting the design and function of some of the components are discussed below.

Outsoles are designed to insulate the foot from wind, wet and cold and to absorb shock. They can be constructed or moulded from rubber assembled from multi-components (offering opportunities to design soles in single or multiple colours), and create or branded lug bases to aid traction and grip, preventing slips and falls. Further stability can be introduced with angled or multi-directional studs for up- and down-hill travel, and the attachment of cleats or ice grippers. Recent developments have incorporated rapid prototype components. Midsoles have similar properties to outsoles. They are lightweight, with zone specific shock absorption, using polyurethane, compression moulded flexible gels and air cushions.

Footpads or insoles provide anatomically shaped cushioning, anti-microbial and anti-friction protection; they can have a moisture membrane or be compression moulded. Liners are the key insulation component, designed to reflect the temperature and the level and duration of activity. Thermal liner systems can be removable, washable and can be designed as a three-component membrane, multi-layered – as many as eight – using felt, foam, wool pile and 'Insulate' or created as an air liner that is pumped up to match the shape of the user's foot (see Fig. 8.13).

Boot uppers need to be waterproof, breathable and durable. They can be designed in leather, a range of nylon deniers, synthetic leather and rubber and in some cases (as with ocean sailing) lined with a membrane and seam sealed. Steel caps, high-density foam or gel padding can provide impact dispersion for the toe, ankle or arch. Heels can be supported with a welt, a full wrap cradle or for some activities (such as snowboarding) an external welded, full-length backstay. Closures for boots include low-tech, single strap and buckles to high-tech, tie lace, speed loop eyelet fastenings and dial-in, individual tight or loose fit. All need to be corrosion resistant and, if laced, to have a bellow or padded tongue. Top edges of boots can be designed with fur cuffs or snow collars with elasticated or tie drawstrings, or with locks to provide a reliable and secure fit to the leg (Fig. 8.13).

Gaiters can be designed as an integral component of an upper or as a separate component. Both are critical to the maintenance of the microclimate of the foot and ankle. Separate gaiters can be designed as a layered insulated thermal unit with a water and wind proofed outer shell. They can be lightweight, with zipper access and a Velcro sealed closure, and are essential if the height of boots is insufficient to protect the foot from deep snow (Fig. 8.14).

8.14 Gaiters. Versatile, durable, dependable and designed to fit virtually any boot. Waterproof and breathable 3 layer Gore-Tex xcr seon upper. Velcro sealed front opening zip for ease of entry (Sub-Zero Boots, 2009).

8.3.4 Stage four: evaluation

A key benefit of user-centred design strategies and three-dimensional, virtual environments is the opportunity to conduct iterative evaluation of design and fit prior to cutting and assembling physical prototypes. Notwithstanding the valuable sustainable approaches and the promise of the evaluation of thermal properties and movement within these three-dimensional environments, a physiological evaluation of the effectiveness of design within the context of user requirements remains an essential criterion.

Little research is reported on the evaluation of cold weather clothing as a stage within a user-centred design approach. Typical models of physiological evaluation of ready-to-wear thermal layered systems follow a schematic structure, developed by Hohenstein, and will be adopted for this chapter (Spindler and Thwaites, 2004; Rossi, 2005; Goldman, 2005). The protocols for the evaluation stage of user-centred design need to be established to reflect the identified user requirements (climatic conditions, activity, personal characteristics, etc.) to ensure that the same parameters are applied in the collection of data across all levels of the physiological system (Fig. 8.15). In addition, checklists or questionnaires need to be compiled with the user to assess

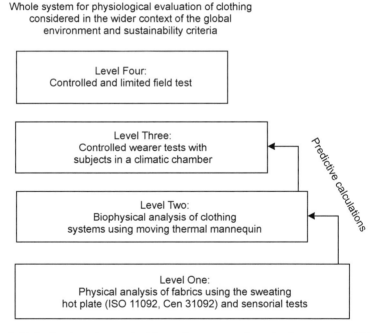

8.15 Evaluation: A sustainable environment: a broader framework within which to consider fabric and clothing tests. Partially adapted from Spindler and Thwaites (2004).

the two practical stages. These will need to identify criteria appropriate to the interrelationships between the:

- body – whole body, head, hands and feet;
- clothing – system as a whole and the efficiency of its component parts, e.g. accessibility of closures, pockets and venting;
- activities – range of movement, posture, tasks, duration within specified terrain;
- climatic conditions – temperature, wind and rain.

A useful framework for a physiological questionnaire has been published by Havenith and Heus (2004).

Evaluation of materials requires the measurement of Rct, Ret, AP, and WP values. In traditional design and development and user-centred design practice, materials testing would be completed prior to selection for design. Although, in the context of evaluation, such information is considered by Shaw (2005) to be of limited value for the prediction of layered clothing properties.

Biophysical analysis uses a life-size, heated and articulated mannequin to measure thermal insulation and calculate evaporative resistance. Measurements taken on mannequins represent an objective clothing function, and take account of factors affecting heat exchange between the body and its environment, including:

- the amount of body covered by textiles and their distribution over the body;
- the effect of design, size, fit and the adjustment of garment features (for example, open or closed fastenings – hoods on or off);
- the variation of body temperature, posture and movement (e.g. walking) (McCullough, 2005).

However, whilst these measurements may be useful for predictive purposes for RTW clothing and regarded as particularly valuable for assessing the effect of design on heat and moisture (Shaw, 2005), they do not necessarily represent real conditions and should be regarded as complimentary to practical testing (Holmér and Nilsson, 1995), suggesting, perhaps, that, for a user-centred design strategy a practical evaluation may be the only requirement. For example, no mannequins were used in the evaluation process for the comparison between new and existing clothing systems for offshore workers (Sandsund et al., 2001). Recommended practical testing include controlled wearer tests using a climatic chamber, and field tests.

Climatic chamber evaluations provide an opportunity not only to verify mannequin measurements but also to create a simulated user environment, where climatic conditions and activity protocols can be replicated. During such tests physiological changes (skin temperature, heart and metabolic rates) and user evaluation of thermal comfort can be recorded. An advantage of user-centred design strategies is that, unlike traditional RTW practices, user knowledge of the process and clothing system (such as aesthetic satisfaction, sensory comfort and fit) are explored before the evaluation could occur.

Field studies are vital to the evaluation of the effectiveness of the design and function of cold weather clothing systems, though it is recognised that it can be difficult to control the variables. The most frequently reported studies of layered military systems suggest that, if user requirements regarding ambient conditions are measured, the duration of activities determined and physiological factors monitored, then a clothing system (and equipment) can be properly evaluated.

The hierarchy shown in Fig. 8.15 is a useful model with which to evaluate the physiology of cold weather clothing. However, with the exception of the selection of materials (BS EN340, 2004), little attention has as yet been given to environmental issues in any of the tests, analyses and standards. The levels have therefore been placed within the wider context of sustainability, where, perhaps the outcomes of an evaluation using such a system would not only be a recommendation for marking and care labelling (as described in the EN340 standard) for protection against foul weather and cold, but would also go some way towards meeting both emerging standards (SMART and ISO 14064) and an ambitious eco-textile labelling system, recently proposed (Mowbray and Wilson, 2009).

8.4 Case studies: motorcycling and climbing

Clothing design for particular forms of activity involving exposure to cold are examined in these case studies of garments for motorcycle road wear and mountain climbing. Both employ the traditional garment layering system, driven by the physiological demands of protection from cold, moisture management and ergonomics. Key differences are considered below.

8.4.1 Rukka motorcycle road wear

This section looks at motorcycle clothing for high-performance road use in a cold environment. The design of a clothing layering system for road use falls into the categories of jackets, trousers, one-piece and divided suits that are ergonomically cut to address a predominant fixed posture. In contrast to the other cold weather activities, recreational motorcycling involves a single predominant posture, a limited range of movements, the risk of high-speed impact and severe wind chill with the need for high visibility. Karlotski (2009) describes how it feels to ride a motorcycle in cold and hostile conditions:

> There is cold, and there is cold on a motorcycle. Cold on a motorcycle is like being beaten with cold hammers while being kicked with cold boots, a bone bruising cold. The wind's big hands squeeze the heat out of my body and whisk it away; caught in a cold October rain, the drops don't even feel like water. They feel like shards of bone fallen from the skies of Hell to

pock my face. I expect to arrive with my cheeks and forehead streaked with blood, but that's just an illusion, just the misery of nerves not designed for highway speeds.

In addition to the clothing layering system described in the previous section consideration has to be given to body protection in high-speed accidents. Hostile environmental conditions demand a high level of personal protection, using body armour or abrasion-resistant materials, strategically positioned within the garment system. Specialist materials, made from fibres such as air-textured nylon Cordura, Kevlar and Lycra have been developed to combine lightness of weight with the injury protection afforded by leather. Standards embrace both protection from ambient weather conditions as well as the inclusion of protective components. The interior design of a motorcycle garment is often as complex as the outer. Personal protection varies in design and composition with a choice of back plates that may be of foam, padding, honeycomb and spacer constructions, hard armour or more recent intelligent shock absorbing polymers, such as Dow Corning Active Protection. Protective materials may be sewn or incorporated into garments (e.g. the Dainese shoulder-mounted airbag).

The concept of cold weather layering system adopted for motor cycling differs from that for climbing in that weight is not as restricted as the machine supports the weight and normally offers some storage and ability to power devices. A one-piece under suit or a co-ordinating top and pants can form a base layer. Close fitting stretchable fabrics help to support the body and reduce muscle fatigue. Functionality may be enhanced by odour control antimicrobial finishes such as Under Armour's Armourblock™ used in base layer products for extreme cold; this manufacturer's ColdGear® branding represents high-density double-sided knit fabrics in two-way stretch constructions with smooth outer faces and soft polyester pile inner surfaces (Under Armor, 2009). Engineered knit fabric structures may provide enhanced breathability at underarm and side panels. The strategic placement of total wind blocking front panels in the base-layer top and leg-wear will help to maintain core body warmth.

Traditional garments include jackets, pants and body warmer vest styles and quilted garments. These consist of lightweight nylon weaves covering polyester thermal wadding, quilted to prevent the migration of the filler. Detachable linings are easily removed by unzipping from the jacket fronts or pants, and unsnapping hook-and-loop connectors at arm and leg cuffs. Mid-layer motor cycling garments now include innovative 'Soft Shell' fabrics, described earlier.

Outer layer garments form a sector of industry that is driven by textile branded fibres. Such fibres are incorporated into sophisticated fabric assemblies. These are often co-branded with garment producers at point of sale, such as BMW (Germany) and Rukka (Finland). These companies work with branded textile producers such as DuPont for fibres, Schoeller and W.L. Gore for textile assemblies and finishes, and with Dow Corning for protection.

Rukka, a Finnish company established half a century ago, seeks out technical textile innovation for motorcycle clothing applications. The ideas of the design development team led by Jasmiine Julin-Aro were triggered by emerging innovations in materials for security and protection, new garment joining techniques and print applications. Based on materials research, Aro begins with hand drawn concept ideas, which are progressed to initial prototypes in collaboration with the experienced in-house technical team. Once garment development has been tested in-house, working prototypes are then verified and refined, through the input of an expert team of riders that includes doctors, biologists, lawyers and engineers. Garment design is tested in a range of conditions, including extreme cold conditions in Finnish Lapland.

The 'SRO Anatomic' Suit, the second generation of Rukka's 'Smart Rider's Outfit' project, is the result of close collaboration with W.L. Gore Inc., and other partners. The stretch Cordura outer material incorporates the GORE-TEX® XCR 3-layer laminate, 'Armacor', which offers flexible safety protectors, achieving a closer fit with minimal aerodynamic drag. The assembly has relatively good waterproofing, with a hydrostatic head of 45 metres, is extremely breathable at <60 R_{et} (BS EN 31092, 1994), and is claimed to be 100% windproof. Kevlar reinforced material is positioned in highly exposed areas, for tear and abrasion-resistance. A 3M Scotchlite reflectively coated Cordura 500 is used strategically to provide visibility at night when the rider is lit up by other road-users' headlights. The suit has a breathable mesh lining.

Other SRO design features include horizontal outer pockets with waterproof zips, two inside pockets, one for a mobile phone, and a readily accessible smart card compartment. The suit has zipped ventilation openings on both shoulders and at lower arm. A Detachable outside Gore-Tex® collar contributes to noise reduction and water protection. The jacket has a high fixed collar with a soft neoprene border lining and a removable, stretch Gore-Tex, draft-proof neck warmer zipped into the jacket on the inside. This material is also used for inner cuffs for a draft and waterproof connection between sleeves and Gore-Tex/ Outlast gloves, with additional short cuffs to fit between inner sleeve and upper material.

The second generation 'SRO Anatomic' body protection suit absorbs impact – using silicone on a flexible spacer warp knit that becomes rigid on impact, but otherwise flexes with body movement. This technology has been developed and patented by Dow Corning (2009). In contrast with rigid and bulky armour components this Active Protection System can be sewn directly into garments and accessories; interestingly, the warp knit spacer structure also contributes to the trapping of still air for insulation. Rukka has adopted this novel Dow Corning textile for the APS air protectors (elbows, shoulders, front, back, knees and hips). Rukka Safety Air Protectors conform to BS EN 1621-2 (2003).

In terms of thermal regulation, Rukka adopts W.L. Gore's AirVantage technology for the SRO's detachable inner jacket. A valve located inside the collar allows the wearer to inflate the jacket to match his or her insulation requirements; for example, it may be fully inflated prior to riding in cold conditions. The clothing system also incorporates OUTLAST® PCM thermal regulation in the jacket and trousers. This is another phase-change material that uses paraffin wax to absorb and release warmth to act as a temperature-balancing buffer to maintain the microclimate, according to the wearer's needs. This technology is suited to quickly changing thermal conditions. The sleeves have 40 g Outlast insulation to offer enhanced thermal protection for the extremities and the trousers have a removable Outlast lining (the lining is not connected to the upper material below the knee so that it may be worn inside purpose-designed Daytona SRO boots). The outer material is worn over boots. Rukka's coordinating SRO trousers feature an AirCushion System as an inner lining in the seat area. This mesh spacer material acts as a climate buffer that reduces condensation and thereby enhances riding comfort. The seat area has an external layer of Keprotec Antiglide fabric to provide enhanced grip and prevent the rider from sliding along the seat. The Rukka SRO (Smart Rider's Outfit) Anatomic clothing system won an award for innovation at Avantex 2007, Frankfurt (Rukka, 2009).

8.4.2 RAB expedition clothing

A climbing layering system consists of the traditional base, mid and outer shell layers. The base layer garment design for cold weather climbing is integral to the selection of fibre types for maximum comfort in fabric structures placed around the contours of the body and are invariably of fine knit structure, cut close to the body with fit and movement enhanced through both elastomeric stretch and mechanical stretch of appropriate knitted stitch constructions. Raglan sleeves and the use of gussets underarm avoid friction. Flatlock seams and zip guard protection at the neck also prevent chafing.

The design of mid layer insulated garments for mountaineering should be simple yet versatile, achieving a relatively fitted garment silhouette. Loose cut garments are less effective in maintaining a clothing microclimate. Ergonomic cut should address the articulation of arms and legs. The cuffs of stretch fleece sleeves may have thumb loops to keep the garment in position while stretching.

The RAB 'Expedition Jacket' is described as of a high specification for high altitude mountaineering, and it is claimed that it is 'the warmest jacket available in the world today'. It has a down filled box wall outer construction with a stitch through baffled inner to eliminate cold spots and maximise warmth. The overall weight is 1,550 g, made up of an abrasion-resistant, breathable coated polyamide outer (at 66 g per square metre) providing weatherproof protection, enhanced by Spandura patches over the key wear points (Pertex, 2009). The filling embraces 450 g of 96% premium white European goose down with 750+EU/850+US fill power. The jacket has a fixed hood for security and warmth in high altitude jet streams. Two huge cargo pockets have internal mesh pockets to accommodate water bottles and food. Two external Velcro closure hand warmer pockets and two zipped pockets provide additional storage. The two-way YKK zip has a double down filled storm flap and chin guard. The waist and hem have adjustable draw cords. The cuffs have adjustable Velcro tabs. The 'Expedition Suit' has, in addition, articulated knees and full-length side zips with multiple pullers that run from ankle to wrist to aid ventilation and access. 'Expedition Salopettes' complement the jacket as a two-part system allowing the user more flexibility than when wearing the suit.

In choosing between synthetic materials and natural down the primary consideration is whether the insulation is to be used in wet or dry conditions. Climbers try to avoid wet conditions but may elect to use synthetic fibres that dry quickly and stay relatively warm when wet. 'Belay' jackets are lightweight and low bulk in relation to fleece constructions and are designed to be worn over a potentially wet shell jacket on belays. The outer fabric should be lightweight, windproof and quick drying, but will be more vulnerable to tearing in comparison with protective shell fabrics. For the mountain environment synthetic down is more practical than natural insulation, as the synthetic will not escape if the garment is torn (with manufacturer guarantees).

Outer protection shell for garments used in climbing should be lightweight and combine strength, durability, impermeability, windproofing with breathability. They are primarily of three-layer construction in a sandwich construction with a membrane laminated between an outer substrate and an inner mesh lining. The choice of layering under the outer shell is of key importance for moisture management. A key consideration in designing the system is the anticipated workload, as any sweat produced will tend to be trapped.

The ergonomic cut should address the predominant posture and range that is characteristic of mountaineering. Cutting for climbing garments should avoid billowing and have sufficient arm lift to ensure that a garment does not ride up and that design features are accessible in relation to the position of a climbing harness or backpack. Pockets and closures should not be restricted in facilitating access to inner layers. Seams may now be welded and bonded, with face fabrics that are lighter in weight but tougher than ever before with strategically placed abrasion resistant 'zones' of lamination. The positioning of design features, such as pockets and closures should facilitate access to the inner layers and not be restricted by a harness or back pack.

A similar jacket that is cut for women in a three-layer Gore-Tex fabric weighs 440–487 g. It has a high-volume mountain hood, is compatible with a helmet, with a single-pull adjustment system. Co-ordinating salopettes have reinforcements (as described earlier), a gusseted crotch, a low-cut bib with front fly, quick adjust suspenders, an elasticised waist, a water-repellent fly zip and side zippers that stop short of the harness/hip belt area. Point of sale material suggests that bright colours will improve posed photographs at the summit!

Headwear insulation is essential. A typical solution would be a cap and balaclava in a stretch fleece such as Polartec's Powerstretch and the hands protected with mitts or gloves filled with down or synthetic fibre (Polartec, 2005). Mitts with inner gloves of fleece are required at high-altitude temperatures. Hoods should be cut to fit over mountain safety items such as helmet and goggles.

Climbing system user feedback

User-centred design strategies produce benefits: here are the thoughts of a practitioner, Dave Taylor, on the design of shell garments for climbers and mountaineers. Taylor acknowledges that these are his 'partisan ideas, gained from personal experience', with which others may not agree.

- Lighter fabric is always better, but not at the expense of durability. It's alright for the professionals, who often have gear sponsors to get through one or two jackets in a season, but mere mortals – who have shelled out £200 for a jacket – want five years' wear, so sometimes a few grams more for a stronger fabric in crucial places is worth it.
- Need stretch or flex under the arms to allow for reaching, especially in these days of the snugger fit. Not convinced by pit zips though – they seem more hassle than they are worth, and really difficult to open and close when wearing a rucksack or having loads of climbing slings, rope and stuff wound round you.
- The bane of my winter mountaineering life is the neck opening. Necks need to be nice and big so that you can zip the front zip right up whilst wearing anything up to four layers under it. (I may be wearing a base layer and two fleeces – all zipped right up and, *in extremis*, even a down or synthetic duvet jacket.) It is a real pain in the neck if the front zip constricts your breathing when everything is fully closed up. Better to have a bigger neck opening, and put in a draw-cord to let the wearer adjust the size?
- I will go for Velcro front closures on the storm flaps every time if I can – if you leave some poppers undone (which happens a lot when you are faffing about with your ventilation). I find that, in extreme conditions, poppers can fill with ice and wind blown snow and it is impossible to fasten them, especially with cumbersome winter gloves on. I once had a close call when I couldn't get my poppers popped (it was important as my zip was also frozen or jammed open, and I was relying on the poppers to close my storm flap).
- The zip pulls on my technical jackets: I used to add my own longer length shock cord pullers so I could grab more easily with gloved hands. (Saves taking gloves off. A friend once lost a glove in strong wind doing this on a winter traverse of the Anoach Eagach Ridge, in Glencoe. He had to spend the next four hours wearing just an inner glove in pretty cold conditions.)

- Please keep outside pockets away from the rucksack waist belt and climbing harness. Or provide decent outside chest pockets so that you don't need to access the lower pockets. I suppose after all you want some lower hand warmer pockets for when you are wandering around town. The inside chest pocket needs to have its access zip between the front zipper and storm flap, so you can get into it without baring your insides to the cold and wet. Also, the inside pocket needs to be made of the same breathable fabric as the outer shell – no good spending all that money on a fancy breathable jacket only to have a 20 by 30 cm patch of plastic glued on the inside chest. I once cut a non-breathable pocket out and replaced it with a homemade mesh pocket.
- Hoods need to be big enough to go on over a variety of headgear from none, to a helmet. But they also need to be adjustable, so that if not wearing any helmet the hood can be pulled in tight to the head (or worn under a helmet). Needs a nice big top 'wired peak' that can be bent to a range of shapes. Needs a good one-handed cord adjuster so that the front hole can be closed right up to just leave the eyes (or goggles) exposed.

8.5 Future trends

8.5.1 Design strategies

User-centred design strategies, three-dimensional technologies and virtual environments all offer opportunities for a more sustainable approach to the design and development of cold weather clothing. Material use, time taken and costs incurred can all be reduced. User potential for satisfaction can be increased as requirements for design, size and shape, fit and function are all more carefully met, helping to reduce garment returns, mark down and subsequent landfill. However, as with the hierarchy of evaluative stages referred to in Fig. 8.15, the impact of these strategies and technologies remains poorly understood.

8.5.2 Wearable technologies

Branded fibres, incorporated into branded fabric assemblies and garments, are beginning to be co-branded with smart textiles and wearable electronics. Relationships have been established between such providers as Polar, the Finnish electronics firm, and the sportswear brand Adidas, while Apple is in collaboration with Nike.

Seamfree Santoni, a knit technology, is being used for high-performance base layer sports garments. Tops, leg wear and full body suits are design engineered by organising of polyester, polyamide, polypropylene and elastomeric yarns in stitch structured zones to provide varying degrees of stretch, strength, protection and wicking. Garment design may be geared to sailing (with a high content of hydrophobic polypropylene content), to climbing (lighter constructions in polyester) and to motorcycle wear exploiting the enhanced strength of polyamide (Tecso, 2009). Padding may be incorporated in the seat area for enhanced comfort, comparable with cycling pants (X-Bionic, 2009). Santoni technology has also been used in the application of textile sensors for vital signs monitoring (Textronics).

8.5.3 Heated clothing

Employees of a Seattle company who rode motorcycles to work year round experienced evident discomfort on cold mornings. In looking for a way to keep these riders warm, owner Gordon Gerbing decided to incorporate heating pads into clothing connected to the machine's battery and established a brand that maintains its prominence in the field (Gerbing, 2009). The styling is slim cut to fit under outer garments, with jacket and trouser liners constructed from a Teflon-coated, wind-resistant, lightweight, durable and soft nylon shell to protect the 100 gram Thinsulate® that stores the heat generated. These garments are highly compressible for storage.

8.6 Acknowledgements

Ray Battersby, motorcyclist, UK
Rab Carrington, RAB, expedition clothing, UK
Rob Croskell, SGS, UK
Olivier Dufau, snowboarder, British Columbia, Canada
Nick Jeffries, motorcyclist, UK
Jasmiine Julin, Aro Rukka, Finland
Paul Oxley, Feridax, UK
Dave Tayler, climber, UK

8.7 Sources of further information and advice

Mass Customization & Personalization Conference (MCPC) conferences
European Outdoor Group Association for Conservation

8.8 References

Abeysekera, J. (1992), The use of personal protective clothing and devices in the cold environment (Lulea University, Sweden), <http://www.luth.se/depts/lib/coldtech/ct92-7.html>, accessed March 2009.
Aihua, M., Yi, L., Roumei, W. and Shuxiao, W. (2008), A CAD system for multi-style thermal functional design of clothing, *Computer-Aided Design*, 40, 916–930.
Allwood, J.M., Laursen, S.E., de Rodriguez, C.M. and Bocken, N.M.P. (2006), *Well Dressed? The present and future sustainability of clothing and textiles in the UK*, University of Cambridge: Institute for Manufacturing, ISBN 1-902546-52-0.

Anderson, K. (2008), Sustainable Fashion. TC2 Bi-Weekly Technology Communicator, November 26 2008. <http://www.tc2.com/newsletter/2008/112608.html>, accessed December 2008.

Anttonen, H., Kinnunen, K. and Niskanen, J. (2005), *Functional glove combinations for cold conditions*, Oulu Regional Institute of Occupational Health, Finland.

Back Country, (2009), commercial website giving details of Alti Mitten, <http://www.backcountry.com/store/ODR0395/Outdoor-Research-Alti-Mitten-Me>, accessed April 2009.

Berghaus (2009a), commercial website explaining the Berghaus layering system, <http://www.berghaus.com/technologies/clothing/berghauslayeringsystem.aspx>, accessed April 2009.

Berghaus (2009b), commercial website showing the Attrition technical jacket, <http://www.berghaus.com/ProductDetails.aspx?ProductID=1836&Gear=2>, accessed April 2009.

Black, A. (2006), User-centred design: the basics of user-centred design. London: Design Council. <http://www.designcouncil.org.uk/en/About-Design/Design-Techniques/User-centred-design-/>, accessed March 2009.

Black, S., Kapsali, V., Bougourd, J., Geesin, F. (2005), Fashion and function factors affecting the design and use of protective clothing, in R. Scott (ed.), *Textiles for Protection*, Cambridge, Woodhead, pp. 60–89.

BS EN 31092 (1994), Textiles. Determination of physiological properties, British Standards Institute, ISBN: 0 580 21444 3.

BS EN 1621-2 (2003), Motorcyclists' protective clothing against mechanical impact. Motorcyclists back protectors. British Standards Institute, ISBN 0 580 42566 5.

BS EN340 (2004), Protective Clothing General Requirements, British Standards Institute, ISBN 0 580 43200 9.

BS EN343 (2003), Protective Against Rain, British Standards Institute, ISBN 978 0 580 62076 8 (incorporates A1:2007).

BS EN342 (2004), Protective clothing ensembles and garments for protection against cold, British Standards Institute, ISBN 978 0 580 62770 5 (incorporates corrigendum March 2008).

BS EN511 (2006), Protective gloves against cold, British Standards Institute, ISBN 0 580 48497 1.

BS EN ISO 9920 (2007), Ergonomics of the Thermal Environment Estimation of Thermal Insulation and Water Vapour Resistance of a Clothing Ensemble, British Standards Institute, ISBN 978 0 580 54807 9.

BS EN ISO 11079 (2008), Ergonomics of the thermal environment – Determination and interpretation of cold stress when using required clothing insulation (IREQ) and local cooling effects. ISBN 978 0 580 54566 5.

Castelani, J.W., O'Brien, C., Baker-Fulco, C., Young, S. and Young, A.J. (2001), Sustaining health and performance in cold weather operations, Technical Note TN/02-2, from US Army Research Institute of Environmental Medicine, Natick, USA.

CWA – Communication Workers of America (2005), Temperature Extremes & the Workplace. BSI (2003), Protective clothing General requirements, BS EN 340: 2003. <http://www.cwa-union.org/issues/osh/articles/page.jsp?itemID=27339145>, accessed March 2009.

DeJonge, J.O. (1984), Foreword: the design process, in S.M. Watkins, *Clothing: The Portable Environment*, Ames, IA, Iowa State University Press, p. vii.

Dow Corning (2009), commercial website, <http://www.activeprotectionsystem.com>, accessed April 2009.

European Standard (2004), EN 342:2004: Protective clothing. Ensembles and garments for protection against cold, ISBN 0 580 44064 8.

Frank, S.M., Raja, S.N., Bulcao, C. and Goldstein, D.S. (2000), Age-related thermoregulatory differences during core cooling in humans. *American Journal of Physiology*, 279, R349–R354.

Gerbing (2009), commercial website, <http://www.gerbing.com/aboutUs.php>, & <http://www.gerbing.com/Products/Liners/heatedJacketLiner.html>, accessed April 2009.

Gill Marine (2009), Respect the Elements, product catalogue (Nottingham: Gill), <http://view.vcab.com/showvcab.aspx?vcabid=njShhehSehcpg>, accessed March 2009.

Goldman, R.F. (2005), Environmental ergonomics: whence-what-whither, proceedings of the Sixth International Manikin and Modeling Meeting, held at the International Research and Training Centre for Information Technologies and Systems in the Ukraine, pp. 39–47.

Gore-Tex (2009), Gore-Tex gloves with X-Trafit (company announcement), see <http://www.gore.com/en_xx/news/x-trafit.html>, accessed April 2009.

Havenith, G. (2002), The interaction between clothing insulation and thermoregulation, *Exogenous Dermatology*, 1, 5, 12.

Havenith, G. and Heus, R. (2004), A test battery related to ergonomics of protective clothing, *Applied Ergonomics*, 35, 3–20.

Held (2009), website of company supplying motorcycle fashion garments and accessories (Burgberg, Germany: Held GmbH), <http://www.held-biker-fashion.de/index.php?plink=index&l=1&fs=&lg=e>, accessed March 2009.

Holmér, I. (2005), Textiles for protection against the cold, in R. Scott (ed.), *Textiles for Protection*, Cambridge, Woodhead, pp. 378–397.

Holmér, I. and Nilsson, H. (1995), Heated manikins as a tool for evaluating clothing, *Annals of Occupational Hygiene*, 39, 6, 809–818.

Holmér, I., Rintamäki, H. (2004), *Barents: Newsletter on Occupational Health and Safety*, Vol. 7, No. 1, Helsinki: Finnish Institute of Occupational Health.

Ice Armor (2009), commercial website for extreme clothing, Minnesota, USA, <http://www.clamcorp.com/icearmor_windcutterhat.html>, accessed April 2009.

IndiaMART (2009), Zipper pullers illustrated on directory of manufacturer at <http://www.indiamart.com/labelcreation/pcat-gifs/products-small/zipper-pullers_10832824.jpg>, accessed April 2009.

Innov-Ex (2008), Conference Report: Innovation in the context of global warming, <http://www.innovation-for-extremes.org/conference/innov_ex_08.html>, accessed March 2009.

ISO 8996 (2004), *Ergonomics of the thermal environment – Determination of metabolic rate*. ISBN: 0 580 45322 7.

ISO 13732-3 (2005), Ergonomics of the thermal environment – Methods for the assessment of human responses to contact with surface – Part 3: Cold surfaces.

Karlotski, D. (2009), The Season of the Bike, 'Motorcycle page', published by the author. <http://www.simpsontaxidermy.com/Motorcycle.htm>, accessed April 2009.

Kasturiya, N., Subbulakshmi, M., Gupta, S. and Raj, H. (1999), System design of cold weather protective clothing, *Defence Science Journal*, 49, 5, 457–464 <http://www.luth.se/depts/lib/coldtech/ct92-7.html>, accessed March 2009.

Lehmuskallio, E. (2001), *Cold Protecting Emollients and Frostbite*. Dissertation, Faculty of Medicine, University of Oulu.

McCann, J. (1999), Establishing requirements for the design development of performance sportswear. Unpublished MPhil thesis, University of Derby.

McCann, J. (2005), Material requirements for design of performance sportswear, in R. Shishoo (ed.), *Textiles in Sport*, Cambridge, Woodhead.

McCullough, E.A. (2005), Evaluation of protective clothing systems using manikins, in R. Scott (ed.), *Textiles for Protection*, Cambridge, Woodhead, pp. 217–232.

Mowbray, J. and Wilson, A. (2009), *Eco-Textile labelling Guide*, Pontefract, UK, Mowbray Communications Ltd, ISSN: 1758-7042.

Musto (2008), Yachting Technical Collection, product catalogue, Laindon, UK, Musto, pp. 60 and 65.

Musto (2009), From Musto HPX ocean jacket and trousers product literature, Laindon, UK, Musto.

Nakamura, K. and Morrison, S.F. (2008), Preoptic mechanism for cold-defensive responses to skin cooling, *Journal of Physiology*, 586, 10, 2611–2620.

Northern Outfitters (2009), Arctic boot from company website, <http://northernoutfitters.com/p-23-arctic-boot.aspx>, accessed April 2009.

O'Hearn, B.E., Bensel, C.K. and Poleyn, A.F. (2005), Biomechanical analyses of body movements and locomotion as affected by clothing and footwear for cold weather climates, Technical Report issued by US Army Research, Development and Engineering Command, Natick, USA.

Optitex (2008), software for apparel design and visualisation: <http://www.optitex.com/en/Apparel_Industry>, accessed April 2009.

Parsons, K.C. (2003) *Human Thermal Environments*, 2nd edn. London, Taylor & Francis.

Parsons, M. and Rose, M. (2008), Innovaton for Extremes: Innovation in the Context of Global Warming, note on proceedings conference held at Lancaster University, <http://www.lums.lancs.ac.uk/files/14334.pdf>, accessed April 2009.

Pertex (2009), manufacturer's website: <http://www.pertex.com/main.php>, accessed April 2009.

Piller, F., Schubert, P., Koch, M. and Möslein, K. (2005), Overcoming mass confusion: Collaborative customer co-design in online communities, *Journal of Computer-Mediated Communication*, 10, 4 (Special Issue: Online Communities Design, theory and practice).

Polartec (2005), press release (New York: Kick Public Relations), <http://www.kickpr.com/polar_1105.html>, accessed April 2009.

RAB (2009), company website, <http://www.rab.uk.com/expedition_gear/clothing/expedition_jacket—31/>, accessed March 2009.

Rossi, R. (2005), Interactions between protection and thermal comfort, in R. Scott (ed.), *Textiles for Protection*, Cambridge, Woodhead, pp. 233–260.

Ruckman, J.E., Murray, R. and Choi, H.S. (1999), Engineering of clothing systems for improved thermophysiological comfort, *International Journal of Clothing Science and Technology*, 11, 1, 37–52.

Rukka (2009), commercial websites:
<http://www.rukka.com/lfashion/rukka/rukkawww.nsf> and
Smart Rider's Outfit:
<http://www.rukka.com/lfashion/rukka/rukkawww.nsf/vwpages/6DE3FE366ED402EAC22572CA0024085B?OpenDocument&Expand=1.6.8>, accessed April 2009.

Ruto, A. (2009), Dynamic Human Body Modelling. Thesis in preparation for submission for degree EngD to London University, 2009.

Sandsund, M., Paasche, A. and Reinertsen, R.E. (2001), Development and evaluation of a clothing system for offshore industry workers in cold environments, in *Proceedings of the Australian Physiological and Pharmacological Society*, p. 135.

SHAPCC (1998), The Health Aspects of Work in Extreme Climates within the E & P Industry: The Cold. Health Subcommittee Report No. 6.65/270, London: E & P Forum.
Shaw, A. (2005), Steps in the protection of clothing materials, in R. Scott (ed.), *Textiles for Protection*, Cambridge, Woodhead, pp. 90–116.
Sizemic (2008), image based on SizUK data provided by Managing Director of Sizemic <sizemic.org>.
Sperling, L. (2005), Ergonomics in user-oriented design, in I. Holmér, K. Kuklane and C. Gao (eds), *Proceedings of the 11th International Conference Environmental Ergonomics*. Thermal Environment Laboratory, Lund University, Lund, Sweden.
Spindler, U. and Thwaites, C. (2004), Modern Military Clothing Systems, International Soldier Systems Conference, Boston, USA.
Sub-Zero Boots (2009), commercial website <http://www.tlcwebs.co.uk/subzero/pages/shooting.htm>, accessed April 2009.
TC2 (2008), news article on commercial website <http://www.tc2.com/newsletter/2008/090308.html>, accessed April 2009.
Tecso (2009), commercial website <http://www.wz-international.com/index.php?option=com_virtuemart&page=shop.browse&category_id=27&Itemid=2>, accessed April 2009.
Thorlo (2009), sales order page <http://www.thorlo.com/socks/snowboard-socks/over-calf/552.php>, accessed March 2009.
Thwaites, C. (2008), Cold weather clothing, in E. Wilusz (ed.), *Military Textiles*, Cambridge, Woodhead, pp. 158–182.
TPC (2008), commercial website describing software for apparel industry, <http://www.tpc-intl.com/software.asp>, accessed April 2009.
Tseng, M.M. and Piller, F.T. (2003), *The Customer Centric Enterprise: Advances in Mass Customization and Personalization*, Berlin, Springer.
Under Armor (2009), commercial website: <http://www.underarmour.com>, accessed April 2009.
Uotila, M., Mattila, H. and Hanninen, O. (2006), Methods and models for intelligent garment design, in H.R. Mattila (ed.), *Intelligent Textiles in Clothing*, Cambridge, Woodhead, pp. 5–18.
Williams, J.D. (2004), *Fashion without limits*, SimplyBe.co.uk, distributed by Chase PR.
X-Bionic (2009), commercial website, <http://www.x-bionic.com/underwear>, accessed April 2009.
Yermakova, I., Tadejeva, J. and Candas, V. (2005), Computer simulator for prediction of human thermal state. Proceedings of the Sixth International Manikin and Modeling Meeting, held at the International Research and Training Centre for Information Technologies and Systems in the Ukraine, pp. 218–221.
Zweld (2009), A description of construction of outerwear using welding technology by Zweld Technology. In *Mountain Hardwear*, Outerwear advanced construction: welding, <www.mountainhardwear.com.au/innovations/MHW_Welding.pdf>, accessed March 2009.

Part II

Evaluation and care of cold weather clothing

9
Standards and legislation governing cold weather clothing

H. MÄKINEN, Finnish Institute of Occupational Health, Finland

Abstract: This chapter recites national and international guidelines on the occupational aspects of cold exposure and necessary protection. It also introduces the development of the legislation and standards for cold protection in the European Union, and deals with the content of current requirement standards as well as testing standards for the evaluation of cold protective clothing and gloves.

Key words: clothing against cold, gloves against cold, legislation, standards.

9.1 Introduction

Cold can be classified as natural or artificial. People living in countries which are near to or north of the 60° North latitude, where the average number of days when the temperature is below zero is over 180 days (Finnish Meteorological Institute, 2008) are particularly exposed to natural cold in their outdoor activities. Because cold seasons have always been a part of these people's lives, they have accepted cold weather as a natural element and have always been used to protecting themselves against the cold. Anttonen (1993) gives examples of tasks performed outdoors. These are work in harbours, shipyards, the forestry and wood industry, agriculture, construction, fishing, traffic and transportation, military and recreation activities. However, exposure to artificial cold indoors has increased due to technical development. Work in the food industry, in cooling plants and cold storages are examples of this type of exposure.

The effects of work in the cold on body heat balance, performance and health, and also on cold protection are quite well known thanks to numerous studies in this area (Rintamäki and Latvala 2003). In addition, several guidebooks on the occupational aspects of cold exposure have been published in different countries (BS 7915, 1998, BS 8800, 1996, DIN 33403-5, 1997 and Hassi *et al.*, 2002). Several thermal standards also provide guidance for the assessment of various aspects of exposure to cold (ISO 15743, 2008, ISO 11079, 2007, ISO/TS 14415, 2005 and EN ISO 9920, 2007). For example, ISO 15743 (2008) 'Working practices in cold: Strategy for risk assessment and management' recommends a three-step approach for the evaluation and management of cold risks. According to ISO 15743 (2008) the first stage is systematic screening of environmental cold stress and individual health effects Standards for protective clothing regarding cold have developed due to the EU Commission legislation on health and safety requirements. This chapter views the legislation and standards which produce requirements and classifications for cold weather clothing.

9.2 Development of legislation and standards

It may be due to the fact that cold weather is a natural element in some countries, that protection against cold has not been seen as a priority during the development of requirements for personal protective equipment (PPE). In Finland, for example, no exact recommendations or regulations existed to regulate cold exposure and the hazardous effects of cold exposure before EU legislations (Anttonen, 1993). Commission directive 89/391 (1989) defines in general the obligatory duty of the employer to ensure the safety and health of workers in every aspect related to the work. The requirements of the general safety regulation (Suomen säädöskokoelma, 2002) in Finland concerning work in cold are as follows:

- At worksites, temperature and humidity must be adjustable without causing harm to employee's health
- If a worker is exposed to humidity, wetness, draught, or cold, protection must be effective
- In general, it is expected that the worker protect him/herself by using sufficient clothing to suit various weather conditions
- The employee must be given protective clothing and ensure that other protective measures are met in temporary outdoor tasks in which the employee is exposed to cold, and for tasks in cold storages.

In other Nordic countries very similar practices were used (Hassi *et al.*, 2002).

In Canada the National Joint Council (2004) has published Personal protective Equipment Directive which is deemed to be part of collective agreements and has been effective from 1 July 1997. It defines some requirements for insulated clothing too; to provide insulation clothing for work in hazardous weather conditions, and to provide insulation clothing designed to prevent hypothermia to individuals when their duties involve significant risk of immersion in cold water.

When employers began to pay attention to worker's protection, the first step in Finland, as well as in Sweden and Norway, was collective labour agreements. In construction work, for example, it was agreed in which conditions workers would be given cold protective clothing, and how many pieces they would receive per year, or how much the employee has to pay for them. The lowest temperature below which the work must be stopped (e.g. under 20 °C) was also agreed (Risikko, 2003).

In 1987, the Single European Act was signed as part of the EU enlargement process. The EU started to develop a single market through a standardized system of laws which apply in all member states, guaranteeing the freedom of movement of people, goods, services and capital. Also included in this was the development of common legislation for work environments, EU directives, and harmonized standards for personal protective equipment (PPE).

9.3 Directives on personal protective equipment

One of the first harmonized Council Directives on goods was the Directive on Personal Protective Equipment (Commission Directive 1989/686/EEC). This defines only the basic safety requirements which PPE must satisfy in order to ensure the health, protection and safety of users. The general requirements applicable to all PPE include, for example, design principles, ergonomics, comfort and efficiency, lightness and design strength.

9.3.1 Requirements for protection against cold

Specific requirements are given for materials and components as well as for complete PPE intended for use in the cold (Commission Directive 1989/686, paragraph 3.7).

> *PPE designed to protect all or part of the body against the effects of cold must possess thermal insulating capacity and mechanical strength appropriate to the foreseeable conditions of use for which it is marketed.*
>
> *Constituent materials and other components suitable for protection against cold must possess a coefficient of transmission of incident thermal flux as low as required under the foreseeable conditions of use. Flexible materials and other components of PPE intended for use in a low-temperature environment must retain the degree of flexibility required for the necessary gestures and postures.*
>
> *This requirement applies to constituent materials and components and not to complete PPE.*
>
> *The manufacturer needs to select materials, components or a combination of them so that in the foreseeable conditions of use:*
>
> - *the thermal flux transmitted through the PPE shall be as low as possible;*
> - *the flexibility remains acceptable to ensure comfort, usability and integrity of the products.*
>
> *The mechanical resistance of materials of components, needs to be, where necessary, appropriate to the impact energy, nature and temperature of the splashes of cold products. PPE materials and other components which may be splashed by large amounts of cold products must also possess sufficient mechanical-impact absorbency.*
>
> *For the complete PPE ready for use under the foreseeable conditions of use is required as follows:*

1. *The flux transmitted by PPE to the user must be sufficiently low to prevent the cold accumulated during wear at any point on the part of the body being protected, including the tips of fingers and toes in the case of hands or feet, from attaining, under any circumstances, the pain or health-impairment threshold.*
2. *PPE must as far as possible prevent the penetration of such liquids as rain water and must not cause injuries resulting from contact between its cold protective integument and the user.*

If PPE incorporates a breathing device, this must adequately fulfil the protective function assigned to it under the foreseeable conditions of use.

The manufacturer's notes accompanying each PPE model intended for brief use in low-temperature environments must provide all relevant data concerning the maximum permissible user exposure to the cold transmitted by the equipment.

Protective clothing for the foreseen risks of cold injuries usually consists of several protective material layers, and protection efficiency depends on insulation capacity as well as proper coverage. The size and model of the cold protective clothing need to be such that cold does not directly harm the user through possible openings. The garments also need to have adequate mechanical strength against abrasion, cuts and tearing.

PPE Guidelines (2006) explains that for whole body protection, it is important that loss of body heat does not cause hypothermia, pain or harm, particularly to the user's extremities (i.e. tips of fingers and toes); and that it prevents the penetration of liquids, such as rain water, that may cause injuries.

With regard to PPE for brief use in cold environments, the manufacturer must provide enough information so that the user can determine, for each of his intended actions, the maximum effective protection time and/or maximum acceptable use time from a physiological point of view.

The second PPE directive (Commission Directive 1989/656/EEC) gives the minimum health and safety requirements for the use of the personal protective equipment by workers at the workplace. The employer is responsible for assessing the risks to the health and safety of the workers and for providing free protective clothing or equipment, where risks are not adequately controlled by other means. The personal protective equipment must (i) be appropriate for the risks involved, without itself leading to any increased risks, (ii) correspond to existing conditions at the workplace, (iii) take into account the ergonomic requirements and state of health of the worker, (iv) fit the wearer correctly after any necessary adjustments.

The conditions of use of PPE, in particular the period for which it is worn, shall be determined on the basis of the seriousness of the risk, the frequency of exposure to the risk, the characteristics of the workstation of each worker, and the performance of the PPE. To assist workplaces in this determination, the standards contain an informative appendix where the measured insulation values have been converted into combinations of ambient air temperature, activity levels and exposure times of eight hours, and one hour (see Section 9.4.9).

9.4 European standards for cold protective clothing

As the above requirements are rather general, the Commission has given a mandate to the European Committee for Standardization (CEN) to develop a European Standard(s), relating in particular to the design and manufacture, and the specifications and test methods applicable to PPE. Standards concerning personal protective clothing and gloves against cold have been prepared under a mandate given to CEN TC 162 by the European Commission and the European Free Trade Association, and support the essential requirements of EU Directive(s).

European Standards are published in order to achieve a common basis in Europe for requirements and test methods for protective clothing ensembles and garments against cold, in the interest of manufacturers in particular, as well as those of test institutes and end-users. The measured properties and their subsequent classification are intended to ensure an adequate protection level under different user conditions. The following PPE standards against cold weather conditions have been published:

- EN 342 Protective clothing (2005). *Ensembles and garments for protection against cold*. Cold environment is defined in this standard as a temperature $< -5\,°C$.
- EN 14058 Protective clothing (2004). *Garments for protection against cool environments*. Cool environment is defined in this standard as a temperature $\geq -5\,°C$.
- EN 343 Protective clothing – Protection against rain (2008). Protection against rain is an essential part of cold protection because wet clothing increases the heat flow from the body, therefore the user can be at risk of hypothermia at as high a temperature as zero. The standard for protection against rain specifies requirements and test methods applicable to the materials and seams of protective clothing against the influence of precipitation (e.g. rain, snowflakes), fog and ground humidity.
- EN 511 Protective gloves against cold (2006). The standard for cold protective gloves specifies requirements and test methods for gloves which protect against convective and conductive cold down to $-50\,°C$.

Parallel to these standards general requirements are given for cold protective clothing in EN 340 (2004) and for cold protective gloves in EN 420 (2004).

Cold protective clothing can be ensembles (i.e. two-piece suits or coveralls) or single garments for protection against a cold environment. There are no specific requirements for headwear and footwear against cold. At present, a standardized requirement to prevent local cooling exists only for cold protective gloves. The existing standards for protective clothing and gloves used in cold or

Table 9.1 Summary of requirements for protective clothing and glove standards for cold work

Type of requirement	EN 342	EN 14058	EN 343	EN 511
Ergonomics • sizes • dimensional change	EN 340	EN 340	EN 340	EN 420
Comfort and efficiency	Water vapour resistance <55 m² PA/W, if required	Water vapour resistance <55 m² PA/W	Water vapour resistance 3 classes	
Design strength	Tear resistance > 25 N		Tensile strength > 450 N Tear resistance > 25 N Seam strength > 225 N	Abrasion resistance Tear resistance Flexibility resistance Extreme cold flexibility resistance
Cold protection 1. Materials		Thermal resistance R_{ct}, 3 classes		
2. Complete PPE	Thermal insulation I_{cle} or I_{cle} $\geq 0.310\,m^2$ (=2 clo) K/W	Thermal insulation I_{cle} or I_{cle}, optional		1. Thermal insulation I_{tr}, 4 classes 2. Thermal resistance R, 4 classes
Air permeability	3 classes	3 classes		
Resistance to water penetration	2 classes, optional	2 classes, optional	After pre-treatments, 3 classes	Whole glove integrity test, class 1 = no leakage, class 0 = not passed

cool environments define performance requirements, explaining the general requirements of the directive as shown in Table 9.1.

9.4.1 Thermal resistance, R_{ct}

Thermal resistance refers to how much the material or material layers insulate against cold. Insulation is tested in accordance with EN 31092 (1993). The

Table 9.2 Classification of thermal resistance R_{ct}

R_{ct} (m²K/W)	Class
0.06–0.12	1
0.12–0.18	2
0.18–0.25	3

thermal resistance R_{ct} of all layers of the garment must be in accordance with Table 9.2. The higher the classification number, the higher the insulation.

9.4.2 Thermal insulation, I_{cle} and I_{cler}

Thermal insulation explains how the whole garment or combination of garments insulates against cold. Effective thermal insulation I_{cle} is measured using a stationary manikin, and the resultant effective thermal insulation I_{cler} is measured using a moving thermal manikin. The minimum value of the resultant effective thermal insulation I_{cler}, or optionally effective thermal insulation I_{cle}, is 0.310 m²·K/W. The test procedure for whole garment testing is given in EN ISO 15831 (2004). The effective, as well as resultant effective thermal insulation of the protective clothing ensemble are measured in combination with standard underwear (B) consisting of undershirt with long sleeves, long underpants, socks (up to the knee), felt boots, thermojacket, thermopants, knitted gloves and balaclava, or optionally with underwear (C) as specified by the manufacturer. Single garments are tested with a standard reference clothing (R) consisting of undershirt with long sleeves, long underpants, socks (up to the knee), felt boots, jacket (one layer), trousers (one layer), shirt, knitted gloves, balaclava.

Thermal insulation for cool protective garments is an optional requirement. The resultant effective thermal insulation I_{cler} must have a minimum value of 0.170 m²·K/W, or the effective thermal insulation I_{cle} must have a minimum value of 0.190 m²·K/W. The effective, as well as resultant thermal insulation of protective clothing are measured in combination with reference clothing as specified in the standard. In the test with underwear B or a single garment, the manikin must not be dressed with any hood if this is not attached to the garment. Only reference items can be used for measurements.

9.4.3 Air permeability (AP)

Good air permeability of the outer layer of the garment means good protection against wind. The test method for air permeability is EN ISO 9237 (1995). Air permeability is classified as shown in Table 9.3.

Table 9.3 Classification of air permeability (AP) in EN 342 and EN 14058

AP (mm/s)	Class
> 100	1
100–5	2
≤ 5	3

9.4.4 Resistance to water penetration (WP)

The resistance to water penetration is optional in cold protective clothing. The test method of resistance to water penetration of the material and seams is EN 20811 (1993). Table 9.4 shows the classification of water penetration of cold protective clothing, and Table 9.5 the classification of protective clothing against rain. For protective clothing against rain, there are classifications for material as new, after pretreatments, and as new for seams.

Table 9.4 Classification of resistance to water penetration of cold protective clothing

Class	WP (Pa)
1	8,000–13,000
2	> 13,000

Table 9.5 Classification of resistance to water penetration of protective clothing against rain

Class	WP (Pa)
1	As new ≥ 8,000
1 2 3	After pretreatment: no requirement ≥ 8,000 ≥ 13,000
1 2 3	As new seams: ≥ 8,000 ≥ 8,000 ≥ 13,000

9.4.5 Water vapour resistance (R_{et})

Water vapour resistance describes the breathability of the material layers. Low resistance means that sweat can escape through the material layers. If water penetration resistance is required for cold protective clothing, it is measured in accordance with EN 31092 (1993), and must be (without underwear) less than 55 m² Pa/W. For protective clothing against rain, water vapour resistance is an obligatory requirement and it is classified as shown in Table 9.6.

Table 9.6 Classification of water vapour resistance of protective clothing against rain

Class	Water vapour resistance, R_{et} (m² Pa/W)
1	> 40
2	20–40
3	< 20

9.4.6 Strength properties

The tearing force of the outer shell material of cold protective clothing (with the exception of vests and excluding elastic and knitted materials) must be at minimum 25 N in both directions of the material. A similar requirement applies to rain protective clothing. In addition, tensile strength of material and seams of rain protective clothing must be measured. The requirement of material is 450 N and of seams is 225 N.

9.4.7 Marking

EN 340 (2004) defines the snowflake in the pictogram indicating that protection against cold is offered and the umbrella pictogram indicates that protection against rain is offered. Figure 9.1 shows how the appropriate performance levels are marked to cold protective clothing according to EN 342 (2005). Figure 9.2 shows marking of cool protective clothing according to EN 14058 (2004). Figure 9.3 shows marking of performance levels of rain protective clothing according to EN 343 (2008).

In addition to levels of performance, the marking must include the care labelling name, trade mark or other means of identification of the manufacturer or his authorized representative; designation of the product type; commercial name or code; and size designation.

208   Textiles for cold weather apparel

EN 342

Performance level	Explanation
Y(B)/Y(C)/Y(R)	I_{cler} in m² K/W of the ensemble (with underwear B and optionally with underwear C of the manufacturer) or of the single garment (with standard reference clothing R)
Y(B)/Y(C)/Y(R)	I_{cle} in m² K/W of the ensemble (with underwear B and optionally with underwear C of the manufacturer) or of the single garment (with standard reference clothing R)
X	Air permeability class
X	Resistance to water penetration class; optional

9.1 Marking of cold protective clothing.

EN 14058

Performance level	Explanation
X	Thermal resistance class
Y	Air permeability class, optional
Z	Resistance to water penetration class; optional
R	Insulation value I_{cler} in m² K/W (optional)
R	Insulation value I_{cle} in m² K/W (optional)

9.2 Marking of cool protective clothing.

EN 343

Performance level	Explanation
X	Level of resistance to water penetration
X	Level of water vapour resistance

9.3 Marking of protective clothing against rain.

9.4.8 Information supplied by the manufacturer

User information is an essential part of the marking. It gives information on the product designation; how to put it on and take it off. It also provides basic information on possible uses, e.g., the recommended temperature values given related to the garment's performance levels, and where detailed information is available, instructions on how to use the information given in the marking as well as information on underwear type, which is used in testing the thermal insulation. It must also give necessary warnings regarding misuse, and washing instructions. The content of user information is defined in EN 340 (2004) for protective clothing and in EN 420 (2004) for protective gloves.

9.4.9 Instructions on how to select levels of performance for use conditions

Cold and cool protective clothing

The standards give guidelines, as an informative annex, where the protective value of measured effective thermal insulation or resultant effective thermal insulation of a garment assembly is converted on the basis of ISO 11079 (2007) into combinations of ambient air temperature and activity level (metabolic heat production), corresponding to a standing wearer and to a wearer moving and performing light or moderate activity. For each level, a minimum temperature is calculated at which the body can be maintained at thermoneutral conditions indefinitely (8 h), and a lowest temperature at which a one-hour period of exposure can be sustained with an acceptable rate of body cooling (Tables 9.7–9.10). Values are based on the conditions that air temperature is equal to mean radiant temperature, relative humidity is about 50%, air velocity is between 0.3 and 0.5 m/s, and walking speed about 1.0 m/s.

Protective clothing against rain

A guide is given for protective clothing against rain (Table 9.11) to illustrate the effect of water vapour permeability on the recommended continuous wearing time of a garment in different ambient temperatures. The table is valid for medium physiological strain $M = 150$ W/m, standard man, at 50% relative humidity and wind speed $va = 0.5$ m/s. With effective ventilation openings and/or break periods, the time for wearing can be prolonged.

9.4.10 Cold protective gloves

Performance levels against convective and contact cold are defined for cold protective gloves (Table 9.12). In addition, requirements and test methods, flexibility behaviour, water impermeability and extreme cold resistance are

Table 9.7 Effective thermal insulation of clothing I_{cle} and cold ambient temperature conditions for heat balance at different exposure durations (EN 420, 2004)

Insulation I_{cle} m²K/W	Wearer standing activity 75 W/m²	
	8 h	1 h
0.310	11	−2
0.390	7	−10
0.470	3	−17
0.540	−3	−25
0.620	−7	−32

Table 9.8 Resultant thermal insulation of clothing I_{cler} and cold ambient temperature conditions for heat balance at different exposure durations (EN 420, 2004)

Insulation I_{cler} m² K/W	Wearer moving activity			
	light 115 W/m²		medium 170 W/m²	
	8 h	1 h	8 h	1 h
0.310	−1	−15	−19	−32
0.390	−8	−25	−28	−45
0.470	−15	−35	−38	−58
0.540	−22	−44	−49	−70
0.620	−29	−54	−60	−83

Table 9.9 Effective thermal insulation of clothing I_{cle} and cool ambient temperature conditions for heat balance at different exposure durations (EN 511, 2006)

Insulation I_{cle} m²K/W	Wearer standing activity 75 W/m²	
	8 h	1 h
0.170	19	11
0.230	15	5
0.310	11	−2

Standards and legislation governing cold weather clothing 211

Table 9.10 Resultant thermal insulation of clothing I_{cler} and cool ambient temperature conditions for heat balance at different exposure durations (EN 511, 2006)

Insulation I_{cler} m² K/W	Wearer moving activity			
	light 115 W/m²		medium 170 W/m²	
	8 h	1 h	8 h	1 h
0.170	11	2	0	−9
0.230	5	−1	−8	−19
0.310	−1	−15	−19	−32

Table 9.11 Recommended maximum continuous wearing time for a complete suit consisting of jacket and trousers without thermal lining (EN ISO 15831, 2004)

Temperature of working environment °C	Class		
	1 R_{et} above 40 min	2 $20 < R_{et} < 40$ min	3 $R_{et} < 20$ min
25	60	105	205
20	75	250	—
15	100	—	—
10	240	—	—
5	—	—	—

— = no limit for wearing time.

Table 9.12 Performance levels against cold of gloves

Cold protection requirement	Levels of performance	Test method
Thermal insulation (convective cold)	Levels: (m²K/W) 1. $0.10 \leq I_{TR} < 0.15$ 2. $0.15 \leq I_{TR} < 0.22$ 3. $0.22 \leq I_{TR} < 0.30$ 4. $0.30 \leq I_{TR}$	EN 511, annex A
Thermal resistance (contact cold)	Levels: (m²K/W) 1. $0.025 \leq R < 0.050$ 2. $0.050 \leq R < 0.100$ 3. $0.100 \leq R < 0.150$ 4. $0.150 \leq R$	ISO 5085-1:1989

given in EN 511 (2006). In informative annex relevant parameters, which should be taken into account in the selection process, are listed as well as required thermal insulation level for three activity levels as a function of ambient air temperature at a wind speed below 0.5 m/s.

9.5 Cold protective clothing standards outside Europe

ASTM has published three testing standards which can be used for evaluation of cold protective clothing.

ASTM-F1291-05 (2005). *Standard Method for Measuring the Thermal Insulation of Clothing Using a Heated Thermal Manikin*

This standard covers the determination of the insulation value of clothing ensembles. It describes the measurement of the resistance to dry heat transfer from a heated manikin.. It provides a baseline clothing measurement on a standing manikin. Therefore the effects of body position and movement are not addressed. The insulation values measured can be given in either clo or SI units.

ASTM F1868-02 (2002). *Standard Test Method for Thermal and Evaporative Resistance of Clothing Materials Using a Sweating Hot Plate*

This method covers the measurement of the thermal resistance and the evaporative resistance under steady-state conditions, of fabrics, films, coatings, foams, and leathers, including multi-layer assemblies, for use in clothing systems. The principle of this method is similar to EN ISO 31092 (1993). The range of this measurement technique according to ASTM F1868-02 for thermal resistance is from 0.002 to 0.2 Km^2/W and for evaporative resistance is from 0.01 to 1.0 $kPam^2/W$.

ASTM F2370-05 (2005). *Standard Test Method for Measuring the Evaporative Resistance of Clothing Using a Sweating Manikin*

This test method covers the determination of the evaporative resistance of clothing ensembles. It describes the measurement of the resistance to evaporative heat transfer from a heated sweating thermal manikin to a relatively calm environment. As ASTM F1291-05 it is a static test providing a baseline clothing measurement on a standing manikin.

9.6 Future trends

The current standards define requirements for outer garments only. For work in cold conditions, quality undergarments and fabrics need to provide good functionality to ensure that underwear keeps the user both warm and dry. Standards for functional underwear could be one new area in traditional standardization.

The standards viewed above define requirements and test methods for traditional cold protective clothing, providing passive protection against cold. Shishoo (2000), Jansen (2008) and Yoo *et al.* (2008) mention new types of textile materials, from which cold protective clothing can be manufactured in the future to respond to changes in their environment, as follows:

- phase-change materials
- shape-memory polymers
- aerogels.

Garments will be available in the future which provide active thermo-regulation with electronic devices embedded or integrated from: implants of conventional electronics and wiring of interactive textiles embodying sensors, actuators and logic circuits built into the structure of the fibres, yarns and fabrics (Kurczewska and Lenikowski, 2008; van Langenhove *et al.*, 2005).

Mattila *et al.* (2006) remind us that in addition to protection and normal textile functions the functionality of electronics, communications, signal monitoring and transfer must also be tested in foreseeable conditions of use as well as the durability, ageing and cleaning, and to secure their compliance with legal requirements. Therefore test methods and requirements are needed to ensure that the products manufactured from these types of materials perform the way they are claimed to. In addition to PPE directive 89/686/EC, the following directives may also contain requirements for new types of cold protective clothing:

- Medical device directive 93/42/EC
- Machinery directive 98/37/EC
- General product safety directive 2001/95/EC
- Waste electrical and electronic equipment 2003/108/EC
- EMC directive 2004/108/EC

To develop an innovation driven economy the European Commission has published a communication *A lead market initiative for Europe*. Protective textiles are identified as one of the six markets which are highly innovative. Standardization is one policy instrument by which the leading market initiatives are supported (European Commission, 2008). In CEN TC 246 a new working group for smart textiles has been constituted. As a first stage the group will prepare a CEN technical report, in which recommendations are made about prioritization of potential work items for this area (CEN/TC 248 WG 31 N 6, 2008).

9.7 Sources of further information and advice

Harmonized standards and access to EU legislation: http://ec.europa.eu/enterprise/newapproach/standardization/harmstds/reflist/ppe.html

9.8 References

Anttonen, H. (1993) Occupational needs and evaluation methods for cold protective clothing, *Arctic Med Res,* 52(9), suppl, pp. 1–76.
ASTM-F1291-05 (2005) Standard method for measuring the thermal insulation of clothing using a heated thermal manikin. ASTM International West Conshohocken PA, www.astm.org.
ASTM F1868 (2002) Standard Test Method for Thermal and Evaporative Resistance of Clothing Materials Using a Sweating Hot Plate. ASTM International West Conshohocken PA, www.astm.org.
ASTM F2370-05 Standard Test Method for Measuring the Evaporative Resistance of Clothing Using a Sweating Manikin. ASTM International West Conshohocken PA, www.astm.org.
BS 7915 (1998) Ergonomics of the thermal environment – guide to design and evaluation of working practices for cold indoor environments, British Standard Institution.
BS 8800 (1996) Guide to occupational health and safety management systems. British Standards Institution.
CEN/TC 248 WG 31 N 6 (2008), CEN/TC 248 WG 31 Smart textiles, report to the plenary meeting of CEN/TC 248. European Commission of standardization.
Commission Directive 1989/391 of June 1989 on the introduction measures to encourage improvements in the safety and health of workers at work. *Official Journal of European Communities* 29.06.1989.
Commission Directive 1989/686/EEC of 21 December 1989 on the approximation of the laws of the Member States relating to personal protective equipment. *Official Journal of European Communities* 30.12.89.
Commission Directive 1989/656 EEC of 30 November 1989 on the minimum health and safety requirements for the use by workers of personal protective equipment at the workplace (third individual directive within the meaning of Article 16 (1) of Directive 89/391/EEC). *Official Journal of European Communities* 30.12.89.
DIN 33403-5 (1997) Deutsche norm. Klima am Arbeitsplatz und in der Arbeitsumgebung. Teil 5. Ergonomische Gestaltung von Kältearbeitsplätzen. (Climate at the workplace and the working environment. Part 5. Ergonomic design of cold workplaces).
EN 14058 (2004) Protective clothing. Garments for protection against cool environment. European Committee for Standardization, Rue de Stassart 36, B-1050 Bruxelles.
EN 20811 (1993) Textiles. Determination of resistance to water penetration. Hydrostatic pressure test. Committee for Standardization, Rue de Stassart 36, B-1050 Bruxelles. International Organization for Standardization.
EN 31092 (1993) Measurement of thermal and water-vapour resistance under steady-state conditions (sweating guarded, hotplate test). Committee for Standardization, Rue de Stassart 36, B-1050 Bruxelles.
EN 340 (2004) Protective clothing. General requirements. Committee for Standardization, Rue de Stassart 36, B-1050 Bruxelles.
EN 342 (2005). Protective clothing. Ensembles for protection against cold. European Committee for Standardization, Rue de Stassart 36, B-1050 Bruxelles.
EN 343 + A 1 (2008) Protective clothing – Protection against foul weather. European Committee for Standardization, Rue de Stassart 36, B-1050 Bruxelles.
EN 420 (2004) Protective gloves. General requirements and test methods. Committee for Standardization, Rue de Stassart 36, B-1050 Bruxelles.
EN 511 (2006) Protective gloves against cold. European Committee for Standardization,

Rue de Stassart 36, B-1050 Bruxelles.
EN ISO 15831 (2004) Clothing – Physiological effects – Measurement of thermal insulation by means of a thermal mannikin. Committee for Standardization, Rue de Stassart 36, B-1050 Bruxelles.
EN ISO 9237 (1995) Textiles. Determination of permeability of fabrics to air. International Organization for Standardization.
EN ISO 9920 (2007) Ergonomics of the thermal environment – Estimation of the thermal insulation and evaporative resistance of a clothing ensemble. International Organization for Standardization.
European Commission, Enterprise and Industry. Lead Market Initiative for Europe. Available at: http://ec.europa.eu/enterprise/leadmarket/leadmarket.htm (accessed 24 November 2008).
Finnish Meterological Institute. Finland's Climate. Available at [http://www.fmi.fi/weather/climate.html] (Accessed 24 November 2008).
Hassi *et al.* (2002) *Opas kylmätyöhön* (Guide for work in cold). Työterveyslaitos (Finnish Institute of Occupational Health), in Finnish.
ISO 5085-1 (1989) Textiles – determination of thermal resistance part 1 – Low thermal resistance. International Organization for Standardization.
ISO 11079 (2007) Ergonomics of the thermal environment – Determination and interpretation of cold stress when using required clothing insulation (IREQ) and local cooling effects. International Organization for Standardization.
ISO 15743 (2008) Ergonomics of the thermal environment – Cold workplaces – Risk assessment and management. International Organization for Standardization.
ISO/TS 14415 (2005) Ergonomics of the thermal environment – Application of International Standards to people with special requirements. International Organization for Standardization.
Jansen, D. (2008) The future of PPE – Smart responsible materials. *1st international conference on personal protective equipment.* 21–23.05.08, Bryges (CD-ROM).
Kurczewska, A., Lenikowski, J. (2008) Garments with variable thermoinsulation with microprocessor temperature controller, *International Journal of Occupational Safety and Ergonomics (JOSE)*, 14(1), 77–87.
Mattila, H. Talvenmaa, P. Mäkinen, M. (2006) Wear Care – Usability of intelligent materials workwear. In H. R. Mattila (ed.) *Intelligent textiles and clothing*. Woodhead Publishing Limited, Cambridge.
National Joint Council (2004) *Personal protective Equipment and Clothing Directive.* Available at http://www.tbs-sct.gc.ca/archives/hrpubs/TBM_119/ppe-eng.rtf (Accessed 24 November 2008).
PPE Guidelines (2006) *Guidelines on the application of Council Directive 89/686/EEC of December 1989 on the approximation of the laws of the member states relating to personal protective equipment*, 17 July 2006.
Rintamäki, H., Latvala, J. (2003) Kylmähaittojen arviointi ja torjunta – standardit ja oppaat ja niiden suosittelemat toimenpiteet (Assesment and prevention of problems due to cold – Standards and guides), *Työ ja ihminen*, 17(1), 5–14, in Finnish with English summary.
Risikko, T. (2003) Tapausselostus kylmän aiheuttamien työterveys- ja –turvallisuus-riskien hallinnasta rakennustyössä . *Työ ja ihminen*, 17(1), 72–80, in Finnish with English summary.
Shishoo, R. (2000) Innovations in fibres and textiles for protective clothing. In K. Kuklane and I. Holmer (eds) *Proceedings of NOKOBETEF 6 and 1st European Conference on Protective clothing*, Stockholm, pp. 79–87.

Suomen säädöskokoelma (2002) No. 738–754. Available at [http://www.finlex-fi/fi/laki/kokoelma2002/20020109.pdf], in Finnish (Accessed 24 November 2008).

Van Langenhove, L., Puers, R., Matthys, D. (2005) Intelligent textiles for protection. In R. Scott (ed.) *Textiles for protection*. Woodhead Publishing Limited, Cambridge, pp. 176–195.

Yoo, S., Yeo, J. S., Hwang, Y.H., Kim, S.G., Hur, Kim, E. (2008) Application of a NiTi alloy two-way shape memory helical coil for a versatile insulating jacket, *Materials Science and Engineering*: A 481, May, 662–667.

10
Laboratory assessment of cold weather clothing

G. HAVENITH, Loughborough University, UK

Abstract: An overview of laboratory tests for cold weather clothing is provided starting from physical measurements on fabrics, and physical measurements on whole garments using thermal manikins. This is extended to human wear trials and climatic chamber experimentation. Insulation and vapour resistance are considered the most relevant parameters followed by wind and waterproofness and moisture absorption properties. The use of test participants in wear trials is considered regarding the information provided by such tests. Tests for innovative fabrics (heated, variable insulation, phase-change materials) are discussed. Finally testing of sleeping bags is considered.

Key words: manikin, field test, wear trials, hot plate, sweat, insulation.

10.1 Introduction

Protecting people from the cold has had a long historical development. With the advancement of textile technology, a shift has taken place from using natural material solutions (fur, down) to manmade fibre and textile solutions or combinations of both. The wide spectrum of material combinations that can be selected at present implies that during clothing development it is impossible to test all possible combinations as real garments in the field, and this puts emphasis on preliminary evaluations at the fibre or textile level to narrow down the choices before going to prototype level. This requires a systematic approach in clothing development which was described by Goldman (1974) and Umbach (1983), presenting a staged approach to the problem. This approach is typically followed only in large development projects as found in the military, given the high cost of going through all the proposed stages. Developments initiated by clothing producers may only use one or two of the stages, and thus often limit themselves to small incremental improvements given the risks involved in making large changes to clothing concepts without extensive prototype evaluations.

The five-stage approach advocated by Goldman and Umbach is in use widely in the clothing research community. It is schematically presented in Fig. 10.1, to which we have added stage zero. The approach should start with a detailed analysis of the task of the intended user of the clothing, thus defining the

218   Textiles for cold weather apparel

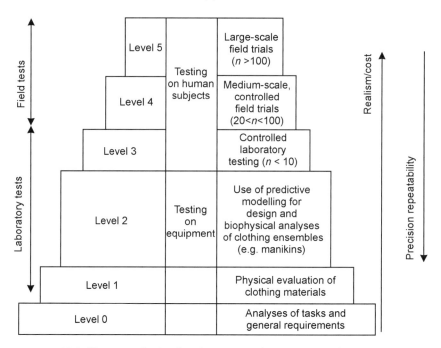

10.1 Six stages in the development and assessment of clothing systems (modified from Goldman, 1974 and Umbach, 1983).

environmental conditions and activities from which the clothing needs to protect. Based on this 'wish list', a pre-selection of fabrics and materials takes place, which are tested for a number of physical properties (stage one), the most important being the thermal protection for the present application. Using the fabric characteristics, a prediction for the overall clothing characteristics can be made, which can be put into the now readily available thermal models that either analyse the balance of heat produced and lost by a hypothetical user (e.g. as done in the Required Clothing Index [EN ISO 11079, 2007]), or actually combine this heat balance approach with a physiological model of human thermoregulations which allows a prediction of the body's responses and the associated risks to the wearers' health or comfort (stage two) (e.g. Lotens, 1993; Wissler and Havenith, 2009).

Typically, this then allows a reduction of the number of materials that could be expected to perform well to a manageable number. These can then be turned into complete garments or ensemble prototypes that can be exposed firstly to manikin (top end of stage two) and secondary to human wear-testing. This moves the testing from 'flat' apparatus to the human shape, increasing the complexity, time and cost of the testing per ensemble but also increasing realism.

In stage three, the clothing will be worn under defined conditions in a climatic chamber and detailed data of the wearer collected. The outcome of this testing may help narrow down the choices to two or three, which can then be tested in small or large-scale field trials. Time required for these two stages as well as cost will increase over previous stages, while detail and precision will suffer. Nevertheless, crucial information will be gained at this stage. This chapter will focus on testing performed in stages one, two (manikin) and three describing tests providing objective data that can be used in development, evaluation and quality control of clothing.

10.2 Clothing properties relevant in cold

Depending on the definition of 'cold', cold protective clothing will be used in temperatures below 5–10 °C, but most likely well below 0 °C (EN standards 342-2004 and 14058-2004 define cold as below −5 °C and cool as above −5 °C). Hence, protection against cold, i.e., heat loss will be the predominant characteristic, with protection against water mainly relevant when temperatures above zero are expected in the work or where other liquids are present in the work environment.

On the other hand, due to fluctuations in the conditions (changing exercise levels, moving in and out of freezer rooms) not only the maximal cold protection is important, but also the range of temperatures in which the clothing can be used. When the workload increases, the clothing may become too warm, the wearer starts to sweat and evaporated sweat would need to be able to pass through the clothing to provide body cooling. Heat stress in the cold, due to protective clothing worn at higher work rates is not uncommon (Rintamäki and Rissanen, 2006). Hence the vapour permeability of the fabrics is important, and for the clothing as a whole the way it can be layered, opened for ventilation, and easily donned and doffed when conditions change. Apart from the actual heat stress, too much moisture accumulation may also affect cold protection once activity levels are reduced again.

A large water content of the clothing system will affect the insulation and can cause the so-called 'after chill', where accumulated moisture evaporates and substantially cools the person when activity is reduced. This increases the risk of hypothermia (reduced body temperature) and therefore is a risk factor. The relevant measures for the above would therefore be:

- heat resistance (convection/radiation)
- vapour resistance/permeability index
- water tightness
- air permeability (affecting heat resistance in wind)
- wicking.

10.3 Material/fabric testing

10.3.1 Heat and vapour resistance

For comparison of different fabrics several tests on heat and vapour resistance are available and many are now defined in ISO (1993), EN (2007) or ASTM (2002, 2003, 2005) standards. For heat resistance measurement, typically a guarded hot plate apparatus of some form is used, which measures the amount of heat lost through a sample at a certain temperature gradient between the plate and the environment. From this, insulation of the fabric can be calculated as:

$$R_{ct} = \frac{\bar{t}_{plate} - \bar{t}_a}{H_{DRY}} - R_0 \qquad 10.1$$

where R_{ct} = heat resistance of fabric sample (m^2.K.W^{-1}); $\bar{t}_{plate}$ = mean hot plate surface temperature (°C); $\bar{t}_a$ = ambient temperature (°C); H_{DRY} = dry heat loss per square metre of plate area (W.m^{-2}); R_0 = heat resistance measured without a sample present (m^2.K.W^{-1}).

Several styles of equipment and different standards are in use for this. Examples are ISO 11092 (1993), CAN/CGSB-4.2 No. 70.1-94 (1994) (now withdrawn), CAN/CGSB 4.2 NO. 78.1-2001 (2001), ASTM D 1518-85 (2003), and BS 4745 (2005). The main principles of these tests are similar, though the design varies in that some clamp the fabric between a hot and a cold plate, others add an air layer between sample and cold plate, or use only a hot plate while the other side of the sample is exposed to ambient air. The method can influence results strongly, e.g., a sleeping bag material will perform worse under compression (though this may be relevant to test), while, e.g., a highly air permeable fabric may be affected by air flows in the equipment.

Similarly, also a number of methods for determination of fabric vapour resistance are available. Examples are: BS7209 (1990), CAN/CGSB-4.2 No. 49 – 99 (1999), ASTM F1868-02 (2002), ASTM F2298 (2003), ASTM E 96 (2005) and ISO 11092 (2003). Test methods can vary dramatically in complexity and cost. For cold weather clothing, where fabrics may be quite thick, not all techniques may be suitable, due to compression, edge effects, etc.

For the guarded hot plate (see Fig. 10.2), the fabric's vapour resistance is measured as:

$$R_{et} = \frac{p_{plate} - p_a}{H - H_{DRY}} - R_{et0} \qquad 10.2$$

where R_{et} = vapour resistance of fabric sample (m^2.Pa.W^{-1}); p_{plate} = mean hot plate surface vapour pressure (Pa); p_a = ambient vapour pressure (Pa); H = total heat loss per square metre of wet plate area (W.m^{-2}); R_{et0} = vapour resistance measured without a sample present (m^2.Pa.W^{-1}); H_{DRY} = dry plate heat loss at t_a (W.m^{-2}) (as, for this measurement, t_a is chosen equal to t_{plate}, this will typically be zero).

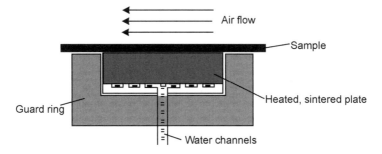

10.2 Schematic drawing of a sweating guarded hot plate according to ISO 11092 (1993).

Other tests provide different results in terms of units, e.g., grams per 24 hours. Comparison of results obtained with different methods should be looked at critically (McCullough *et al.*, 2003; Huang and Qian, 2008) as discrepancies exist due to the techniques used. Methods, e.g., involving desiccants tend to create circumstances that do not exist in the real world. Most of such methods are mainly used for quality control purposes. For real life performance, the sweating guarded hot plate may be one of the better predictors, though cheaper alternatives may also provide useful information to the development process.

The results of the heat resistance measurements then need to be judged against the insulation required, while in relation to the vapour resistance one may work on the principle, the lower the better. However, when fabrics differ in heat resistance, they usually differ in thickness and this thickness will also increase vapour resistance. Comparing the vapour resistance *per se* may therefore not be meaningful and it is more relevant to compare the ratio between heat and vapour resistance (the higher the better). This is usually expressed as the water vapour permeability index:

$$i_{mt} = S \cdot \frac{R_{ct}}{R_{et}} \qquad 10.3$$

where $S = 60\,\text{Pa.K}^{-1}$.

This number varies between zero (totally impermeable) and one (air) (though the typical upper limit is due to its definition around 0.5), and as mentioned before to avoid moisture accumulation the highest number is best here. With some water vapour permeable membranes, moisture transfer can be affected by the humidity at the membrane and by the local temperature (Havenith *et al.*, 2004). In such cases, measurements at actual ambient temperatures can be relevant.

10.3.2 Air permeability

When wind speed increases, a fabric's heat and vapour resistance may decrease in relation to its air permeability. This can be quantified via the fabric air permeability that can be determined, e.g., by EN ISO 9237. For cold protection, a low number is better in order to provide wind protection. For cold protective clothing, air penetration can be beneficial when heavy work is performed in the clothing (inducing heat stress). In that case, design features like variable ventilation openings, allow the microclimate ventilation to be increased. We will return to this later in the garment-testing phase.

10.3.3 Waterproofness/water penetration

The material's resistance against water penetration can be measured, e.g., by EN 20811 (1992). The higher the number, the better. New tests have been developed that test the garment as a whole (EN 14360, 2004), which include the effect of design features. It should be noted that for fabrics meant to be used in extreme cold (e.g. cold storage at $-40\,°C$) no liquid will be present, and as waterproofing (coating, membranes) typically reduces vapour permeability, the criterion for this test should be adapted to the application range of the fabric. If liquid protection is not required, leaving out the waterproof layer/membrane/coating will increase the temperature range of application of the clothing, as the vapour resistance decreases. Nevertheless, such layers also contribute to windproofness of the clothing and a balance may need to be struck here. A good example here are, e.g., wind-stopper® fabrics where windproofness is increased with a limited penalty for vapour permeability.

10.3.4 Wicking/buffering

Where the protective clothing ensemble includes underwear or other next-to-skin materials, wear comfort can be affected by the wicking ability of the fabrics. Discomfort is linked to the presence of liquid on the skin, and the removal of this, either by optimised evaporation or by wicking the moisture away from the skin, is thus a relevant factor. Wicking fabrics can benefit comfort and cooling in two ways. When the person starts to sweat in their cold protective garments, this sweat can be absorbed by the fabric, spread over a bigger area and thus facilitate evaporation. Also, by removing the liquid from the skin and transporting it away from the skin–fabric interface, clinging of clothing with its associated discomfort is reduced. Also this reduces 'after-chill', the continued cooling of sweat on the skin that occurs after exercise if moisture remains on or close to the skin and promotes fast drying of fabrics. When the person is very active, and thus produces a lot of sweat, the comfort effect related to wicking ability is limited, as with saturation of the clothing the spreading becomes less relevant, and comfort is low in all fabric types.

The wicking/buffering effect can be measured on the sweating guarded hot plate, allowing the liquid water to touch the fabric and looking at overall moisture loss, or by measuring the microclimate response to a short sweating burst (Mecheels and Umbach, 1976). Other methods look at the wicking of liquid into vertically hanging strips of fabric after fixed time periods (BS 3424-18, 1986 and DIN 53924, 1997), or looking at the dispersion of a drop of liquid on a fabric (visual test). Trying to make this more objective, electrical conductivity of fabrics has been used to define the water absorption speed (Van Langenhove and Kiekens, 2001).

10.3.5 Thermal character

One parameter that may have an important role in the buying behaviour of consumers is the initial sensation when the garment touches the skin (the feeling when the garment is touched/gripped in the shop). This may not affect the functional performance of the garments and hence it is only briefly mentioned here. Examples of instruments for assessing parameters related to this sensation can be found in the Kawabata system and, e.g., in the Alambeta tester, the latter looking at the 'warm-cool feeling' in the first two seconds of skin contact.

10.4 Garment and ensemble testing: physical apparatus

As the step from fabric testing to human subject lab and wear trials is quite big, a need was present to have an intermediate step to test complete garments and ensembles on a physical test apparatus, providing the precision and repeatability that is limited in human wear trials, but still incorporating all effects that differentiate garments from fabrics like fit, drape, ventilation, etc. This method was developed from the 1940s onwards in the form of thermal manikins, which are now available in a variety of sizes (baby–adult), shapes (male–female) and technologies (metal, resin shells, water filled fabric shells, etc.), of which many now can produce 'sweat' (uniformly wet skin or regional variation in sweat output).

Manikins tend to be regulated in a number of individual body zones, so that local insulation can be studied in addition to whole body insulation. This allows a better analysis of 'cold spots' in the clothing design. Most current manikins have between 16 and 30 body zones that can be individually assessed. Measurements are typically performed with the manikin static, as well as with the manikin performing walking movements. Also measurements without wind (static) and with wind can be performed. In this way, effects of pumping, ventilation and air penetration can be performed which is crucial for cold weather garment assessment (Havenith and Nilsson, 2004, 2005).

10.4.1 Insulation

The principle of the manikin (see Fig. 10.3) is the same as that of the (sweating) hot plate, albeit now in a human shape (EN ISO 9920 (2007), ASTM F1291

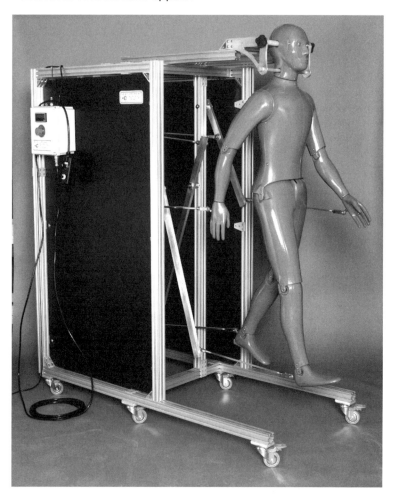

10.3 Example of a thermal manikin (model type 'Newton', MTNW, Seattle).

(2005) and ASTM F2370 (2005), EN ISO 15831 (2004)). For dry heat insulation (the prime parameter for cold protection) the manikin surface is controlled at a set temperature and the energy (heat) required keeping this set temperature is then directly related to the manikin's insulation:

$$I_T = \frac{\bar{t}_{sk} - t_a}{H_{sk}} = \frac{\sum \alpha_i t_i - t_a}{\sum (\alpha_i \cdot H_i)} = \frac{\sum \alpha_i \cdot (t_i - t_a)}{\sum (\alpha_i \cdot H_i)}$$ 10.4

where α_i = (surface area of segment i)/(total surface area of manikin); I_T = insulation of complete ensemble including enclosed and surface air layers (m^2.K.W^{-1}); $\bar{t}_{sk}$ = average skin temperature (°C); t_i = temperature of segment i (°C); t_a = ambient temperature (°C) (if radiation is present, this is replaced by operative temperature); H_i = heat loss of segment i (W).

If the measurement is performed with a uniform skin temperature over the body, i.e., $t_i = t_{sk}$ = constant, then equation 10.4 becomes more simple and follows the so-called parallel method:

$$I_T = \frac{\bar{t}_{sk} - t_a}{\sum(\alpha_i \cdot H_i)} \qquad 10.5$$

For other skin temperature settings (e.g. the comfort mode where skin temperature differs over the body) equation 10.4 is the only correct calculation. In the literature, another calculation method ('serial') can be found, but the required boundary conditions for this (uniform heat flux over the body) are almost never met.

10.4.2 Vapour resistance

For the measurement of vapour resistance, usually the ambient temperature is set equal to skin temperature to eliminate any dry heat loss (isothermal conditions). Then the calculation becomes:

$$R_{e,T} = \frac{\bar{p}_{sk} - p_a}{H_{sk}} = \frac{\sum \alpha_i \cdot p_i - p_a}{\sum(\alpha_i \cdot H_i)} = \frac{\sum \alpha_i \cdot (p_i - p_a)}{\sum(\alpha_i \cdot H_i)} \qquad 10.6$$

If the measurement is performed with a uniform skin vapour pressure over the body, i.e., $p_i = p_{sk}$ = constant, then equation 10.6 becomes:

$$R_{e,T} = \frac{\bar{p}_{sk} - p_a}{\sum(\alpha_i \cdot H_i)} \qquad 10.7$$

If, instead of measuring heat flux, a measurement of mass loss due to sweating is performed, no regional resistances can be calculated. The calculation becomes:

$$R_{e,T} = \frac{\bar{p}_{sk} - p_a}{H_{sk}} = \frac{\bar{p}_{sk} - p_a}{\text{mass loss} \cdot \lambda} \qquad 10.8$$

where $R_{e,T}$ = vapour resistance of complete ensemble including enclosed and surface air layers (m².Pa.W^{-1}); $\bar{p}_{sk}$ = average skin vapour pressure (Pa); p_i = vapour pressure of segment i (Pa); p_a = ambient vapour pressure (Pa); H_i = heat loss of segment i (W); mass loss = amount of moisture evaporated from ensemble per unit of manikin surface area per second (g.m^{-2}.sec^{-1}); λ = latent heat of evaporation at skin temperature (approx. 2430 J.g^{-1}).

One of the main problems with vapour resistance measurements on the various manikin types is to ensure a uniform vapour pressure on the skin, and to measure this. In manikins that have a pre-wetted skin before dressing, this is deduced from the skin temperature (assuming saturation). In manikins with uneven sweat distribution (e.g. simulating certain scenarios with realistic sweat distributions (Havenith et al., 2008a), this is problematic, especially in terms of ensuring that a segment is evenly wetted over its surface. In most manikins, the wetted surface lies on top of the temperature controlled surface, which may introduce additional errors as the wet surface's temperature will then be lower due to the evaporative cooling, the more evaporation takes place, and if not taking this into account an error is made.

If measurements are not done in isothermal conditions, the heat flows need to be corrected for dry heat losses. Also, the cooler the air in which the measurement is done, the higher the likelihood of condensation taking place in the clothing. It has been shown, that this has a major effect on heat loss, especially in cold weather garments with limited vapour permeability (Havenith *et al.*, 2008b) and thus non-isothermal measurements can provide highly relevant information for cold weather garments. This aspect can only be measured if the actual heat losses are measured directly. Manikins that deduct heat loss from mass loss may make substantial errors here, as this condensation heat loss is not included in the mass loss calculation.

Before the advance of sweating manikins, vapour resistance was often deduced from the combination of dry insulation data from the manikin and fabric vapour resistance data from the sweating hot plate (Mecheels and Umbach, 1976). At the moment, mostly direct measurements are performed. A more in-depth discussion on manikin measurements is provided in Chapter 11.

10.4.3 Clothing ventilation

Ventilation of the clothing microclimate is particularly important in cold weather clothing. For cold protection, high insulation values in extreme conditions require windproof clothing that can be made rather airtight at openings (collars, zips, etc.) too. This, however, means that when activity levels increase and more heat is generated than lost, heat stress may develop and the clothing needs to be adaptable to the higher heat loss requirement. A high permeability index is the starting point for this allowing sweat evaporation, but given the tightly woven textiles mostly with coatings or membranes used as outer layer to ensure windproofness, this is quite limited in scope. The alternative is incorporation of ventilation openings in the clothing that can be adjusted to the need for microclimate ventilation (Havenith *et al.*, 2003). This provides a direct convective pathway from the skin to the environment, without hindrance in terms of fabric heat and vapour resistances (see Fig. 10.4).

Manikin pumping factors

In order to evaluate the effectivity of microclimate ventilation this needs to be measured either directly (tracer gas methods) or indirectly (manikin heat loss). Umbach (1983) and Mecheels and Umbach (1976) describe a series of manikin measurements from which they deduct the clothing ventilation. By measuring heat loss from the manikin dressed in the test clothing while static, while

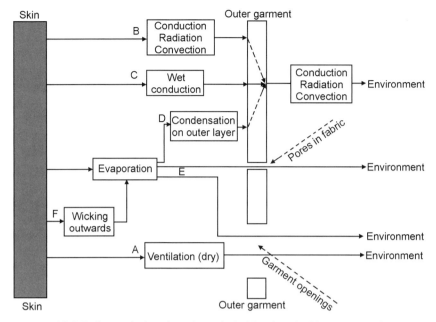

10.4 Pathways for heat loss through clothing (used with permission from JAP, Havenith *et al.*, 2008b).

walking with all openings tightly closed, and while walking with all openings maximally open, they were able to break down the manikin heat loss into a maximal value (resting) and an active value with and without ventilation. Dry ventilation can be teased out as:

$$\text{Ventilation conduction} = \frac{1}{I_{\text{ventilation}}}$$

$$= \frac{1}{I_T \text{ (walking, open)}} - \frac{1}{I_T \text{ (walking, closed)}} \quad 10.9$$

It should be noted that in their approach wind speed was not changed (they focus mainly on the pumping effect of walking), but this could easily be added. They applied the same approach to evaporative ventilation heat loss.

Prediction equation of ventilation effect of wind and movement

The effect of wind and movement on clothing insulation was extensively studied by Nilsson *et al.* and by Havenith *et al.*, who integrated and summarised their findings producing prediction equations for the size of these effects for different clothing types including cold weather (Havenith and Nilsson, 2004, 2005). For standard cold weather clothing they provide the following equation:

$$I_{T,r} = \left[e^{\{[-0.0512*(\nu_{ar}-0.4)+0.794*10^{-3}*(\nu_{ar}-0.4)^2 -0.0639*w]*p^{0.1434}\}} \right] \cdot I_{T,\text{static}} \quad 10.10$$

where $I_{T,static}$ = insulation measured on static manikin at 0.4 m.s^{-1} wind speed; $I_{T,r}$ = resultant insulation including effects of ventilation (wind and walking); 0 < walk < 1.2 m.s^{-1} and 0.4 < wind < 18 m.s^{-1} and 1 < p < 1000 lm^{-2}.s^{-1}; r^2 = 0.968; SEE = 0.048; with p = clothing air permeability (typical permeability values that can be used are 1 for garments with impermeable membranes; 50 for densely woven workwear and 1000 for highly permeable garments (e.g. fleece without membrane)).

For the lower area of the wind range better results were obtained in a separate analysis:

$$I_{T,r} = e^{((-0.0881*(\nu_{ar}-0.4)+0.0779*(\nu_{ar}-0.4)^2-0.0317*(w))*p^{0.2648})} \cdot I_{T,static} \quad 10.11$$

with r^2 = 0.931, SEE = 0.023; 0 < walk < 1.2 m.s^{-1} and 0.4 < wind < 1 m.s^{-1} and 1 < p < 1000 lm^{-2}.s^{-1}. Though Havenith et al. produced similar equations for vapour resistance (latest version published in ISO 9920), no such equation was developed specifically for cold weather clothing.

Direct measurement of ventilation

As ventilation is about the exchange of microclimate air with 'fresh' environmental air, this can be measured using tracer gases that, mixed in the microclimate air, will follow such air movement. Two methods for this have been developed that can be used on human subjects, but also on manikins (the manikin does not have to be a thermal one, though the controlled surface temperature would add more realism in terms of convection patterns).

The first method, originally developed by Birnbaum and Crockford (1978) blows in a gas (usually Nitrogen) under the clothing that displaces the oxygen present in the microclimate. When this gas supply is switched off, oxygen from the ambient air re-enters the clothing by ventilation and diffusion, and the time constant of the oxygen concentration change in the microclimate can be used to quantify the ventilation (Ueda and Havenith, 2002). The number produced is the number of air exchanges that take place per unit of time. To translate this into ventilation (in volume units), this needs to be multiplied with the microclimate volume. The latter was traditionally difficult to assess, but with 3D scanning equipment this can now be done more reliably (Daanen et al., 2002).

The second method, developed by Lotens and Havenith (1988), uses a diluted tracer gas (usually argon), that is blown into the microclimate in a continuous stream, and the steady state microclimate concentration is sampled, from which the ventilation (volume per time) can be calculated directly (Havenith et al., 1990, 2003). This method has been shown to be repeatable and accurate, if the tracer gas is distributed equally over the microclimate regions. Efforts are under way to measure this ventilation for different regions of the body separately (Ueda et al., 2006).

Infra-red image analysis

Especially for cold weather clothing, the analysis of cold spots is important to identify design flaws. As mentioned above, this can be done on manikins that have a detailed measurement system with many individual zones. An alternative method is the use of infra-red imaging. With the manikin at a fixed surface temperature or the clothing worn by test subjects, and the clothing worn long enough to create a stable temperature distribution, Infra-red analysis of the clothing surface temperature may reveal local 'hot spots' which are manifests of location where body heat finds little resistance to come to the surface, and which may be experienced as cold spots by the wearer. Clothing fit can play an important role here as where clothing gets tight, insulation may be lost as the insulative materials get squashed. Also seams that leak heat can be easily identified in this way.

10.4.4 Gloves and footwear manikins

For the special applications of gloves and footwear, equipment is often used functioning as a manikin, but consisting only of an arm with a hand for the glove testing and a leg with a foot for shoes (Kuklane *et al.*, 1997, 1999). These devices usually have more detail in these areas than whole body manikins have, though this is not a requirement for EN 511 (2006) which describes glove testing for cold environments.

10.5 Garment and ensemble testing: human subjects

As discussed by many authors (Goldman, 1988; Lotens, 1988; Umbach, 1983; Havenith and Heus, 2004), wear trials of garments provide invaluable information that cannot be gained from lab tests on test equipment. Apart from wear trials in the field, many laboratories carry out laboratory trials with humans in climatic chambers, where most outdoor conditions can be set up accurately, and more importantly, reproducibly. Keeping the conditions constant, it is possible to compare a number of garments on the same test subjects and get a good comparison of their properties in real use.

10.5.1 Heat and vapour resistance

Using indirect calorimetry, it is possible to measure most components of human heat exchange with the environment. Such an analysis (see Holmér and Elnas, 1981; Havenith *et al.*, 1990; Havenith, 2002a,b for details) allows the calculation of the effective heat and vapour resistance of the clothing, including all effects of movement and wind. Unfortunately, this is a very time-consuming measurement, and therefore costly. Also the repeatability is lower than that of manikin testing, and the measurement needs to be repeated on several test participants to become reliable and allow statistical comparisons (Havenith and Heus, 2004).

10.5.2 Freedom of movement

The freedom of movement test (Havenith and Heus, 2004) takes account of the design and fit of the garments in relation to their effect on task performance. For this purpose, the participants need to perform task-related activities where possible, but where this, e.g., requires equipment that cannot be brought into the lab, one can consider 'simplified' tasks instead that nevertheless have a relation with actual task performance. For example, for rescue workers who need to be able to move fast and freely in their cold protective clothing, an obstacle course can be devised that challenges the clothing design in this respect. Movements that will 'stretch' the garments, like bending, kneeling, crawling can be incorporated, and the time needed to finish such an obstacle course is a good indication of the clothing's performance in this respect.

The clothing's performance is measured in relation to the time to do the obstacle course in a training suit. In addition various small tests (time needed for donning, 80 m sprint, running in 8 shaped pattern underneath a lowered bar, sit and reach and stand and reach test, Sargeant's jump, can be added to evaluate the way in which the clothing hampers movement. Tests follow a within-subject design, in which each participant wears all garments and a control. Tests need to be spread out in time sufficiently to avoid them influencing each other (fatigue, time of day, heat/cold accumulation), and performed in a balanced order over the different participants to avoid order effects. Such tests should preferably be validated for the profession studied (Lotens and van de Linde, 1982; Havenith and Heus, 2004). In addition to the performance measurements, questionnaires taken during these tests provide essential information on the ergonomic properties of the clothing.

For clothing for extreme cold, the clothing weight together with the stiffness, friction between multiple layers, fit, etc., may cause substantial performance loss. They can be measured with timed trials as discussed here, but in extreme cases it may be relevant to quantify the extra energy usage when wearing the clothing. This is done via a measurement of oxygen consumption in fixed work trials (walking, obstacle course, etc.). An example of this is given in Fig. 10.5, where the effect on energy consumption was measured for two historical protective ensembles: replica clothing as worn by Robert Falcon Scott and Roald Amundsen on their respective 1911/1912 journeys to the south pole and for a modern-day ensemble used for the same type of expedition. The figure shows the dramatic difference in energy consumption increase for the different ensembles compared to a track suit control and indicates the expected differences in progress speed, nutritional requirements and fatigue for different clothing

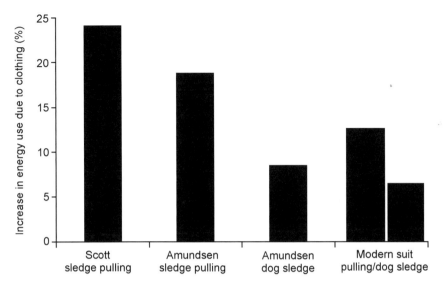

10.5 Increase in energy consumption due to wearing protective clothing versus a track suit control for clothing as worn by Robert Falcon Scott and Roald Amundsen on their expedition to the South Pole in 1911–12, and for modern expedition type clothing (Havenith and Dorman, 2007). Values for different activities: sledge pulling (Scott team) and dog sledge travel (Amundsen team).

designs. Recent studies have shown that many cold protective garments cause increases in energy consumption of the wearer in the order of 8 to 17% (Dorman *et al.*, 2006; Dorman and Havenith, 2009) with a concomitant increase in heat production during these tasks.

10.5.3 Ergonomical design assessment

As the clothing is worn for only brief periods during the tests described above, not all ergonomic problems may surface. Hence, an assessment by an ergonomist and/or expert on the profession can provide additional information. When all the clothing and equipment worn during a certain task is combined the clothing should still be functioning properly and providing the basic functions (protection, storage, etc.) to the wearer. For this aspect an expert panel (ergonomics and topic specialist) assess the clothing. While the clothing is worn by participants of different statures and builds, and importantly also by panel members, the clothing is evaluated for freedom of movement, proper design (overlap between jacket and trousers, arm length), compatibility with other equipment (e.g. gloves, boots). The clothing performance is measured using consumer evaluation type tables (scores: much better, better, equal to, lower/worse, etc., than average or than old suit or a reference suit) (Havenith and Heus, 2004).

10.5.4 General considerations for testing on human participants

Test selection

The test battery described in Havenith and Heus (2004) represents a series of tests that in the past have been shown to discriminate between clothing ensembles in a meaningful manner. The test battery is not exhaustive, however. With this testing one should keep in mind that a large investment of time and cost is involved, and therefore one should always make a selection of tests that is most relevant to the specific clothing that is under investigation. For this reason, it is essential that the experimenter tries to anticipate what kind of results he or she may get from a test and actually consider the usability of such results in the evaluation. To give an example about selecting the conditions: when one wants to test whether clothing has sufficient 'breathability' (i.e. high vapour permeability) one should allow vapour transport in the selection of the climate.

If one were to test in an environment with vapour pressure equal to that of the skin, one would never find any differences between garments due to the absence of a vapour pressure gradient and results could not be interpreted for other conditions. The test then is superfluous as results could have been predicted. Trying to differentiate between breathable membranes for fire-fighter clothing in a firehouse test at extreme temperatures, where vapour pressure of the environment will be higher than that on the skin, and where thus no vapour will leave the skin is an example of choosing the wrong test conditions. On the other hand such conditions are good for testing the heat protection of suits; however, that is a different question.

Another point related to test selection is that the tester or the representative for the profession should in advance consider what they would do with the results and what priority they would give them. For example, testing suits at a low temperature and vapour pressure will generally give the opposite ranking to testing these suits at very high temperatures and vapour pressures due to their insulative properties. If one is not able to give priority to one of these conditions for the task performance, there is no point in doing the tests as no overall conclusions could be drawn.

Short-term versus long-term wear trials

The tests described here are short-term tests. Long-term user trials (Behman, 1988, Lotens, 1988) would provide additional information on ageing and wear and tear of materials and haberdashery. Considering the cost implications of such tests, they are feasible only in very large procurement projects (e.g. military), and go beyond the scope of the present chapter. In such wear trials a lot of information is gathered by questionnaires and regular inspections of the clothing. In the tests described here, part of that information can be collected by questionnaires filled in by the participants after each test. The actual test procedures, apart from providing objective data, have an important second function in this respect. This is to get the participants in the test clothing through task-related activities that will bring to light shortcomings in wearability, compatibility, design, sizing, production quality, quality of zippers, fasteners, seams, etc.

Test reproducibility: human subjects as evaluation tools

The tests discussed above are geared towards precise and reproducible results. In the tests proposed in Section 10.5, measurements are performed using human subjects as 'evaluation tools'. This implies that the experimenter will have to expect human variability to influence the results. While inter-participant variability is desirable in order to evaluate the cold protective clothing for use by different populations (e.g. sizing effect), intra-participant variation (differences between days for the same participant in the same condition, etc.) is undesirable and efforts should be made to minimise this. Pre-test conditioning of clothing and participants are thus important, as is the participants' motivation level, which needs to remain high. Though this variability will not produce bias in the data if the experimental design is properly balanced, it will increase noise levels in the data. Experience shows that the tests described here can discriminate well between different clothing ensembles for various cold protective clothing types, and that they are very sensitive in exposing typical weak spots in the clothing design or manufacturing process.

10.6 Special applications

10.6.1 Smart/innovative fabrics

Several types of smart or innovative fabrics can be found in cold weather clothing. The main types at this moment are fabrics with integrated heating, fabrics/garments with variable insulation, and fabrics incorporating phase-change materials.

Heating

Actively heated garments, using a power supply and woven in or intrinsically conductive fibres can be used to generate extra heat and improve wearer comfort. Used in, e.g., gloves, they can extend allowable exposure time of workers with maintained manual dexterity substantially. Testing of such equipment in the lab situation can be performed on thermal manikins, or for gloves and shoes on special thermally controlled manikin body sections. These would have to be tested in realistic climates to be able to discriminate between the passive and active insulation parameters, and to see how much of the power usage actually benefits the wearer.

Variable insulation

Garments with variable insulation, which may be changed automatically using controllers, are mostly based on changing the thickness of the enclosed air layer in the garment, e.g., by spacer fabrics in combination with shape-memory materials pushing the fabric layers 'open' when cooled, or by inflating bladders integrated in the garment. These types of garments can easily be evaluated on a standard manikin by measurement of the insulation provided in the different states the garment can take. Apart from dry insulation, in these cases it is important to evaluate vapour resistance as well, as the introduction of systems described here (e.g. air bladders) may have a penalty in terms of increased vapour resistance.

Phase-change materials

One of the more recent additions to cold weather clothing systems is that of phase-change materials (PCM). The classical phase-change material is water, which freezes (changes phase) at 0 °C, and while freezing releases a vast amount of heat, i.e., slows down the temperature drop. The main problem is that 0 °C is not often reached inside clothing, certainly not close to the body and thus water is outside the useful range for a PCM. Modern PCMs, consisting of different molecular chain length paraffin waxes, can be produced with a range of PC temperatures. Most are set between 20 and 35 °C. The idea is that when a sudden temperature drop or increase occurs, the micro-encapsulated PCMs are pushed through their phase change and thus will buffer (slow down) the sudden drop so the wearer remains more comfortable for longer. When re-heated, the phase change in the other direction will buffer any sudden temperature increases.

Micro-encapsulated PCMs are typically integrated into the fabric or the fibre or added as a coating or integrated in a foam. This implies that only small amounts are present (mostly less than 80 grams per square metre) and only a small amount of heat can be absorbed or released in most normal garments. The effect is measurable on fabric samples (Bo-an *et al.*, 2004, 2005), and, e.g., ASTM standard D7024-04 (2004) is very much fine-tuned to demonstrate the effect.

The presence of PCMs in garments is often noticed by consumers when gripping/feeling the fabric, though this initial grip is rather different in terms of effect from what happens during actual wear. Overall there is little evidence for a positive impact on people wearing these garments (Shim and McCullough, 2000). Most studies suggesting an effect have either measured only the fabrics or have methodological problems, in that their PCM garment differs from their control in several aspects, not just the presence of PCM. PCMs tend to make the garment heavier (increasing metabolic rate of the wearer and thus heat generation) and reduce the vapour permeability (Pause, 2000) thereby reducing evaporative heat loss. Shim and McCullough (2000), e.g., observed more moisture accumulation in the garments with PCM, i.e., a negative effect.

Another problem with PCMs is that their PC temperature needs to be fine tuned to the conditions in order to be activated in use. This means that if the climate gets, e.g., 10 °C colder, the PCMs may go out of their active range and not be activated. An impact of PCM was demonstrated by Reinertsen *et al.* (2008), but in this case the PCM was present in large amounts (several pounds) in a situation similar to an ice vest. In this case enough heat storage capacity was present and the PCM allows a more comfortable action temperature of the substance than in the case of water/ice.

In terms of lab testing, as mentioned before ASTM D 7024 (2004) provides a methodology that is geared towards high sensitivity for PCM effects. Its test conditions are modified for each specific PCM fabric to ensure it catches the precise PCM temperature change for the sample. Also heat capacities of the equipment are chosen extremely low in order to detect the effects. This test may not have much relevance for the real-life situation, however. A sample with good test results in this test may not be activated at all during actual use in garments, as it may not be placed in the precise location where it would reach the activation temperature.

Several attempts have been made to use manikins for such testing, either by moving the manikin between climates (Shim and McCullough, 2000) or by changing the power to the manikin in analogy to the ASTM test (Havenith, unpublished data). In both instances it has been difficult to see effects, and those observed showed only small changes in heat loss (below 10 W) which may in part have been due to other differences between the PCM and control clothing (difference in conductivity due to PCM). Effects lasted around 15 minutes in the changing climates test. In the changing power test, results showed slightly reduced amplitude of manikin surface temperature with PCM, which supports the positive PCM effect, but the mean surface temperature with PCM was lower suggesting a lower insulation with PCM during the transient, i.e., a negative effect.

Though benefits for wear situations with a single phase change (stepping out into the cold) may be unproven, positive effects may be expected where multiple changes take place in succession, e.g., for forklift truck drivers driving in and out of cold stores. Test methods with real-life value for this are not yet fully developed.

10.6.2 Sleeping bags

Sleeping bags may be seen as an ultimate form of cold weather clothing. Especially expedition and mountaineering bags aim to provide maximal cold protection at minimal weight. In the assessment of sleeping bag performance, the main problem is the translation of the physical data to a realistic range of use of the sleeping bags. This is mostly due to the immense variation between individual users in their experience when using a bag, with experienced mountaineers, e.g., tolerating some level of cold stress while still sleeping on one hand and the common anecdotal response of female recreational users suffering severe discomfort in the same bag at much higher temperatures at the other.

The first step in sleeping bag assessment is the determination of the physical parameters: heat and vapour resistance. Heat resistance for cold protection, vapour resistance for looking at the higher 'warm tolerance' threshold for use. Moisture accumulation during prolonged use can also be an important parameter to determine as moisture will affect the insulation (Havenith and Heus, 1989; Havenith, 2002a).

It is important to define a number of factors when measuring/calculating sleeping bag insulation:

- the clothing worn while sleeping
- the model of the bag
- the floor insulation
- the insulation of any mattress used.

All these will influence the range of use of the bags.

Thickness

The dry insulation of sleeping bags is strongly linked to its thickness. Figure 10.6 shows an overview of flat plate measurements of sleeping bag fillings, showing that thickness is a very good predictor of insulation of a bag filling (assuming appropriate design and manufacture). Thickness measurements are easy and fast.

A simple approach to predicting sleeping bag performance or for quality control can therefore be a measurement of the thickness of the filling or the bag, ideally in two conditions: uncompressed and compressed (Havenith and Heus, 1989). Uncompressed represents the upper side of the bag, which is measured with a low pressure (typically 10 to 20 Pa for reproducibility, with lower values often used for down bags). The compressed side represents that for the lower section of the bag underneath the user and this is thus compressed to a pressure equivalent to that of a person (1500 Pa). This measurement also works well to look at the washing behaviour of bags, which is especially important for non-personal bags that are washed regularly (Havenith and Heus, 1989; Figs 10.6 and 10.7).

A first-order approximation model of the sleeping bag insulation based on these thicknesses was developed by Havenith (1994), based on typical specific insulation values of sleeping bag fillings compressed (30 mKW^{-1}) and uncom-

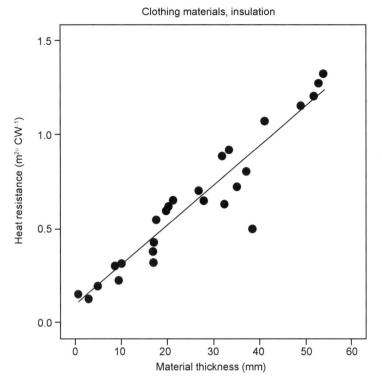

10.6 Heat resistance of insulative fabrics and battings in relation to their thickness (redrawn from Havenith, 2002b).

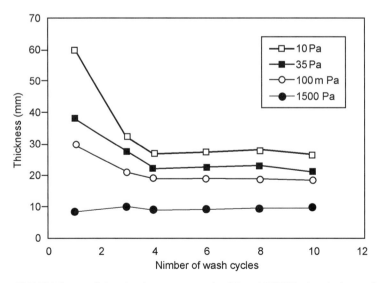

10.7 Thickness of sleeping bags measured at 20 and 1500 Pa, in relation to the number of washing cycles performed (Havenith, unpublished data).

pressed (20 mKW^{-1}) and on the covered and exposed body areas (exposed head: 7% of body area for blanket model bag; 3% for mummy model) and the body area lying on the floor (38%), compressing the filling and the remainder insulated by the uncompressed filling.

For a blanket model bag:

$$I_T = \frac{1}{\left(\dfrac{0.57}{I_{up}} + \dfrac{0.36}{I_{down} + I_{mat}} + \dfrac{0.07}{I_{face}}\right)} = \frac{1}{\left(\dfrac{0.57}{20 \cdot d_{up}} + \dfrac{0.36}{30 \cdot d_{down} + I_{mat}} + \dfrac{0.07}{I_{face}}\right)}$$

10.12

For a mummy bag:

$$I_T = \frac{1}{\left(\dfrac{0.61}{I_{up}} + \dfrac{0.36}{I_{down} + I_{mat}} + \dfrac{0.03}{I_{face}}\right)} = \frac{1}{\left(\dfrac{0.61}{20 \cdot d_{up}} + \dfrac{0.36}{30 \cdot d_{down} + I_{mat}} + \dfrac{0.03}{I_{face}}\right)}$$

10.13

where I = insulation (m^2.K.W^{-1}), d = thickness (m) and I_{face} = 0.105 m^2.K.W^{-1}.

A second-order approximation is then to measure the actual insulations uncompressed and compressed on a flat plate and use the measured numbers in equations 10.12 and 10.13, rather than the thicknesses.

Manikin measurement

Manikins can be used to measure sleeping bag insulation. EN 13537 (2002) describes such a measurement and defines the use of 'a rigid support (12 mm wood board)' to simulate the ground, with air passing underneath; 'a mat, representative of the user group' and 'a two piece track suit and knee-long socks' as clothing. Given the large impact of the mat, representing more insulation towards the floor than the actual underside of the bag in many cases, this is an important factor to specify and keep constant in comparisons. Based on the heat resistance measurement they specify that a utility range shall be reported, including a comfort, a limit and an extreme temperature.

A problem with many manikins for sleeping bag measurements is that their weight is much lower than that of a human being, causing too little compression of the filling underneath the person. Hence, these manikins would need to be weighed down for the test.

Human subject testing

Similar to the insulation testing of clothing with human participants, this can also be done for sleeping bags (Havenith and Heus, 1989) by measuring heat production and heat losses, or by trialling bags at different temperatures and asking for subjective feedback. In general, it is the temperatures of the feet and toes that when measured provide the most sensitive discrimination between different bags. Once a person cools, feet cool fastest and are most sensitive to the insulation provided (Havenith and Heus, 1989).

Minimal temperature for sleep

After determining the insulation of a bag there often is a need to estimate the minimal temperature for sleep. Numerous models have been produced for this, with a variety of assumptions.

A simple first-order calculation is provided by (Havenith, 1994):

$$T_{a,min} = 33 - \frac{70 \cdot I_T}{1.8} = 33 - 39 \cdot I_T \qquad 10.14$$

Assuming a mean skin temperature of 33 °C and a heat production of the body of 70 W with a 1.8 m² body surface area (38.9 W.m⁻²). These calculations, given the low heat production while asleep, are highly sensitive to the estimation of that heat production. Several authors have therefore also done calculations where they 'accept' a certain amount of body cooling within a certain period of sleep. Some examples of calculations are, for steady state comfort sleep:

$$T_{a,min} = 32 - 34.8 \cdot I_T \quad \text{(Goldman, 1988)} \qquad 10.15$$

$$T_{a,min} = 32.8 - 31.1 \cdot I_T \quad \text{(McCullough, 1994)} \qquad 10.16$$

For limited time sleep with some cooling:

$$T_{a,min} = 32 - 43.2 \cdot I_T \quad \text{(6 hours; Goldman, 1988)} \qquad 10.17$$

$$T_{a,min} = 31.4 - 37.3 \cdot I_T \quad \text{(8 hours; McCullough, 1994)} \qquad 10.18$$

In contrast to these simple calculations, the ISO standard model contains a higher number of assumptions regarding the user's posture, gender, etc. A table of sleep temperature limits in relation to measured insulations is provided in the standard.

10.7 Future trends

A number of developments in lab testing are anticipated. In the area of manikin development, the sweating features will become more detailed, and with emerging knowledge on whole body sweat patterns (Havenith *et al.*, 2008a; Machada-Moreira *et al.*, 2008), manikins are expected to show differentiated sweating over different zones. Though this may not be relevant to the general determination of ensemble vapour resistances, given the introduction of more sources of uncertainty and error by this method, it will enable comparison of garment systems by going through various realistic scenarios for climate, work load and associated physiological responses (i.e. sweat production).

The other major development is expected in the area of assessment of smart clothing. New methods need to be developed to assess the real world value of smart clothing with conditions representative for actual work situations.

10.8 References

ASTM F1868 – 02 (2002) 'Standard Test Method for Thermal and Evaporative Resistance of Clothing Materials Using a Sweating Hot Plate', American Society for Testing and Materials, West Conshohocken, PA.

ASTM D1518 – 85 (2003) 'Standard Test Method for Thermal Transmittance of Textile', American Society for Testing and Materials, West Conshohocken, PA.

ASTM F2298 (2003) 'Standard Test Methods for Water Vapor Diffusion Resistance and Air Flow Resistance of Clothing Materials Using the Dynamic Moisture Permeation Cell', in *Annual Book of ASTM Standards 11.03*, American Society for Testing and Materials, West Conshohocken, PA.

ASTM D7024 (2004) 'Standard Test Method for Steady State and Dynamic Thermal Performance of Textile Materials', American Society for Testing and Materials, West Conshohocken, PA.

ASTM F1291 (2005) 'Standard Test Method for Measuring the Thermal Insulation of Clothing Using a Heated Manikin', American Society for Testing and Materials, West Conshohocken, PA.

ASTM F2370 (2005) 'Standard Test Method for Measuring the Evaporative Resistance of Clothing Using a Sweating Manikin', American Society for Testing and Materials, West Conshohocken, PA.

ASTM E96/E96M – 05 (2005) 'Standard Test Methods for Water Vapor Transmission of Materials', American Society for Testing and Materials, West Conshohocken, PA.

Behman F W (1988) Chapter 16; 'Field evaluation methods'. In: *Handbook on clothing RSG 7, Panel VIII*, Nato. (available at www.environmental-ergonomics.org).

Birnbaum R R and Crockford G W (1978) 'Measurement of the clothing ventilation index', *Applied Ergonomics* 9(4), 194–200.

Bo-an Y, Kwok Y L, Li Y, Yeung C Y, Song Q (2004) 'Thermal regulating functional performance of PCM garments', *International Journal of Clothing Science and Technology* 16(1/2), 84–96.

Bo-an Y, Kwok Y L, Li Y, Zhu Q and Yung C (2005) 'Assessing the performance of textiles incorporating phase change materials', *Polymer Testing* 23(5), 541–549.

BS 3424-18 (1986) *Testing coated fabrics. Methods 21A and 21B. Methods for determination of resistance to wicking and lateral leakage*, British Standards Institute, London.

BS 7209 (1990) *Specification for water vapour permeable apparel fabrics leakage*, British Standards Institute, London.

BS 4745 (2005) *Determination of the thermal resistance of textiles. Two-plate method: fixed pressure procedure, two-plate plate method: fixed opening procedure, and single-plate method*, British Standards Institute, London.

CAN/CGSB 4.2 No. 70.1-94 (1994) [withdrawn] *Textile Test Methods – Thermal Insulation Performance of Textile Materials*, Canadian General Standards Board.

CAN/CGSB-4.2 No. 49 – 99 (1999), *Textile Test Methods Resistance of Materials to Water Vapour Diffusion*, Canadian General Standards Board.

CAN/CGSB 4.2 No. 78.1-2001 (2001) *Textile Test Methods – Thermal Protective Performance of Material for Clothing*, Canadian General Standards Board.

Daanen H A M, Hatcher K and Havenith G (2002) 'Determination of clothing microclimate volume', *Environmental Ergonomics X*, The Organizing and International Program Committees of the 10th International Conference on Environmental Ergonomics (eds), The 10th International Conference on Environmental Ergonomics, Fukuoka, Japan, 66-5668.

DIN 53924 (1997) -03 *Prüfung von Textilien – Bestimmung der Sauggeschwindigkeit von textilen Flächengebilden gegenüber Wasser (Steighöhenverfahren)*, German Standardisation Institute, Berlin.

Dorman L and Havenith G (2009) 'The effects of protective clothing on metabolic rate', *European Journal of Applied Physiology* 105(3), 463–470.

Dorman L, Havenith G, Brode P, Candas V, den Hartog E, Holmer I, Meinander H, Nocker W and Richards M (2006) *Modelling the Metabolic Effects of Protective Clothing*, Central Institute for Labour Protection – National Research Institute, 3rd European Conference on Protective Clothing (ECPC), Poland.

EN 20811 (1992), ISO 811 (1981) *Textiles. Determination of resistance to water penetration. Hydrostatic pressure test*, European Committee for Standardization, Brussels; International Organisation for Standardisation, Geneva.

EN ISO 9237 (1995) *Textiles. Determination of the permeability of fabrics to air*, International Organisation for Standardisation, Geneva.

EN 13537 (2002) *Requirements for sleeping bags*, European Committee for Standardization, Brussels.

EN 14058 (2004) *Protective clothing. Garments for protection against cool environments*, European Committee for Standardization, Brussels.

EN 14360 (2004) *Protective clothing against rain. Test method for ready made garments, Impact from above with high energy droplets*, European Committee for Standardization, Brussels.

EN 342 (2004) *Protective clothing. Ensembles and garments for protection against cold*, European Committee for Standardization, Brussels.

EN ISO 15831 (2004) *Clothing. Physiological effects. Measurement of thermal insulation by means of a thermal manikin*, International Organisation for Standardisation, Geneva.

EN 511 (2006) *Protective gloves against cold*, European Committee for Standardization, Brussels.

EN ISO 11079 (2007) *Ergonomics of the thermal environment. Determination and interpretation of cold stress when using required clothing insulation (IREQ) and local cooling effects*, International Organisation for Standardisation, Geneva.

EN ISO 9920 (2007) *Ergonomics of the thermal environment. Estimation of thermal insulation and water vapour resistance of a clothing ensemble*, International Organisation for Standardisation, Geneva.

Goldman R F (1974) 'Clothing design for comfort and work performance in extreme thermal environments'. *Transactions of the New York Academy of Sciences*, series ii, 36(6).

Goldman R F (1988) In: *Handbook on clothing RSG 7, Panel VIII*, Nato. (available at www.environmental-ergonomics.org).

Havenith G (1994) 'Classification for sleeping bags model "mummy"' (in Dutch), *Report TNO-TM 1994 C-33*, TNO Human Factors Research Institute, Soesterberg.

Havenith G (2002a) 'Moisture accumulation in sleeping bags at sub-zero temperature; effect of semipermeable and impermeable covers', *Textile Research Journal* 72(4), 281–284.

Havenith G (2002b) 'Clothing and thermoregulation', *Allergologie* 25(3), 177.

Havenith G and Dorman L (2007) 'Race to the South Pole – Scott and Amundsen's clothing revisited', *Proceedings of the 12th International Conference on Environmental Ergonomics*, I B Mekjavic, S N Kounalakis, N A S Taylor, Biomed, Piran, Slovenia, August, pp. 150–152.

Havenith G and Heus R (1989) 'Insulation, comfort and moisture accumulation of six prototypes of sleeping bags for the Royal Netherlands Army' (in Dutch). *Report IZF 1989-40*, TNO Institute for Perception, Soesterberg.

Havenith G and Heus R (2004) 'An ergonomic test battery of protective clothing', *Applied Ergonomics* 35/1, 3–20.

Havenith G and Nilsson H (2004) 'Correction of clothing insulation for movement and wind effects, a meta-analysis', *Eur J Appl Physiol* 92, 636–640.

Havenith G and Nilsson H (2005) 'Correction of clothing insulation for movement and wind effects, a meta-analysis', *Eur J Appl Physiol* 93, 506 (erratum to previous publication).

Havenith G, Heus R, Lotens W A (1990) 'Clothing ventilation, vapour resistance and permeability index: changes due to posture, movement and wind', *Ergonomics* 33/8, 989–1005.

Havenith G, Ueda H, Sari H and Inoue Y (2003) 'Required clothing ventilation for different body regions in relation to local sweat rates', *Proceedings 2nd European Conference on Protective Clothing (ECPC), Challenges for Protective Clothing*, 21–24 May, Montreux, Switzerland.

Havenith G, den Hartog E and Heus R (2004) 'Moisture accumulation in sleeping bags at −7 and −20 °C in relation to cover material and method of use', *Ergonomics* 47(13), 1424–1431.

Havenith G, Fogarty A, Bartlett R, Smith C and Ventenat V (2008a) 'Male and female upper body sweat distribution during running measured with technical absorbents', *European Journal of Applied Physiolog* 104(2) 245–255.

Havenith G, Richards M, Wang X, Broede P, Candas V, den Hartog E, Holmer I, Kuklane K, Meinander H and Nocker W (2008b) 'Apparent latent heat of evaporation from clothing: attenuation and "heat pipe" effects', *J Appl Physiol* 104, 142–149.

Holmér I and Elnas S (1981) 'Physiological evaluation of the resistance to evaporate heat transfer by clothing', *Ergonomics* 24, 63–74.

Huang J and Qian X (2008) 'Comparison of test methods for measuring water vapor permeability of fabrics', *Textile Research Journal* 78, 342–352.

ISO 11092 (1993); EN 31092 (1994) *Textiles. Determination of physiological properties. Measurement of thermal and water-vapour resistance under steady-state conditions (sweating guarded-hotplate test)*, International Organisation for Standardisation, Geneva; European committee for standardization, Brussels.

Kuklane K, Nilsson H, Holmér I and Liu X (1997) 'Methods for handwear, footwear and headgear evaluation', Nilsson H and Holmér I eds. *Proceedings of a European Seminar on Thermal Manikin Testing*, Arbetslivsrapport 1997:9, Solna, Sweden.

Kuklane K, Afanasieva R, Burmistrova O, Bessonova N and Holmér I (1999) 'Determination of heat loss from the feet and insulation of the footwear', *Int J Occup Saf Ergon* 5(4), 465–476.

Lotens W A (1988) 'Optimal design principles for clothing systems', *Handbook on clothing* RSG-7 (Nato) (Chapter 17).

Lotens W A (1993) *Heat Transfer from Humans Wearing Clothing*, PhD Thesis Technische Universiteit Delft.

Lotens W A and Havenith G (1988) 'Ventilation of rainwear determined by a trace gas method'. In: Mekjavics, Bannister, Morrison (eds), *Environmental Ergonomics*.

Taylor and Francis, Philadelphia, pp. 162–175.
Lotens W A and van de Linde F J G (1982) 'A comparison of three firefighter suits,Part II: Ergonomics and practice tests (Een vergelijking van drie brandweerpakken II: Ergonomie en praktijkproeven)', *Report Institute for Perception* IZF 1982 C-14.
McCullough E A (1994) 'Determination of the insulation value and temperature rating of sleeping bags', *Proc. Outdoor Retailer Coalition, Reno, NV*, 20 August.
McCullough E A, Kwon M and Shim H (2003) 'A comparison of standard methods for measuring water vapour permeability of fabrics', *Meas. Sd. Technol* 14(8), 1402–1408.
Machado-Moreira C A, Smith F M, van den Heuvel A M J, Mekjavic I B and Taylor N A S (2008) 'Sweat secretion from the torso during passively-induced and exercise-related hyperthermia', *Eur J Appl Physiol* 104(2), 265–270.
Mecheels J, Umbach K H (1976) 'Thermophysiologische Eigenschaften von Kleidungssystemen (Thermophysiological properties of clothing systems)', *Melliand Textilberichte* 57(12), 1029–1032.
Pause B H (2000) 'New heat protective garment with phase change material', *Performance of Protective Clothing: Issues and Priorities for the 21st Century*, Seventh Volume, ASTM STP 1386.
Reinertsen R E, Faerevik H, Holbo K, Nesbakken R, Reitan J, Royset A and Thi M S (2008) 'Optimising the performance of phase change materials in personal protective clothing systems', *International Journal of Occupational Safety and Ergonomics*, in press.
Rintamäki H and Rissanen S (2006) 'Heat strain in the cold', *Industrial Health* 44(3), 427–432.
Shim H and McCullough EA (2000) 'The effectiveness of phase change materials in outdoor clothing', *Ergonomics of Protective Clothing, Proceedings of nokobetef 6 and 1st European Conference on Protective Clothing*, Stockholm, Sweden, 7–10 May.
Ueda H and Havenith G (2002) 'The effect of fabric air permeability on clothing ventilation', *Environmental Ergonomics X*, The Organizing and International Program Committees of the 10th International Conference on Environmental Ergonomics (eds), 10th international Conference on Environmental Ergonomics, Fukuoka, Japan, September, 621–624.
Ueda H, Inoue Y, Matsudaira M, Araki T and Havenith G (2006) 'Regional microclimate humidity of clothing during light work as a result of the interaction between local sweat production and ventilation', *International Journal of Clothing Science and Technology* 18(4), 225–234.
Umbach (1983) 'Evaluation of comfort characteristics of clothing by use of laboratory measurements and predictive calculations', *Research Inst. of National Defence Intern. Conf. on Protective Clothing Systems* (SEE N83 26454 15 54), Sweden, 141–149.
Van Langenhove L and Kiekens P (2001) 'Textiles and the transport of moisture', *Textile Asia* 32–43.
Wissler E H and Havenith G (2009) 'A simple theoretical model of heat and moisture transport in multi-layer garments in cool ambient air', *European Journal of Applied Physiology* 105(5), 797–808.

11
Evaluation of cold weather clothing using manikins

E. A. McCULLOUGH, Kansas State University, USA

Abstract: This chapter discusses the use of thermal manikins to measure the insulation value and evaporative resistance of cold weather clothing ensembles. Manikin data can be used in heat loss models to predict the environmental conditions for comfort. The pumping effect in clothing can be measured with the manikin walking, and the effect of phase-change materials in cold weather clothing can be evaluated by moving the manikin between chambers at different environmental temperatures.

Key words: manikins; thermal insulation; evaporative resistance; cold weather clothing; thermal comfort.

11.1 Introduction

The body loses heat through conductive, convective, and radiant heat exchange with the environment and by the evaporation of sweat. The heat is lost from the body surface and through respiration (convection and evaporation). Cold weather protective clothing systems are designed to provide significant thermal resistance (insulation) while minimizing resistance to the evaporation of sweat. This chapter will discuss the use of heated manikins to measure the insulation and evaporative resistance of cold weather clothing systems.

11.2 Manikin tests vs. fabric tests

Sweating guarded hot plates can be used to measure and compare the thermal resistance and evaporative resistance of different types of fabrics and multi-component systems (McCullough *et al.*, 2004). The most commonly used methods for this purpose include ASTM F 1868, Standard Test Method for Thermal and Evaporative Resistance of Clothing Materials Using a Sweating Hot Plate (ASTM, 2006) and ISO 11092, Textile B Physiological Effects B Measurement of Thermal and Water Vapor Resistance Under Steady State Conditions (Sweating Guarded Hot plate Test) (ISO, 1993). These standards limit the thickness of materials tested to about 2–3 cm because the guard sections around the test section of the plate are not large enough to prevent heat loss from the edges of thick samples. Many cold weather fabric/batting/fabric systems are thicker than that.

Although hot plate data can be important information to use in comparing and selecting materials for cold weather garments, manikin measurements account for many additional factors that affect the thermal and evaporative resistance of clothing systems. These include:

- the amount of body surface area covered by textiles and the amount of exposed skin
- the distribution of textile layers and air layers over the body surface (i.e., non-uniform)
- looseness or tightness of fit
- the increase in surface area for heat loss (i.e., clothing area factor) due to the textiles around the body
- the effect of product design
- the adjustment of garment features (i.e., fasteners open, hood up, etc.)
- variation in the temperature (and heat flux) on different parts of the body
- the effect of body position (i.e., standing, sitting, lying down)
- the effect of body movement (i.e., walking, cycling).

Therefore, manikin measurements are realistic, in that they quantify the effect of a clothing system on the heat exchange between the whole body and the environment. However, manikins, environmental chambers, and computer control and data acquisition systems are expensive to acquire and complex to maintain.

11.3 Thermal manikins

Thermal manikins have been used by many researchers to measure the thermal resistance (insulation) and evaporative resistance of clothing systems. These resistance values are used in biophysical models to predict the comfort and/or thermal stress associated with particular environmental conditions and the activity of the wearer. Wyon (1989) described the historical development of manikins, and Holmér (1999) continued this work, describing standard test methods for the use of thermal manikins and standards that require data from manikin tests. It is impossible to count the number of heated manikins in use because some laboratories have published numerous scientific articles on data collected with the same manikin, and researchers from labs without manikins have collaborated with researchers with manikins. However, manikins are in use in the United States, Canada, France, Sweden, Finland, Norway, Denmark, Germany, United Kingdom, Switzerland, Hungary, China, Korea, and Japan. A recent trend has been the development of specialized heated body parts such as a head, hand, and foot/calf so that the thermal effectiveness of the design and materials used in head gear, gloves/mittens, and footwear can be determined with more precision (Kuklane *et al.*, 1997).

It is important to remember that manikins do not simulate the human body physiologically. They are thermal measuring devices in the size and shape of a human being that are heated so that their surface temperatures simulate the local and/or mean skin temperatures of a human being. They do not respond to changes in the environment or clothing like the human body does.

11.3.1 Segmented thermal manikins

Most manikins are divided into body segments with independent temperature control and measurement (even though they are intended to be used to quantify the heat transfer characteristics of total body systems). The segments can all be controlled at the same temperature (i.e., 34 °C), or a skin temperature distribution where the extremities have lower temperatures than the head and trunk can be achieved. These manikins can indicate the relative amounts of heat loss from different parts of the body under specific environmental conditions and/or measure the insulation value or evaporative resistance value of each segment. However, most segmented manikins have some internal heat transfer from one segment to another and the movement of heat within the clothing layers between segments – both of which will lead to inaccurate local results. In addition, the use of segmented manikins has led to the use of two different methods for calculating clothing resistance values: the parallel (total) method where all heat losses, temperatures (area-weighted), and areas are summed before the total resistance is calculated, and the serial (local) method where the individual resistances for each body segment are calculated and then summed (Nilsson, 1997). The serial calculation usually produces higher values and more variable results due to the uneven distribution of insulation over the body (Nilsson, 1997; McCullough *et al.*, 2002). When a body segment (e.g., abdomen) is well insulated relative to the others, its heat loss may be very low or zero, causing the measured insulation value of the segment to be too high, and the resulting serial calculation for the body to be high.

11.4 Measuring the thermal resistance of cold weather clothing systems

The first thermal manikin was developed by military researchers in the United States in the 1940s (Belding, 1949). A manikin was needed to measure the insulation properties of protective clothing and sleeping bag systems because measurements on pieces of fabric could not be related to whole-body systems with accuracy.

11.4.1 Standards

The first thermal manikin test method was developed in 1996: ASTM F 1291, Standard Test Method for Measuring the Thermal Insulation of Clothing Using a Heated Manikin (ASTM, 2006). Recently, ISO 15831, Clothing – Physiological

Evaluation of cold weather clothing using manikins   247

Effects – Measurement of Thermal Insulation by Means of a Thermal Manikin was approved also (ISO, 2004). The ASTM method calls for a stationary manikin test, whereas, the ISO standard includes a walking test as well.

11.4.2 Method

To measure the thermal resistance, a manikin is dressed in the clothing system and placed in a cool/cold environmental chamber. Then the amount of electrical power required to keep the manikin heated to a constant skin temperature (e.g., 33–35 °C) is measured under steady-state conditions. The power input is proportional to body heat loss (see Fig. 11.1).

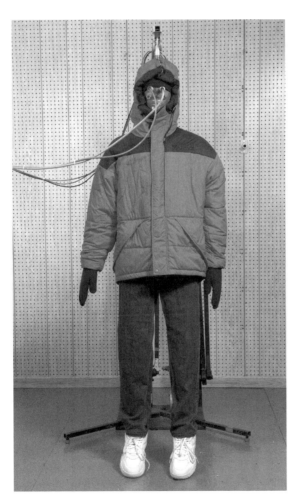

11.1 The thermal manikin at Kansas State University dressed in cold weather clothing.

The total thermal insulation value (R_t) is the total resistance to dry heat loss from the body surface, which includes the resistance provided by the clothing and the air layer around the clothed body. R_t is measured directly with a manikin and is calculated by:

$$R_t = (T_s - T_a) \cdot A_s / H \qquad 11.1$$

where R_t = total thermal insulation of the clothing plus the boundary air layer (m²·C/W), T_s = mean skin temperature (°C), T_a = ambient air temperature (°C), A_s = manikin surface area (m²), and H = power input (W).

A clo unit is normally used for expressing clothing insulation since it is related to commonly worn ensembles. A warm business suit ensemble provides an average of approximately 1 clo of insulation for the whole body, where 1 clo of insulation is equal to 0.155 m²·°C/W (Gagge et al., 1941). When insulation values are reported in clo units, the symbol I is usually used instead of R. Therefore,

$$I_t = 6.45 \cdot (T_s - T_a) \cdot A_s / H \qquad 11.2$$

Because R_t or I_t includes the resistance at the surface of the clothed body, it is influenced by air velocity and temperature level (as it relates to incident radiation). These factors can be easily dealt with in physiological models which predict how much heat a person will lose or gain under a specific set of conditions. However, in some models and applications, it may be preferable to separate the resistance of the clothing from the resistance of the air layer.

Intrinsic clothing insulation (R_{cl}, I_{cl}) indicates the insulation provided by the clothing alone and does not include the insulation provided by the surface air layer around the clothed body. R_{cl} is defined by

$$R_{cl} = R_t - (R_a / f_{cl}) \qquad 11.3$$

and I_{cl} is defined by

$$I_{cl} = I_t - (I_a / f_{cl}) \qquad 11.4$$

where R_{cl} = intrinsic clothing insulation (m²·°C/W), R_t = total thermal insulation of the clothing plus the boundary air layer (m²·°C/W), R_a = resistance of the boundary air layer around the nude manikin (m²·°C/W), f_{cl} = clothing area factor (unitless), I_{cl} = intrinsic clothing insulation (clo), I_t = total thermal insulation of the clothing plus the boundary air layer (clo), I_a = resistance of the boundary air layer around the nude manikin (clo).

11.4.3 Clothing area factor

The clothing area factor (f_{cl}) is the ratio of the projected area of the clothed body area to the projected area of the nude body. Photographs of the manikin or a person are taken from three azimuth angles: 0° front view, 45° angle view, and

90° side view and two altitude angles: 0° and 60°. The distance between the camera and the manikin and the focal length must stay the same for all photographs. The size of the projected areas of the nude manikin and the clothed manikin can be determined in several ways. A planimeter can be used to trace around each silhouette and determine the projected area for each view. Alternatively, the areas may be determined by cutting along the outside edges of the silhouette (i.e., projected area) in each photograph and weighing them (assuming the same type of paper was used for printing all photographs). The relative size of the areas may also be determined by manipulating the photographs using computer software such as Adobe Photoshop to blacken the silhouettes and compare the relative number of pixels used in each area. The sum of the projected areas of the clothed manikin is then divided by the sum of the projected areas of the nude manikin; the resulting f_{cl} is a dimensionless index greater than 1 (McCullough et al., 2005).

It is time-consuming to take all of the photographs and determine the relative size of the projected areas. Therefore, the clothing area factor of clothing ensembles can be estimated from tables in standards such as ISO 9920 and ASTM F 1291 (ISO, 1995; ASTM, 2006) and clothing databases. The f_{cl} varies from 1.00 (nude) to a maximum of about 1.70 for some protective clothing ensembles with a breathing apparatus. The f_{cl} for winter clothing ranges from 1.3 to 1.5 – depending upon how thick the garments are.

The term (R_a/f_{cl} or I_a/f_{cl}) is the resistance provided by the air layer around the clothed body. It is smaller than the air layer resistance for the nude body because the clothing increases the surface area and thus provides a greater area for heat transfer. As the f_{cl} increases, the intrinsic clothing insulation also increases (for a given level of total insulation).

11.5 Measuring the evaporative resistance of cold weather clothing systems

There are relatively few sweating manikins available for measuring the evaporative resistance or vapor permeability of clothing (McCullough et al., 2002). Some manikins are covered with a cotton knit suit and wetted out with distilled water to create a saturated sweating skin. However, the skin will dry out over time unless tiny tubes are attached to the skin so that water can be supplied at a rate necessary to sustain saturation. Other manikins have sweat glands on different parts of the body (Holmér et al., 1996). Water is supplied to each sweat gland from inside the manikin, and its supply rate can be varied. A new type of sweating manikin uses a waterproof, but moisture-permeable fabric skin, through which water vapor is transmitted from the inside of the body to the skin surface (Fan and Qian, 2004). Some manikins keep the clothing from getting wet by using a microporous membrane between the sweating surface and the clothing, but this configuration may increase the insulation value of the nude manikin.

The evaporative resistance of cold weather clothing is not as important as the insulation it provides. If a person gets too warm in cold weather clothing (i.e., when generating additional body heat through exercise), he/she can adjust garment openings or remove garment layers to dissipate the extra heat. Therefore, the sweating manikin test is not routinely conducted on cold weather ensembles except for those worn by people who work in the cold (e.g., the military).

11.5.1 Standards

The first sweating manikin standard was developed in 2005: ASTM F 2370, Measuring the Evaporative Resistance of Clothing Using a Sweating Manikin (ASTM, 2006). It specifies procedures for measuring the evaporative resistance of clothing systems under isothermal conditions – where the manikin's skin temperature is the same as the air temperature (i.e., there is no temperature gradient for dry heat loss). An alternative protocol in the standard allows the clothing ensemble to be tested under environmental conditions that simulate actual conditions of use; this is called the non-isothermal test. The same environmental conditions are used for the insulation test and the non-isothermal sweating manikin test. The air temperature is lower than the manikin's skin temperature, so dry heat loss is occurring simultaneously with evaporative heat loss, and condensation may develop in the clothing layers. The evaporative resistance determined under non-isothermal conditions is called the apparent evaporative resistance value. The apparent evaporative resistance values for ensembles can only be compared to those of other ensembles measured under the same environmental conditions.

11.5.2 Method

To conduct a sweating manikin test, the surface of the manikin is heated to skin temperature and saturated with water. The manikin is dressed in the clothing, and the evaporative resistance of the clothing system is determined by measuring the power consumption of the heated manikin. Even under isothermal conditions, it will take electrical power to keep the manikin heated because the process of evaporating moisture on the surface removes heat. More power is needed under non-isothermal conditions.

The equation for calculating the total resistance to evaporative heat transfer provided by the clothing is

$$R_{et} = (P_s - P_a) \cdot A_s / [H - ((T_s - T_a) \cdot A_s / R_t)] \qquad 11.5$$

where R_{et} = resistance to evaporative heat transfer provided by the clothing and the boundary air layer (m²·kPa/W), P_s = saturated water vapor pressure at the skin surface (kPa), and P_a = the water vapor pressure in the air (kPa), R_t = total thermal insulation of the clothing plus the boundary air layer (m²·°C/W), T_s = mean skin temperature (°C), T_a = ambient air temperature (°C), A_s = manikin surface area (m²), and H = power input (W). The mean R_t value from the dry tests on the ensemble is needed to calculate R_{et}.

The equation for calculating the intrinsic evaporative resistance provided by the clothing alone (R_{ecl}) is analogous to that for dry resistance:

$$R_{ecl} = R_{et} - (R_{ea}/f_{cl}) \qquad 11.6$$

where R_{ea} = the resistance to evaporative heat transfer for a still air layer (m²·kPa/W).

11.5.3 Moisture permeability index

The permeability index (i_m) indicates the maximum evaporative heat transfer permitted by a clothing system as compared to ideal maximum from an uncovered surface (i.e., a slung psychrometer). It was defined by Woodcock (1962) as

$$i_m = (R_t/R_{et})/LR \qquad 11.7$$

where i_m = permeability index and LR = the Lewis relation, commonly given the value of 16.65 °C/kPa. The permeability index usually ranges from about 0.50 for a nude manikin to about 0.05 for an impermeable single-layer ensemble with a low thermal resistance and high evaporative resistance.

11.6 Moving manikins

Most of the time, manikins are used in the standing position, but more and more researchers are attaching their manikins to external locomotion devices and measuring clothing insulation with the manikin walking (McCullough and Hong, 1994; Kim and McCullough, 2000; Nilsson et al., 1992; Olesen et al., 1982). Body motion increases convective heat loss and decreases the insulation value of clothing. This value has been referred to as resultant or dynamic insulation. ISO 15831 (ISO, 2004) gives a protocol for using a walking manikin to measure resultant insulation. Few laboratories have used a sweating manikin while it was walking (Richards and Mattle, 2001). Examples of how the insulation provided by an ensemble decreases as a function of body movement is shown in Table 11.1.

11.7 Using manikins under transient conditions

Manikins are designed for steady-state measurements, so their control systems do not work well during transients (i.e., changing environmental conditions). The changes in heat loss over time do not simulate the thermal responses of

252   Textiles for cold weather apparel

Table 11.1 Static and dynamic insulation values for cold weather clothing ensembles[a]

Ensemble description	Clothing area factor f_{cl}	Intrinsic clothing insulation I_{cl} (clo)		
		Static	Dynamic	% Change
1. Extreme cold weather expedition suit with hood (down-filled, one-piece suit), thermal long underwear top and bottoms, mittens with fleece liners, thick socks, insulated waterproof boots	1.50	3.67	3.21	12
2. One-piece ski suit, thermal long underwear top and bottoms, knit head/ear band, goggles, insulated ski gloves, thin knee length ski socks, insulated waterproof boots	1.28	1.60	1.13	30
3. One-piece fiberfill ski suit with hood, thermal long underwear top and bottoms, goggles, insulated ski gloves, thin knee length ski socks, insulated waterproof boots	1.27	1.97	1.53	22
4. Extreme cold weather down-filled parka with hood, shell pants, fiberfill pants liner, thermal long underwear top and bottoms, sweat shirt, mitten shell with inner fleece gloves, thick socks, insulated waterproof boots	1.47	3.28	2.53	23
5. Knee length down-filled coat, thermal long underwear bottoms, jeans, T-shirt, long sleeve flannel shirt, hat with fleece liner and ear flaps, insulated ski gloves, thick socks, low cut leather work boots	1.52	2.45	1.50	39
6. Fiberfill jacket, jeans, T-shirt, thermal long underwear bottoms, long-sleeve flannel shirt, baseball cap, thick socks, low cut leather work boots	1.40	1.68	1.30	23
7. Fleece long sleeve shirt, fleece pants, briefs, athletic socks, athletic shoes	1.29	1.19	0.86	28

[a]Air layer thermal resistance around the nude manikin was 0.68 clo while standing and 0.49 clo while walking. These values can be used with data in the table to calculate total insulation values (I_t).
Source: Kim and McCullough, 2000

human beings; they depend upon the power capacity and control system (i.e., time constants) of the manikin. The thermal capacitance of a manikin is different from the human body, so even if the manikin's program allows the skin temperature to change during a transient, the net effect on energy balance for the manikin is not the same as it is for the human body. However, manikins have been used on a limited basis to quantify and compare the impact of thermal changes in clothing during step changes in the environmental conditions (Shim et al., 2001). For example, the temperature of a garment containing phase-change materials may change temporarily as the result of a change in air temperature. To study these transient effects, the manikin must be moved quickly from one environmental chamber to another while heat loss from the manikin is recorded over time.

11.8 Temperature ratings

Many manufacturers of cold weather clothing want to indicate the amount of warmth (i.e., insulation) their products will provide to consumers at the point of sale. In the United States, this is often expressed as a temperature rating on product labels and in product descriptions in catalogs. A temperature rating is commonly understood to mean the lowest air temperature at which the average adult person will be thermally comfortable when using the product. Although it is not always stated on labels or in catalogs, manufacturers are assuming that consumers will wear the appropriate amount of clothing with the outdoor garments. The use of these auxiliary products can greatly increase the insulation provided by the clothing ensemble, and therefore, lower the temperature for comfort.

ASTM Technical Committee F23 on Protective Clothing recently developed ASTM F 2732, a new standard practice on determining the temperature ratings for cold weather protective clothing. The standing manikin test is used to measure the insulation value of a jacket, insulated pants, or a coverall worn over a standard set of garments. The insulation value is used in a whole-body heat loss model to determine the air temperature for comfort under a certain set of conditions (e.g., activity level, wind speed, relative humidity, etc., are specified at one level). Although evaporative heat losses are included in the model prediction, the evaporative resistance value for the ensembles is estimated as opposed to being measured with a sweating manikin. The temperature ratings are simply guidelines as to the amount of warmth an outdoor garment will provide. The other garments worn with the cold weather garment and the activity level of the wearer vary widely during use and greatly affect the air temperature for comfort.

11.9 Conclusions

Thermal manikins will continue to be used to measure the thermal resistance (i.e., insulation value) and evaporative resistance of cold weather clothing systems. They can be used to evaluate and compare clothing ensembles under steady-state and transient conditions, and their data can be used to predict the thermal comfort of the wearer.

11.10 References

American Society for Testing and Materials (2006), *2006 Annual book of ASTM standards, Vol. 11.03*, American Society for Testing and Materials, West Conshohocken, PA.

Belding H S (1949), Protection against cold. In Newburgh L H, *Physiology of Heat Regulation and the Science of Clothing*. Philadelphia, PA, Saunders, 351–367.

Fan J and Qian X (2004), 'New functions and applications of Walter, the sweating fabric manikin', *European Journal of Applied Physiology*, 92, 641–644.

Gagge A P, Burton A C and Bazett H D (1941), 'A practical system of units for the description of heat exchange of man with his environment', *Science*, 94, 428–430.

Holmér I (1999), 'Thermal manikins in research and standards', 3rd int conf *Thermal Manikin Testing*, Sweden, 1–7.

Holmér I, Nilsson H and Meinander H (1996), 'Evaluation of clothing heat transfer by dry and sweating manikin measurements in performance of protective clothing', in Johnson J S and Mansdorf S Z, *Performance of Protective Clothing: Fifth Volume, ASTM STP 1237*, West Conshohocken, PA, ASTM, 360–366.

International Organization for Standardization (1993), ISO 11092, *Textiles – physiological effects – measurement of thermal and water vapour resistance under steady-state conditions (sweating guarded hotplate test)*, International Organization for Standardization, Geneva.

International Organization for Standardization (2004), ISO 15831, *Clothing – physiological effects – measurement of thermal insulation by means of a thermal manikin*, International Organization for Standardization, Geneva.

International Organization for Standardization (1995), ISO 9920, *Ergonomics of the thermal environment – estimation of the thermal insulation and evaporative resistance of a clothing ensemble*, International Organization for Standardization, Geneva.

Kim C S and McCullough E A (2000), 'Static and dynamic insulation values for cold weather protective clothing', in Nelson C N and Henry N W, *Performance of Protective Clothing: Issues and Priorities for the 21st Century: Seventh Volume, ASTM STP 1386*, West Conshohocken, PA, ASTM, 233–247.

Kuklane K, Nilsson H, Holmér I and Liu X (1997), 'Methods for handwear, footwear and headgear evaluation', European seminar *Thermal Manikin Testing*, Sweden, 23–29.

McCullough E A, Barker R, Giblo J, Higenbottam C, Meinander H, Shim H and Tamura T (2002), 'Interlaboratory evaluation of sweating manikins', 10th int conf *Environmental Ergonomics*, Japan, 467–470.

McCullough E A and Hong S (1994), 'A data base for determining the decrease in clothing insulation due to body motion', *ASHRAE Transactions*, 100 (1), 765–775.

McCullough E A, Huang J and Kim C S (2004), 'An explanation and comparison of sweating hot plate standards', *Journal of ASTM International*, 1 (7), (online journal at ASTM.org).

McCullough E A, Huang J and Deaton S (2005), 'Methods of measuring the clothing area factor', in Holmér, I Kuklane K and Gao C *Environmental Ergonomics XI: Proceedings of the 11th International Conference*, Ystad, Sweden, May, 433–436.

Nilsson H (1997), 'Analysis of two methods of calculating the total insulation', European conf *Thermal Manikin Testing*, Sweden, 17–22.

Nilsson H, Gavhed D and Holmér I (1992), 'Effect of step rate on clothing insulation measurement with a moveable thermal manikin', 5th int conf *Environmental Ergonomics*, The Netherlands.

Olesen B, Sliwinska E, Madsen T and Fanger P (1982), 'Effect of body posture and activity on the thermal insulation of clothing: measurements by a moveable thermal manikin', *ASHRAE Transactions*, 88 (2), 791–805.

Richards M and Mattle N (2001), 'Development of a sweating agile thermal manikin (SAM)', 4th int conf *Thermal Manikins*, Switzerland.

Shim H, McCullough E and Jones B (2001), 'Using phase change materials in clothing', *Textile Research Journal*, 7 (6), 495–502.

Woodcock A H (1962), 'Moisture transfer in textile systems, Part I', *Textile Research Journal*, August.

Wyon D P (1989), 'Use of thermal manikins in environmental ergonomics', *Scandinavian Journal of Work, Environment and Health*, 15 (1), 84–94.

12
Human wear trials for cold weather protective clothing systems

I. HOLMÉR, Lund University, Sweden

Abstract: Human wear trials provide valuable information on the immediate responses of persons to exposure conditions. Wear trials may be designed to study subjective responses, physiological responses or conditions for human heat balance. Valuable information is gathered on individual responses as well as individual variation. Wear trials are necessary for evaluation of products but also for validation of, for example, prediction models of human responses. Experiments with subjects are expensive and time consuming and require careful planning and design.

Key words: heat balance, physiological responses, subjective perception, methods.

12.1 Introduction

The prime purpose of clothing in a cold environment is to admit control of heat fluxes around the human body and protect it against excessive heat losses endangering heat balance. The net heat balance of the body is a function of environmental climatic factors, metabolic heat production and heat transfer through clothing. Heat transfer is governed by physical laws and thermal properties of clothing. The individual, however, has significant degrees of freedom to adapt by behavioural measures such as changing activity and adjustment of clothing. The success of such adaptations can only by judged by the individual himself or by physiological measurements.

Much of the clothing physiology research today is still based on wear trials in laboratories or in the field. However, researchers soon become interested in the development of thermal or clothing indices in order to better describe and explain how humans responded to complex thermal environments and the role of clothing (Gagge et al., 1941; Woodcock, 1962; Givoni and Goldman, 1972, 1973). Much of the early work is reported in Newburgh (1949) and Burton and Edholm (1955). With the development and advancement of computers predictive modelling became popular and several more or less advanced mathematical models have been proposed. Some of them are focusing on the human thermoregulatory system and its interaction with the thermal environment and use simplified clothing models (Wissler, 1985; Fiala et al., 2001; Werner, 1989).

Other models attempt to describe in detail the dynamics of clothing heat transfer (Lotens, 1993; Li and Holcombe, 1998; Danielsson, 1990; Barnes and Holcombe, 1996).

Physiological wear trials in clothing research serve several purposes:

- evaluation of comfort and strain
- assessment of energetic requirements
- assessment of freedom of movement and body motions
- assessment of compatibility with other equipment
- determination of individual variability in responses
- basis for development of empirical models or indices
- validation of mathematical prediction models.

12.2 Types of human wear trials

Human wear trials may be grouped into three categories based on the type of measurements taken and information gathered.

1. Thermal comfort assessment by subjective judgements.
2. Physiological wear trials.
3. Heat balance evaluation.

12.2.1 Thermal comfort assessment

Assessment of the performance of clothing using the wearer's own subjective judgements, provides valuable information about its function and behaviour and the variability in perception between persons. Typically, individual variation is great and many subjects may be required in order to get statistically significant results if differences between clothing ensembles are small. Although relatively easy to perform, these kind of wear trials often become time consuming and expensive. Their value lies in the fact that it shows how people really perceive the clothing. The efficiency of subjective wear trials can be improved if the number of test ensembles is kept low and has been tested for their properties with, for example, a thermal manikin.

Measurements

Standardized scales for human perception of thermal factors are readily at hand (ISO-10551, 1995). Typically, the subject is asked to rate his perception of thermal sensation, comfort and acceptability. Table 12.1 shows the standard scale for assessment of thermal sensation and comfort. Numerous other scales and questions may be asked depending on purpose of the wear trial. It is important that the wording and the scales are relevant, easy to understand, and unequivocal.

Table 12.1 Rating scales for assessment of subjective responses to thermal stress. Modified from (ISO-10551, 1995)

(a) **Thermal sensation** How are you feeling now in your: body, hands, feet, face, thighs, arms? −4 Very cold −3 Cold −2 Cool −1 Slightly cool 0 Neutral +1 Slightly warm +2 Warm +3 Hot +4 Very hot	(b) **Thermoregulatory responses** Are you shivering? No Yes (c) **Pain sensations** 0 no pain 1 slight pain 2 moderate pain 3 considerable pain 4 severe pain

(d) **Acceptance**
Do you find your thermal state acceptable?
    0 Yes, in the daily exposure
    1 Yes, but only at a single exposure
    2 No

Procedures

Factors to consider in the design of subjective wear trials include the number of subjects, type of exposure, type of activity and length of exposure. In addition scales must be chosen and time for voting and number of votes must be decided. Most commonly the same persons are subject to repeated exposures with different ensembles. Ensembles and exposures must be randomly assigned and the votes of the subjects must be independent and confidential to the other participants.

Evaluation

Results of subjective wear trials are evaluated by statistical analysis. If several votes are taken as a time series, they must be statistically treated as repeated measures for each individual. As previously mentioned, large number of subjects may be required in order to detect significant differences between ensembles.

12.2.2 Physiological wear trials

In laboratory studies quite sophisticated physiological measurements can be done during standardized exercise tests. In field trials miniaturized sensors and data loggers are required in order not to interfere with body movements and work tasks.

Measurements

From a clothing point of view the temperatures of skin and body cores, sweating and oxygen consumption are of interest. ISO 9886 (ISO-9886, 2004) defines critical physiological variables for assessment of thermal strain, specifies requirements for their measurement and gives guidance on limit values.

Core temperature

Representative measures of deep core temperature require measurements at sites that are more or less convenient for the subject. Rectal temperature, measured 10 cm or more beyond the anal sphincter, is often used as a measure of core temperature in laboratory as well as in field conditions. Esophageal and tympanic temperatures are also good indicators of core temperature but more difficult to measure. Core temperatures obtained by measurements in the ear canal or in the mouth are less reliable and must be interpreted with caution. In a cold environment they are more easily affected by surrounding cold air.

Core temperature represents different volumes of tissues depending on the heat balance of the body. In a warm environment it merely represent the temperature of the whole body except for the most superficial parts (skin). In the cold peripheral circulation reduces, tissue temperatures fall and the core temperature may represent only the deeper and more central tissues.

Skin, layer and surface temperatures

The surface temperature of the skin may vary considerably (0 to 45 °C), at least for short times, without apparent injury to the tissue. The skin temperature is a function of the heat fluxes at the body surface and as such influenced by deep tissue temperature and ambient temperature. At rest in thermoneutral conditions ('comfort') mean skin temperature is 33–34 °C, with an even distribution over the body surface. This is about 3–4 °C lower than core temperature. This internal temperature gradient diminishes in the heat and increases in the cold.

The fall in skin temperature is more pronounced the more peripheral are the tissues. The greatest drop is seen in hands and feet (fingers and toes). Central skin temperatures show a much smaller drop. Mean skin temperature drops primarily as a consequence of the larger reductions in extremity temperatures. Skin temperature is measured with sensors taped to the surface. It is important that sensors are small and in good contact with the surface to eliminate influence of the ambient conditions, particularly in very cold environments. Suitable sensors are thermistors and thermocouples. Skin temperature can also be measured without contact by infrared sensors. The sensor detects the infrared radiation from the surface and converts it into a temperature. Special cameras are available for infrared imaging and can provide photos of the temperature distribution over the focused area.

Mean skin temperature is determined as an area weighted mean value of measurements at representative sites on the surface. Several formulas exist based on different numbers of measuring points. The following formulas, based on 4 and 8 points respectively, are common in laboratory investigations and proposed in ISO 9886. For cold conditions it is recommended to use at least 8 points and include additional, peripheral sites, for example finger, foot and toe.

$$\bar{t}_{sk} = 0.28 \cdot t_{neck} + 0.28 \cdot t_{back} + 0.16 \cdot t_{hand} + 0.28 \cdot t_{shin} \qquad 12.1$$

$$\bar{t}_{sk} = 0.07 \cdot t_{forehead} + 0.175 \cdot t_{back} + 0.175 \cdot t_{chest} + 0.07 \cdot t_{upperarm}$$
$$+ 0.07 \cdot t_{forearm} + 0.05 \cdot t_{hand} + 0.19 \cdot t_{thigh} + 0.2 \cdot t_{calf} \qquad 12.2$$

Surface temperature and layer temperatures are measured in order to study temperature and moisture gradients within a clothing ensemble during a wear trial.

Surface temperatures can also be measured with infrared sensors, for example, in a camera. This kind of camera has become cheaper and more readily available in recent years. Resolution has also been improved. The pictures or videos taken with the camera provide detailed information about the distribution of temperatures over the observed area.

Mean body temperature

For heat balance analysis an average temperature of the body tissues is required and the simplest way to define the mean body temperature is by calculating a mean value of core and skin temperature. The following two formulas apply to conditions when the body is either neutral/warm or cold. The latter may apply when mean skin temperature falls below 32 °C.

$$\bar{t}_b = 0.8 \cdot t_{re} + 0.2 \cdot \bar{t}_{sk} \qquad \text{warm} \qquad 12.3$$

$$\bar{t}_b = 0.65 \cdot t_{re} + 0.35 \cdot \bar{t}_{sk} \qquad \text{cold} \qquad 12.4$$

Sweat rate and humidity

In cold exposures sweating should be avoided due to the limitations for evaporation and moisture transfer in multilayer clothing and cold air. Clothing must be kept dry in order to preserve good insulation and warmth. Under certain circumstances, such as heavy work and impermeable clothing, sweating can often not be avoided. Measurements of sweating includes weighing of the person and all pieces of garments immediately before and after the exposure on a balance of sufficient accuracy (1 g).

Small humidity sensors are available for placement at discrete positions in the layers. Most sensors measure relative humidity and must be accompanied by a temperature measurement at the same point in order to give relevant and reliable information related to the heat and moisture fluxes.

Oxygen consumption and heart rate

Measurements of the oxygen consumption and heart rate during wear trials provide information about the energy cost and work load of the activity. By these measures the effects of clothing weight, bulk and friction on physical workload can be evaluated. Heart rate, however, is less sensitive as it is influenced also by other factors.

12.2.3 Procedures

Wear trials for the evaluation of, for example, cold protective clothing are mostly carried out in climatic chambers with accurate control of air temperature, humidity and wind. Subjects must comply with certain rules regarding meals, drinks and activity before the trial. Subjects are equipped with sensors and dressed with experimental clothing at room temperature for at least 20–30 min before exposure. Clothing should be stored at the same room temperature and humidity. Subjects may rest or exercise on a bicycle or a treadmill for a sufficiently long time for the physiological responses to stabilize and reach a steady state. This often requires exposure times of an hour or longer. During this time clothing and exercise intensity are kept constant. In certain applications clothing or activity may be changed if the purpose is to test the dynamics of the physiological responses itself. When protection properties are evaluated some conditions may lead to unacceptable physiological strain and exposure must be interrupted when critical values are reached.

In physiological wear trials fewer subjects need to be studied, since the measurements are more accurate and the variability within and between subjects in physiological responses is smaller. Nevertheless, at least 8–10 subjects are often required to find significant differences between experimental conditions that were meaningful and valuable for the study.

Evaluation

The protective value of cold weather clothing was studied in 8 subjects performing 2 hours of walking at 5 km/h in a climatic chamber at 0, −9, −12 and −19 °C, respectively (Gavhed and Holmér, 1989). The clothing ensemble was composed of long underwear, light thermal underwear with long arms and legs, polyester overall, socks, boots, gloves and cap. Basic insulation was 1.70 clo and total insulation 2.15 clo. All subjects attained steady state levels for all physiological variables that were continuously monitored (rectal and skin temperatures, heart rate). Metabolic energy and heat production calculated from measurements of

Table 12.2 Mean values for the last 30 minutes of a 2-hour exposure (walking) to different ambient temperatures of 10 subjects dressed in a clothing ensemble with a total insulation of 2.15 clo (Gavhed and Holmér, 1989)

Variable	0 °C	−9 °C	−12 °C	−19 °C
T_{sk}, °C	32.9	31.8	31.1	29.9
T_{re}, °C	37.7	37.6	37.6	37.6
M-W, W/m^2	174	174	181	178
E, W/m^2	34	11	11	8
$R+C$, W/m^2	128	143	157	159
S, W/m^2	−6	1	−4	−9
$I_{T,r}$, clo	1.63	1.85	1.77	1.91
Thermal sensation	2.0	0.5	−0.2	−1.0

oxygen uptake were almost the same or 177 W/m^2. Table 12.2 lists the measured and calculated variables during the last 30 min of exposure. Rectal temperature was maintained slightly elevated at about 37.6 °C during all climatic conditions. The increasing cold stress by lowered ambient temperature was seen in significantly lowered mean skin temperature and sweat rate. The subjects were able to maintain heat balance over the whole temperature interval by thermoregulatory responses (sweating and vasoconstriction (lowered skin temperature)). The perceived thermal sensation of conditions was significantly related to air temperature. At 0 °C conditions are felt warm, at −9 and −12 °C conditions are close to thermoneutral and at −19 °C they are felt slightly cool (compare Table 12.2).

Figures 12.1 and 12.2 show the dynamic temperature responses of subjects to light exercise at three cold climatic conditions (−6, −14 and −22 °C) (Gavhed and Holmér, 1998). Ten subjects walked at 2 km/h (110 W/m^2) for 90–100 minutes wearing a three-layer winter ensemble comprising briefs, long-sleeved underpants and shirt, fibre pile jacket and trousers, overall and unlined jacket, socks, boots, mittens and balaclava. Total insulation of the ensemble was 2.7 clo (basic insulation 2.3 clo).

At 60 minutes all subjects achieved an apparent steady state in rectal temperature at about 37.3 °C. Mean skin temperature continuously dropped over time and reached a level at 50 minutes that was related to air temperature and was significantly different between conditions. From 60 to 100 minutes at −6 and −14 °C rectal temperature tended to fall and mean skin temperature continued to drop at a slower rate.

After 60 min at −22 °C five subjects started to walk faster (6 km/h) for 40 minutes and five subjects continued at 2 km/h but put on an additional insulated parka. Both changes resulted in abated mean body cooling rates (Fig. 12.2). Mean body temperature was similar in both conditions. However, the rectal temperature continued to decrease slightly despite added insulation, while the increased metabolic heat production resulted in a 0.5 °C higher rectal tempera-

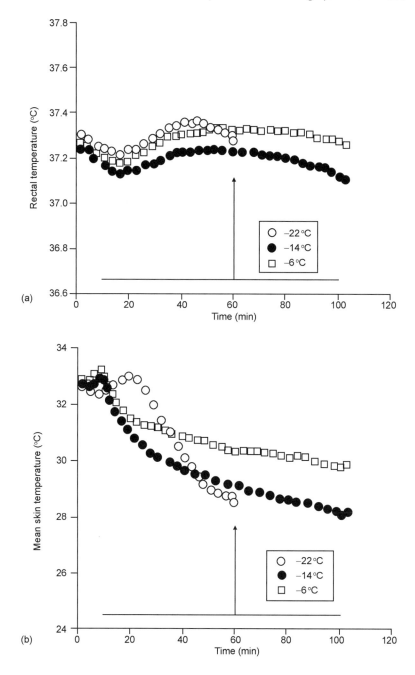

12.1 (a) and (b) Thermal effects of slow walking (2 km/h) for 90 minutes in cold conditions in clothing with a basic insulation of 2.3 clo (Gavhed and Holmér, 1998). Average response of 10 subjects. Work and exposure period was 50 min (−22°C) and 100 min, respectively.

264 Textiles for cold weather apparel

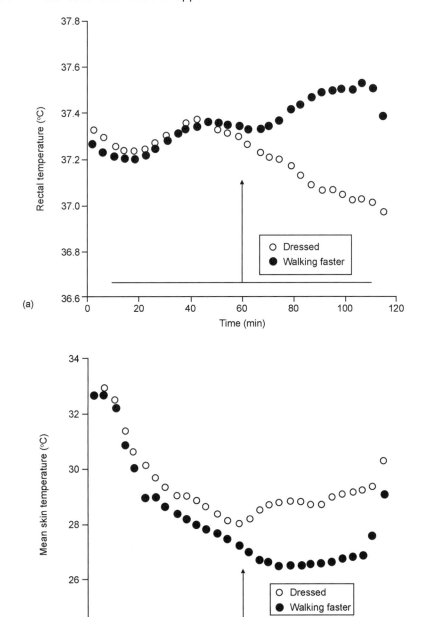

12.2 (a), (b), (c) and (d) Thermal effects of light work on 10 subjects (walking 2 km/h) for 110 min at −22 °C in clothing with a basic insulation of 2.3 clo. After 50 min five subjects increased walking speed to 6 km/h and five subjects donned a parka (Gavhed and Holmér, 1998). Compare with Fig. 12.1.

Human wear trials for cold weather protective clothing systems 265

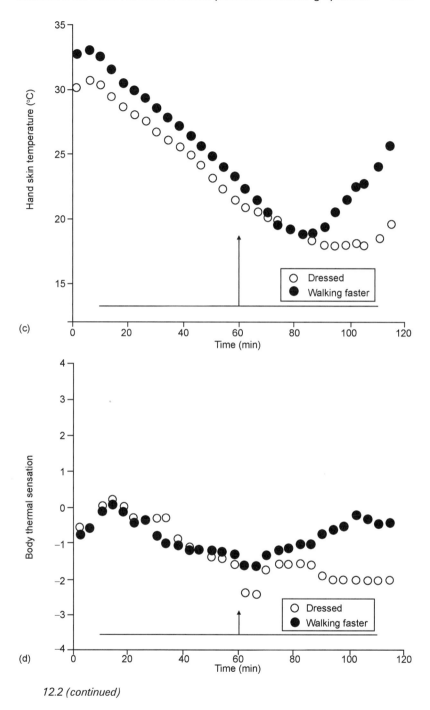

12.2 (continued)

Table 12.3 Suggested criteria for evaluation of strain. Modified from (ISO-11079, 2007) and (Afanasieva, 1994)

Level/effect	Thermal sensation	Whole body cooling		Extremity cooling
		Mean skin temperature °C	Body heat debt kJ/m²	Finger skin temperature °C
1 light strain	Neutral, 'comfort'	$T_{sk} = 35.7-0.0285*M$	−145	24
2 high strain	Cool	$T_{sk} = 33.34-0.0354*M$	−290	15
3 very high strain	Cold	$T_{sk} = 30.36-0.0310*M$	−435	7

ture. On the other hand, mean skin temperature increased with more clothing and decreased with faster walking (Fig. 12.2). Hand and foot temperatures tended to increase and thermal sensations tended to be warmer with increased exercise intensity. The added insulation raised the temperature mainly of the covered body parts, but the hands and finger temperatures levelled off or continued to cool (Table 12.3).

12.2.4 Heat balance analysis

In addition to determination of physiological strain, wear trials can be used for the evaluation of the body's heat balance. This kind of analysis is also known as partitional calorimetry. From the heat balance analysis it is possible to calculate the actual (resultant) thermal properties of the clothing ensemble.

Heat balance equations

Equation 12.5 is a mathematical description of the heat balance of the body.

$$S = M - W - RES - E - R - C - K \qquad 12.5$$

Metabolic energy production (M) minus the effective mechanical energy (W) is the internal heat production. Heat exchange takes place in the respiratory tract (RES), on the skin by evaporation (E), radiation (R), convection (C), and conduction (K). For our applications K can be considered very small and is omitted from the analysis. Tissue heat content (S) may change depending on the values of the equation.

From a clothing point of view the interesting variables to determine are E and $R + C$. E provides information about the probable evaporative heat exchange through clothing and $R + C$ gives information about the convective and radiative heat transfer. E and $R + C$, respectively, can be determined by solving equation 12.5 for known values of the other variables.

Human wear trials for cold weather protective clothing systems   267

M is determined from measurements of oxygen uptake (V_{O_2} in l/min) during the wear trial by

$$M = 5.873 \cdot (0.23 \cdot RQ + 0.77) \cdot V_{O_2} \cdot 60/A_{Du} \qquad 12.6$$

where M is in W/m², 5.873 is the energy equivalent of oxygen in Wh per l O_2, RQ is the respiratory quotient and A_{Du} is the body surface area in m². It is useful to use W/m² (Watts per square metre body surface area) as almost all heat exchange takes place via the body surface area. This can be calculated from body weight (Wt in kg) and body height (H in m) by the DuBois formula.

$$A_{Du} = 0.202 \cdot Wt^{0.425} \cdot H^{0.725} \qquad 12.7$$

RES has a dry and a humid component. Both are related to M and calculated by

$$RES = E_{res} + C_{res} \qquad 12.8$$

$$E_{res} = 0.0173 \cdot M \cdot (p_{ex} - p_a) \qquad 12.9$$

$$C_{res} = 0.0014 \cdot M \cdot (t_{ex} - t_a) \qquad 12.10$$

$$t_{ex} = 29 + 0.2 \cdot t_a \qquad 12.11$$

where E_{res} is evaporative heat exchange via respiration, C_{res} is convective heat exchange via respiration, p_{ex} is the saturated water vapour pressure at expired air temperature (t_{ex}), p_a is ambient water vapour pressure and t_a is ambient air temperature. It is assumed that expired air is saturated and has a temperature (t_{ex}), that is related to inspired (ambient) temperature (t_a).

S can be calculated on the basis of the mean body temperature (equations 12.3 and 12.4). More useful in heat analysis is the calculation of the rate of change in body heat content or body heat storage, S.

$$S = 0.97 \cdot Wt \cdot \frac{dt_b}{dt} \cdot \frac{1}{A_D} \qquad 12.12$$

where 0.97 is the specific heat of the body tissues in Wh/kg, Wt is body weight in kg and the derivative is the rate of change in mean body temperature. Values of S for certain levels of strain are given in Table 12.3.

The external mechanical work, W, is equal to 0 during walking and running on a plane surface, for example a treadmill. Walking uphill contains mechanical work in increasing the body's potential energy. This can be calculated on the basis of body weight (Wt in kg (including clothes and weights)), inclination (α in degrees) and walking or running speed (v_w in m/s) by

$$W = Wt \cdot g \cdot v_w \cdot \sin\left(\alpha \, \frac{\pi}{180}\right) \Big/ A_{Du} \qquad 12.13$$

Cycling on a bicycle ergometer also contains a mechanical component in overcoming the friction of the swing wheel. This amount is normally known, as it can be regulated and controlled by the experimenter.

The evaporative heat exchange is calculated on the basis of measurements of body and clothing weights immediately before and after the wear trial. The weight change of the body is caused by sweating and moist expired air. Sweat production is determined by

$$m_{sw} = m_{sb} - m_{sa} - m_{res} \qquad 12.14$$

where m_{sw} is sweat mass (g), m_{sb} is the subject's weight before test (g) and m_{sa} is the subject's weight after test (g).

Accumulated sweat mass in clothing is calculated by

$$m_{acc} = m_{ca} - m_{cb} \qquad 12.15$$

where m_{acc} is the accumulated sweat mass (g), m_{cb} is the clothing weight before test (g) and m_{ca} is clothing weight after test (g).

Sweat evaporation (m_e in g) is determined by

$$m_e = m_{sw} - m_{acc} \qquad 12.16$$

Weight loss due to humidifying inspired air (m_{res} in g) is calculated by

$$m_{res} = \frac{0.0173 \cdot M \cdot (p_{ex} - p_a)}{0.68 \cdot 60 \cdot A_{Du}} \qquad 12.17$$

where p_{ex} is the saturated water vapour pressure at expired air temperature and p_a is the ambient water vapour pressure; both in kPa.

Finally, evaporative heat exchange (W/m²) is calculated by

$$E = \frac{2425 \times m_e}{D \times A_{Du}} \qquad 12.18$$

where 2425 is the enthalpy of water evaporating at 32 °C (J/g), and D is the test duration (s).

Convective and radiative heat exchange $(R + C)$ is calculated by solving equation 12.5 with the other variable known. $R + C$ is also equal to

$$R + C = \frac{t_{sk} - t_a}{I_{T,r}} \qquad 12.19$$

Since both t_{sk} and t_a are measured in the experiment, the value of the total, resultant insulation of the clothing ensemble ($I_{T,r}$ in m²°C/W) under the prevailing conditions can be calculated. The clo-unit is equal to 0.155 m²°C/W.

$$I_{T,r} = \frac{t_{sk} - t_a}{R + C} \qquad 12.20$$

Measurements

Calculation of heat fluxes described in the previous section can be made on the basis of the physiological measurements reported in Section 12.2.2, e.g., skin and core temperature, metabolic rate and sweat loss. Heat fluxes by convection

Human wear trials for cold weather protective clothing systems 269

and radiation can also be measured with small heat flux sensors attached to the skin at representative sites.

Procedures

Procedures are similar to those described in Section 12.2.3. Exposure time should be long enough for the physiological adjustment to the conditions and to achieve a steady state period at the end of the exposure. This period could either represent constant values of the variables or values that change at a relatively constant rate (Figs 12.1 and 12.2). Usually, metabolic rate for the selected activity is reached within 5–10 minutes and then remains constant. Temperatures are usually measured at least every minute and values for the final 10–20 min considered as steady state values.

Evaluation

Evaluation of heat balance can be made for the whole exposure period or for the steady state period. If the whole period is analyzed it is usually the amounts of heat that are calculated and the net heat storage in the body is analyzed. The purpose of this kind of evaluation is to get an overall view of the heat exchange during the whole exposure period.

In many studies the steady state period is of more interest to evaluate as it gives information about the conditions for the maintenance of heat balance. Negative values for body heat storage rate (S) indicates a negative heat balance (Table 12.2). This information can be used for the assessment of the risks of body cooling and hypothermia associated with this kind of exposures. The steady state period must also be used, for example, when resultant clothing insulation is calculated (equation 12.20).

In the first study reported in Section 12.2.2 the metabolic rate was almost the same or about 177 W/m^2 (Table 12.2). Respiratory heat exchange increases slightly from 17 to 20 W/m^2 in the coldest condition. Evaporative heat exchange becomes less in the cold conditions, whereas convection and radiation increase. The heat balance is well maintained as the value of S only varies between 0 and −9 W/m^2. As already reported, rectal temperature is the same and constant at the end of exposure. The subjects adjust to the colder conditions by a drop in mean skin temperature. This drop is also recorded in the thermal sensation votes (Table 12.2). The total insulation value of clothing, as measured with a thermal manikin, was 2.15 clo (EN 342, 2004). Using equation 12.20 the resultant total insulation ($I_{T,r}$) for the actual experimental conditions can be calculated.

Values vary from 1.63 to 1.91 clo. Values are lower than the manikin values because of the pumping effect and the relative air velocity caused by walking. The lowest value at 0 °C most likely also can be attributed to some moisture absorption in clothing layers as a result of a higher sweat rate. Conditions at −12 °C appear to provide a neutral thermal balance (thermal sensation vote is close to 0). The value predicted with I_{REQ} (ISO-11079, 2007) including the boundary air layer is 1.8 clo and quite close to the measured $I_{T,r}$ of 1.77 clo for this condition.

Figures 12.1 and 12.2 report data from a second study that also can be used for a heat balance analysis. Skin temperature shows an almost linear drop after about 40 minutes, whereas rectal temperature remains almost constant during the 60 to 100 minutes of exposure. The rate of change in body heat content can be calculated by equation 12.12. The period for the analysis is minute 40 to 60 for $-22\,°C$ and minute 80 to 100 for the others. The rate of change in S was -16, -25 and $-43\,W/m^2$, respectively and significantly different for the three air temperatures. Metabolic rate was 107, 108 and 118, respectively and not significantly different between temperatures. Respiratory heat losses were 12, 13 and $14\,W/m^2$, respectively, and evaporative heat loss about $5\,W/m^2$ in all conditions. Accordingly, 106, 115 and $142\,W/m^2$ was lost by convection and radiation for the three conditions. By using equation 12.20, we find that $I_{T,r}$ for the three temperatures would be 2.2, 2.4 and 2.3 clo, respectively, at -7, -14 and $-22\,°C$. This compares well with the total insulation value of the ensemble at 2.7 clo obtained with a thermal manikin (EN 342, 2004). As described above, the main reason for this reduction is the increased air velocities around the body and within clothing due to body movements.

Heat balance could not be maintained in any of the three conditions due to negative heat storage. Similar negative values for heat storage were calculated with I_{REQ} (ISO-11079, 2007) for the actual conditions (-12, -25 and $-47\,W/m^2$, respectively). The time to lose $40\,Wh/m^2$ (Table 12.2) was calculated at 3.3, 1.6 and 0.85 hours, respectively. This recommended exposure time for light strain appears to fit well with the responses of the subjects.

12.3 Discussion

Human wear trials can add significant information regarding function and performance of a cold weather protective clothing system. First of all they provide a measure of the actual response of the subjects to exposure conditions and, thus, information about how well clothing protects from cooling. This information can also be used for estimates of how well clothing protects at temperatures slightly higher and lower than test conditions. Too large extrapolations of results are not recommended. Secondly, wear trials add significant information about individual variation. This variation is not only caused by clothing fit but also by variation in the thermoregulatory responses. People also perceive thermal conditions differently due to variation in experience, expectations and preferences. Thirdly, wear trials (as do thermal manikins) comprise a three-dimensional, dynamic heat and moisture transfer process that is different from that of fibres and fabrics. Therefore, promising new properties found at this level may become small or even negligible when a whole clothing ensemble is tested.

The time and costs associated with wear trials are the main problems. Due to the individual variation, a large number of subjects may be required in order to obtain significant statistical differences between test samples or conditions. A combination of methods can help to improve the relevance, accuracy and sensitivity of tests. Fabrics and textiles for garment manufacturing can be tested for critical properties. Only fabrics that are assumed to differ in critical properties are selected for making test clothing. Prototypes of clothing are tested with a thermal manikin for prediction of their effects on human heat exchange. Only prototypes with significant differences in critical properties are selected for wear trials. Conditions for the wear trials should be chosen so that the thermal stress on the subjects is large enough to evoke significant thermoregulatory responses.

Heat balance analysis provides a useful tool to validate technical tests of clothing, for example, with manikins or predictive models. The partitional calorimetry that is used for determination of heat fluxes is based on the assumptions that fluxes are constant and no heat or moisture is stored during the passage through clothing. Therefore, such experiments must be designed so that sweat can freely evaporate and pass through layers. This may be difficult to avoid during heavy work in the cold or when special protective clothing is used. Under such circumstances, the evaporative heat loss, in particular, can be erroneously calculated and so the convection and radiation (Havenith *et al.*, 2008). Nevertheless, the overall heat balance can be evaluated, but a correct differentiation between evaporation on one hand and convection and radiation on the other cannot be made.

Wear trials are necessary in developing and validating predictive thermal models. More or less empirical models derived from one or several sets of data, should be validated with independent data. Pure mathematical models require validation in wear trials. Validations should be made by independent researchers. Any modification of the model based on the results of the validation must be reported.

As in many other parts of industry much of the development work is done with computers. Advanced mathematical models allow the analysis and evaluation of heat and mass transfer in fibres, textiles, garments, layers and clothing systems. However, in the end the models are no better than the quality of the data and assumptions on which they are based. Validating wear trials are still necessary in many contexts, but can be limited and controlled by careful selection of test samples, design of exposure conditions and choice of relevant, reliable and accurate measurements.

12.4 Sources of further information and advice

Scott R (ed.) *Textiles for protection*. Woodhead Publishing, Cambridge, 2005.
Shishoo R (ed.) *Textiles in sport*. Woodhead Publishing, Cambridge, 2005.
Parsons K. *Human thermal environments*. Taylor & Francis, London, 2003.

Hassi J, Mäkinen T, Holmér I, Påsche A, et al. (2002) *Handbook for Cold Work*, Oulu Regional Institute of Occupational Health.

12.5 References

Afanasieva, R. (1994) Physiologic and hygienic clothing requirements for protection against cold. In Holmér, I. (ed.) *Work in cold environments*, Solna, National Institute of Occupational Health.

Barnes, J. C. and Holcombe, B. V. (1996) Moisture sorption and transport in clothing during wear. *Text Res J*, 66, 777–786.

Burton, A. C. and Edholm, O. G. (1955) *Man in a cold environment*, New York, Edward Arnold.

Danielsson, U. (1990) Convective heat transfer measured directly with a heat flux sensor. *J Appl Physiol*, 68, 1275–1281.

EN 342 (2004) Protective clothing – Ensembles for protection against cold.

Fiala, D., Lomas, K. J. and Stohrer, M. (2001) Computer predictions of human thermoregulatory and temperature responses for a wide range of environmental conditions. *Int J Biometeorol*, 45, 143–159.

Gagge, A. P., Burton, A. C. and Bazett, H. C. (1941) A practical system of units for the description of the heat exchange of man with his environment. *Science*, 94, 428–430.

Gavhed, D. and Holmér, I. (1989) Protection against cold by a standard winter clothing during moderate activity. *Third Scandinavian Symposium on Protective Clothing against chemicals and other Health Risks*. Gausdal, Defense Research Establishment.

Gavhed, D. and Holmér, I. (1998) Thermal responses at three low ambient temperatures: Validation of the duration limited exposure index. *International Journal of Industrial Ergonomics*, 21, 465–474.

Givoni, B. and Goldman, R. (1972) Predicting rectal temperature response to work, environments and clothing. *J Appl Physiol*, 32, 812–822.

Givoni, B. and Goldman, R. (1973) Predicting heart rate response to work, environment, and clothing. *J Appl Physiol*, 34, 2, 201–204.

Havenith, G., Richards, M., Wang, X., Bröde, P., Candas, V., Den Hartog, E., Holmér, I., Kuklane, K., Meinander, H. and Nocker, W. (2008) Apparent latent heat of evaporation from clothing: attenuation and 'heat pipe' effects. *J App Physiol*, 104, 142–149.

ISO 10551 (1995) Ergonomics of the thermal environment – Assessment of the influence of the thermal environment using subjective judgement scales.

ISO 9886 (2004) Ergonomics – Evaluation of thermal strain by physiological measurements.

ISO 11079 (2007) Ergonomics of the thermal environment. Determination and interpretation of cold stress when using required clothing insulation (IREQ) and local cooling effects.

Li, Y. and Holcombe, B. V. (1998) Mathematical simulation of heat and moisture heat transfer in a human–clothing–environment system. *Text Res J*, 68, 389–397.

Lotens, W. A. (1993) *Heat transfer from humans wearing clothing*. TNO – Institute for Perception, Soesterberg.

Newburgh, L. H. (Ed.) (1949) *Physiology of heat regulation and the science of clothing*, Philadelphia, London, W. B. Saunders Company.

Werner, J. (1989) Thermoregulatory models. *Scand J Work Environ Health*, 15 suppl 1, 34–46.

Wissler, E. H. (1985) Mathematical simulation of human thermal behavior using whole body models. In Shitzer, A. and Eberhart, R. C. (eds) *Heat transfer in medicine and biology*. Plenum Press.

Woodcock, A. H. (1962) Moisture transfer in textile systems, Part II. *Text Res J*, 719–723.

13
Care and maintenance of cold weather protective clothing

N. KERR, J. C. BATCHELLER and E. M. CROWN,
The University of Alberta, Edmonton, Canada

Abstract: Cleaning procedures have undergone many changes in the last decade, many of which greatly affect the maintenance of cold weather clothing. In this chapter, both traditional and newer cleaning processes and equipment are reviewed and applied to cleaning cold weather clothing. Problem areas for maintenance of cold weather clothing are identified, described and illustrated.

Key words: contaminated cold weather clothing, cleaning, home laundering, dry cleaning, wet cleaning, protective clothing.

13.1 Introduction

There are many issues that make the care and maintenance of cold weather clothing difficult and which therefore require special consideration. The fact that cold weather clothing usually comprises more than one layer of often different materials is a key aspect of this problem. What is appropriate maintenance for one component may be either ineffective or harmful for another component. This is especially problematic when cold weather work clothing becomes highly contaminated as is often the case in some sectors.

Another important consideration is that some traditional cleaning procedures have undergone, and are undergoing, drastic changes in response to human health and environmental concerns of both consumers and governments, related to the chemicals and processes used. To this end, fabric care companies and their raw material suppliers are trying to make sustainable products that satisfy the public's new environmental sensibility (McCoy, 2007a).

The many aspects of clothing care that have changed in the past decade will affect how cold weather clothing is maintained. Some of these changes include the formulation of detergents to make them 'greener', the design and operation of home laundering equipment, new organic solvents for dry cleaning, and the development of professional wet cleaning equipment so that organic solvents need not be used. These developments are reviewed as three potential methods for cleaning cold weather clothing are discussed: home or domestic laundering, professional dry cleaning, and professional wet cleaning. In each case, issues of particular relevance to cold weather clothing are highlighted. This review is followed by a discussion of care and maintenance issues relevant to specific materials for outer layers, insulation layers and membranes, linings and trim, as well as cold weather underwear and socks. Known problems encountered with several of these materials are discussed and illustrated. The chapter closes with a brief consideration of new and expected future trends. Sources of further information on this complex subject are provided.

13.1.1 Regular maintenance, shelf life and removal from service

A primary physical reason for discarding fashion clothing is change in appearance that has occurred during wear, cleaning, or storage. For protective clothing or other industrial garments, appearance generally may not be as important as function. Each cold weather garment has one or more primary functions such as protecting the wearer from specific hazards in the workplace or environment, keeping the wearer warm and dry, or ensuring the wearer is visible from a distance. The decision to discard a garment is normally therefore based on presumed loss of a specified amount of functionality or failure to meet a specific standard for that type of garment. Apparent loss of cleanliness may not be a reason to discard an item unless the dirt remaining after cleaning is deemed to affect the properties defined in the relevant performance standard; for example, residual oil in a coverall worn by an oilfield worker may cause it to fail to meet flame-resistance criteria. Existing guidelines for removal from service tend to be general and based on observation, and usually cannot be strictly performance based because performance testing is usually destructive. Decisions to retire garments are therefore often based on experience in interpreting observations. For cold weather clothing, this practice is problematic because problems with inner layers may not be noted.

If clothing is allowed to become very dirty or heavily contaminated, it will be difficult to clean and may be hazardous to wear. If oily stains and embedded particulate dirt are not removed promptly, oxidation of the oils will make them less soluble and perhaps impossible to remove (Obendorf, 2004). Oil-stained protective clothing may lose its flame resistance (Crown *et al.*, 2004). All items of cold weather clothing should be cleaned regularly and stored in a ventilated area away from heat and should never be stored without cleaning. Gloves, which are an important component of cold weather clothing systems, should have a cleaning protocol to avoid infections.

There have been incidences reported of methicillin-resistant *Staphylococcus aureus* (MRSA) infections among groups involved in competitive sports, and protective clothing and equipment have been implicated in the transmission of *Staphylococcus aureus*. Thus, cleaning sportswear, including cold weather clothing, after each use is recommended (Barrett and Moran, 2004). No evidence of similar contamination specific to users of cold weather protective clothing has been reported; nonetheless, the potential for gloves and helmets to be shared amongst users could provide some risk. Particularly in winter climates where low humidity upsets skin hydration, abraded and chaffed skin could be suspectible to *Staphylococcus* infection.

13.1.2 Preparation for cleaning

Before cleaning, garments should be examined for mechanical damage and repaired. It is important to determine the presence of various types of dirt or contamination. Soils can be water-based, oil-based or insoluble such as metal oxides, and particulates (clays, carbon black). Dirt on a garment is a complex mixture of the three components. It is characteristic of an individual's body chemistry, the closeness of garments to the skin and activities while wearing the clothing. There may be a build up of soil residues from previous wear and washing, some residues deeply penetrating cracks in fibers. Particulate soil deposits on fibre surfaces, in the crevices between fibres and on yarn surfaces. Large particulates, and clusters of smaller particles are usually coated with oil. Oily deposits often appear as streaks because the oil has wicked along yarns. In very dirty garments, the oil encapsulates yarns and because of its sticky texture, collects particulate soils (Obendorf, 2004). As soils age, they oxidize and become less soluble in water or organic solvents. Pre-treatment of stains before washing or dry cleaning facilitates their removal. Very dirty areas on garments should be sprayed or flushed with an appropriate pre-treatment spray or spotting liquid. Pre-treatment sprays used on dirty areas of a garment before aqueous cleaning contain organic solvents and detergent to soften and solubilize or dissolve oily stains. If a garment is to be dry cleaned, stains are treated with pre-spotters containing water and detergent and the garment is dried before it enters the solvent bath.

In many jurisdictions, a manufacturer's care label in cold weather clothing should indicate appropriate procedures for aqueous or dry cleaning, including cleaning temperature, solvent, drying method and heat sensitivity. The symbols on the label may conform to those shown in Fig. 13.1. Deviating from the suggested method for cleaning is not recommended: cold weather clothing often contains hidden layers, trims or finishes that are not identified on the label and may be sensitive to heat or a particular solvent.

13.2 Home (domestic) laundering procedures

To conserve energy, consumers are urged to use both a low volume of water and low wash temperatures. Easter and Ankenman (2006) evaluated the effect of both factors on laundering dark-coloured cotton and cotton blends. The reader

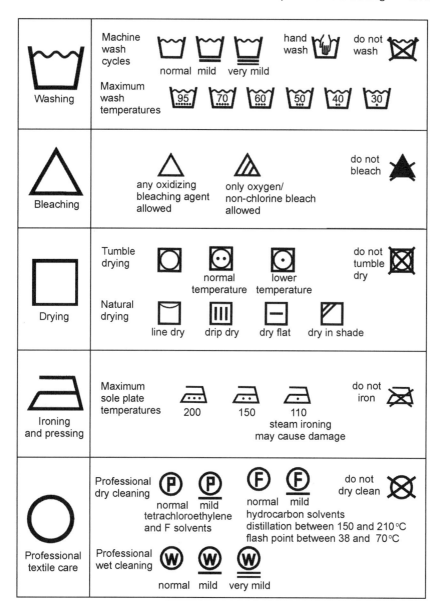

13.1 Care labelling symbols based on ISO 3758:2005.

should note, however, that they used a traditional North-American top-loading washer and a traditional liquid detergent rather than machines or detergent designed for high efficiency, low volume laundry. Washing fabrics in a low volume of water did not noticeably affect colour retention, smoothness, pilling, or stain repellency; however, there was an increase in dye transfer, perhaps because of the higher detergent concentration in the low volume wash. Washing with cold water (21 °C and 15.5 °C) was not as effective as warm water (32 °C) in removing motor oil from fabric that had a stain release/stain repellent finish. The oil was removed most effectively when the fabric was washed in warm water with no fabric softener added. Washing with very cold water (15.5 °C) negatively affected colour retention, smoothness and stain release of the fabrics. Thus, if traditional vertical axis machines are used to launder cold weather clothing, it may be preferable to wash in warm water (30 °C) rather than cold water.

13.2.1 High efficiency domestic washers with low water platforms

To save water and reduce energy consumption, the laundry appliance industry has introduced front- and top-loading high efficiency (HE) washers that are gradually replacing traditional deep-water, top-loading vertical-axis washers in North America. In Europe, front-loading, horizontal-axis washers have been the preferred home machine for many years. As well as front-loading HE washers, there are top-loading HE washers with a vertical axis or rotating plates/discs (Ankeny et al., 2006). HE washers use a water volume 20 to 66% of that of traditional vertical-axis washers, with front-loading machines using less water than top-loading washers (Soap and Detergent Association, 2005). Low water use may mean that garments are not rinsed as effectively as when a larger volume of water is used. Also there is a higher concentration of detergent in an HE wash load than in a traditional deep-water washer (Thiry, 2008), although it should be low-sudsing. Especially in such cases, cold weather clothing may benefit from extra rinses to ensure that detergent residues (wetting agents) do not remain in garments and compromise water repellent or other finishes and/or breathable microporous membranes.

The cleaning action of both types of HE washers differs from that in the traditional vertical-axis washer. In vertical-axis HE machines there is a small agitator or wheels, plates or discs that move the laundry as they spin or rotate. Re-circulated water may also spray onto the laundry during the wash and rinse cycle. Rather than immersion in deep water and agitation back and forth, the laundry in a front-loading machine is repeatedly lifted out of the low volume wash solution and tumbled gently as the drum rotates on its horizontal axis, clockwise and counter clockwise. This gentle action may be preferable for multi-component fabric assemblies used in cold weather clothing. The fact that HE washers require the use of an HE detergent that is low sudsing, however, may be a limitation in cleaning cold weather clothing if a garment manufacturer specifies use of a specific detergent tailored for that garment.

Ankeny et al. (2006) evaluated the effect of the three washing machines/laundry platforms (traditional vertical-axis, vertical-axis HE, and horizontal-axis

HE) on colour retention, surface appearance (pilling) and shrinkage of 100% cotton knit fabric finished with a cationic and a nonionic softener (microemulsion silicone). To assess differences in mechanical action of the washers, four Danish Mechanical Action pieces of fabric were added to each wash load for five runs. The water level was high in the traditional vertical-axis washer and low in both HE washers. Results indicated that the traditional vertical-axis washer washed fabrics with greater mechanical action and caused greater shrinkage than did either of the HE machines. Cleaning efficiency of the three machines was not assessed. Further developments in home laundry equipment are discussed in Section 13.6.

13.2.2 Detergent formulations

Detergents with the nonionic surfactants nonylphenol ethoxylates (NPEs) and nonylphenols were banned by the European Union beginning in 2005 (Anon., 2005). In the USA, NPEs have not been banned. Canadian facilities, however, are required to plan for the reduction of NPEs in waste water (McCoy, 2008).

Detergent formulations (Table 13.1 and the Appendix) have changed considerably as manufacturers have attempted to make sustainable products wholly or in part from natural starting materials such as alcohols or oils from plant sources (e.g., palm, corn). As these detergent innovations are described, the reader should consider the performance that is expected of a cold weather garment such as a parka and whether new types of detergent or additives to boost its performance are necessary or even desirable; for example, some additives may clog the micropores of a breathable layer, or change the water repellency, flame resistance or anti-microbial properties of a garment.

As noted above, energy can be saved by laundering in cold water. Newly formulated cold water detergents give improved cleaning results over earlier ones and are designed to work effectively at temperatures between 20 °C and 30 °C. Detergent formulators have found builders, enzymes and bleach activators that work well at low temperatures (e.g., sodium percarbonate bleach and a manganese-based bleach activator). However, in cleaning cold weather clothing that is heavily contaminated with oil or oil-based dirt, cold-water formulations are unlikely to be the most effective choice. In cold water washing, the oils are not likely to be softened sufficiently to be removed by the roll-up mechanism. Rather, the hydrophobic portion of the surfactant molecules is attracted to the oil and solubilizes it. Enzymes that are effective in cold water are an important additive to help break down other insoluble dirt residues.

Newer detergents often have surfactants with a natural base such as a palm or coconut oil-based alcohol modified with ethylene oxide. Vegetable oil alcohols are mostly 12 carbon atoms long and foam well. Methyl esters from palm oil are combined with ethylene oxide to make methyl ester ethoxylates that have a strong performance in cleaning. Nonionic detergents made from alkyl poly-

Table 13.1 Components of a detergent: traditional and contemporary alternatives

Component	Traditional	Contemporary
Surfactant	Anionic – soap, fatty alcohol sulfates, linear alkyl benzene sulfonate Nonionic – alcohol ethoxylates, alkyl phenol ethoxylates; e.g., nonylphenol ethoxylate (NPE), ethylene oxide/propylene oxide, alkyl polyethers	Anionic – methyl ester sulfonate (MES) Nonionic – methyl ester ethoxylate alkyl polyethoxide (APE) Blends – alcohol ethoxylates/ethanol amines
Foam depressants		Silicones, soap for HE detergents
Builders	Tripolyphosphates, carbonates, silicates, zeolites, citrates, organophosphonates	Polycarboxylates, soda ash/silicate Polyacrylates, silicate/carbonate cogranule Hydroxyethyl iminodiacetic acid (CEN 08)
Anti-redeposition agents	Carboxymethyl cellulose	Acrylate polymers Carboxymethylinulin from chicory root
Enzymes		Amylases for starch-based stains, cellulases for cotton fuzz removal, proteases to break down protein stains, lipases for oil-based soils, mannases for carbohydrates of the mannan family (e.g., guar gum)
Bleach	Perborates, percarbonates + activator	Perborate + cold water activator tetraacetyl ethylene Percarbonate with manganese-based activator
Colour protection	Polyvinyl pyrrolidone	Dye transfer inhibitor – imidazole derivative
Optical brighteners Fluorescent whitening agents (FWA)	Stilbene, azole, coumarin, pyrazoline derivatives,	0.05–0.3% FWA in most commercial detergents
Soil release agents		Soil release polymers (SRP) delivered by the detergent coat low polarity fibre surfaces with a very thin layer of amphiphilic (hydrophilic) polymer (Obendorf, 2004)
Filler	Sodium sulfate	Reduction in amount of filler
Miscellaneous	Fragrance, colour beads, opacifiers, anti-caking agents to improve flow of powders	Many are fragrance free and/or have no colour beads.

glucosides (APGs) from a natural source have been more successful in hard surface cleaners than in laundry detergents (McCoy, 2007b). Alkylpolyethoxide (APE) nonionic detergents can be made from palm or coconut-based detergent alcohols modified with ethylene oxide.

HE detergents for use in high efficiency (HE) machines are formulated for low-liquor washing machines and are essential if using HE machines with bulky cold weather clothing. According to the Soap and Detergent Association (2005) they are low-sudsing and disperse soils quickly. Many HE detergents are also up to three times more concentrated than traditional liquids, requiring less packaging and lower energy for transportation, but having no effect on laundry effectiveness.

Two-in-one detergents, or 'softergents' contain detergent ingredients plus fabric softener, thus saving the purchase of a separate softener. Such detergents may not be ideal for cold weather clothing with microporous breathable membranes that are often rendered ineffective in the presence of fabric softeners.

Because some clothing and household items like sports wear and towels may retain odour after laundering, detergent formulators have found additives that help reduce odour in laundered goods. Odour control in spray products such as Febreeze is achieved with cyclodextrins, cage-shaped glucosides that trap odour molecules but do not work well in laundering. To achieve odour control in laundry detergents, zinc ricinoleate is used because it is able to react chemically with nitrogen and sulfur containing compounds associated with body odour (McCoy, 2007b). Such detergents may be useful for cold weather clothing items like underwear and socks that are worn next to the body.

Enzymes such as amylases, cellulases and lipases are added to detergent formulations to enhance stain removal and are tailored for specific classes of stains. A newly added enzyme, mannase, is directed at stains that reappear on garments after they appear to have been washed away. Food stains containing carbohydrate residues such as starch tend to adhere to cotton and not be removed by detergents alone. The residues on washed clothing are colourless, and do not become visible until they pick up dirt during wear or washing. Dirt removed from other garments during washing can adhere to the carbohydrate residues and a stain appears in a location where one was not previously visible. A particularly tenacious polysaccharide from the mannan family is guar gum, found in many foods and household products. Mannase in a detergent formulation will help to remove guar gum (McCoy, 2001).

13.3 Professional textile care

13.3.1 Dry cleaning procedures

Professional dry cleaning is recommended by some manufacturers of cold weather clothing when laundering at home is considered unsafe because of water sensitive components, dyes or finishes. The long standing tradition of cleaning with perchloroethylene (PCE) or hydrocarbon solvents (Stoddard solvent, 140F solvent) is changing because of environmental and health concerns over the safety of PCE, a potential occupational carcinogen. Perchloroethylene is strictly regulated in many countries (EPA, 1998a). It is being phased out in California and will be banned in Los Angeles by 2020 (McCoy, 2005).

Environmental regulations related to clothing care are often included in clean air and water legislation (for example, such regulations are similar in Germany, the United States and Canada). Dry cleaning machines and condensates in the machine are still regulated. The handling of aqueous wastes, contact water treatment and the design of drains are also normally regulated (Kurz, 1996). Both manufacturers' recommendations and consumer practices for dry cleaning cold weather garments may need to change as new dry cleaning solvents and cleaning processes are introduced. Because PCE is still the most widely used solvent worldwide, a typical dry cleaning procedure with PCE is outlined below and potential problems for cold weather garment assemblies are discussed. New solvents that may replace PCE are described briefly in Section 13.6, although information about their cleaning effectiveness is lacking or anecdotal.

Dry cleaning with perchloroethylene (PCE)

The following description is typical of a cleaning cycle at a traditional dry cleaning establishment, though some details may vary (National Institute of Drycleaning, 1959; EPA, 1998a).

- Preparation: Garments are classified into three groups: wearing apparel, household items and special items (furs, leathers). Garments may be further sorted by fibre type, weight and colour. They are examined and poor condition recorded: rips, tears, lost buttons, major stains. The care label may be reviewed if present. Pockets are cleaned. Fragile buttons are covered or temporarily removed. Fragile garments may be cleaned in a net bag. Noticeable stains are often spotted after cleaning rather than before because oil based stains may be readily removed in the cleaning cycle.
- Washing: in a dry-to-dry machine, the garments are loaded in the wash drum and the cleaning conditions are selected. 'Charged' PCE at a temperature of approximately 21–32 °C enters the drum and garments tumble in the wash solvent for a selected time. To facilitate removal of water-based soils, the PCE is charged with 0.5 to 4% (vol/mass) detergent and with water sufficient to give the solvent a relative humidity (SRH) of 70 to 75% (NID, 1959).
- Rinsing: if the detergent charge is low (0.5 to 2%), there is usually no rinse and detergent residues remain in the clothing. If a higher charge is used (2 to 4%) that might darken light-coloured fabrics, a rinse in clear solvent may be used to remove detergent residues.
- Drying: solvent is extracted from garments by whirling of the wash wheel, then any remaining solvent is removed during tumbling of garments in a

stream of warm air. When most of the solvent is removed, a stream of fresh air helps to strip away any remaining solvent vapour.
- Finishing: garments are checked for residual soils or stains, and spotted and re-cleaned if necessary. Steaming and pressing follow.

Water repellent and waterproof finishes are applied in three ways: by immersion in a PCE/water repellent solution that is pumped from a separate storage tank into the machine's cleaning cylinder; a commercial repellent mixture (non-PCE based) is sprayed onto the garments by hand; and cleaned garments in a wire basket are immersed in a dip tank with the water repellent solution, drained, then transferred to a dryer (EPA, 1998a, 2-8).

Potential problems in dry cleaning cold weather clothing with PCE

The detergent and additives to the solvent used by a dry cleaner are generally not known to the client. The dry cleaner should be consulted to determine whether garments are rinsed in clean solvent. If they are not rinsed, residues remaining in the garments will affect the water repellency. Residues may also clog pores in microporous breathable membranes. Some textiles materials such as olefins, rubber and polyurethanes and adhesives swell in PCE and should be cleaned in hydrocarbons or other solvents. Coatings may separate and olefin piping in corded seams may shrink. Elastane (spandex) fibres swell in PCE; although they return to their original size when the solvent evaporates, stabilizers are extracted in the process (Boliek and Jensen, 1994). Poly(vinyl chloride) will stiffen and shrink if exposed to heat or if solvent soluble plasticizers are removed during cleaning.

13.3.2 Professional wet cleaning procedures

Professional wet cleaning is a relatively new process that uses water as the solvent to clean fabrics. Many professional cleaners now have equipment for wet cleaning that they operate in tandem with their solvent cleaning machines, although a growing number of shops only do wet cleaning (EPA, 1998a).

Wet cleaning of cold weather clothing at a commercial establishment may be more appropriate than dry cleaning in organic solvents. It is appropriate for very dirty work clothing, sportswear, rainwear, garments with microporous membranes, and items with components that dissolve or soften in PCE (Hasenclever, 1996). For many years, dry cleaners have wet cleaned a small percentage of incoming garments, sometimes by hand or in home-style washing machines. Large commercial washers for industrial cleaning have also existed. In discussing European research related to the development of care labels for wet cleaning, den Otter (1996) stated, 'it is absolutely necessary to distinguish between washing and wet cleaning' and to convince consumers that garments must be treated by a professional wet cleaner if so labelled. Consumers cannot assume that a garment labelled with a 'wet clean' symbol may be safely cleaned in a home washer. The wet cleaning process is much gentler than laundering because of newly designed washers and dryers. The equipment is computer controlled to give multiple programs with such variations as load capacity, mechanical action of the drum, length of the wash and dry cycle, number of rinses, water level, temperature, and detergent and additives such as conditioners or sizing (Star and Vasquez, 1999).

Special detergents that are milder than home laundry detergents have been developed for machine wet cleaning (EPA, 1998b). Detergent formulations are proprietary; however, detergents for wet cleaning appear to contain similar ingredients to regular detergents: anionic and non-ionic surfactants, builders, anti-redeposition agents, pH adjusters and fragrance (EPA, 1998a). Finishing of wet cleaned garments is more time consuming than that for dry cleaned items. Tensioning equipment is often used to reduce shrinkage.

The gentleness of wet cleaning procedures is evident from the description of 'gentle' and 'very gentle' cleaning procedures that have been associated with the wet cleaning symbols used in Europe (Kruessmann, 1996) and now incorporated into ISO 3758 – 2005, Textiles – Care labelling code using symbols (ISO, 2005) as 'mild' and 'very mild'. A wet cleaning symbol with a 'normal' classification indicates that the textiles should be professionally wet cleaned because of their structure or finish. A 'gentle/mild' classification is for items such as soft woollens that should not be home laundered. A 'very gentle/very mild' classification is for items such as angoras and silks that should not be laundered because they are very sensitive to mechanical action.

When dry cleaners began wet cleaning items that were labelled 'Dry Clean only', they found that dimensional change was a problem that led to customer dissatisfaction (Patton, 1996). If care label instructions are not followed, there is a greater risk of garment failure such as excessive shrinkage. Wentz (1996) indicated that shrinkage, both felting and relaxation, can occur during cleaning, drying or finishing, although 75 to 95% occurs during the washing part of a wet cleaning cycle (den Otter, 1996). The new wet cleaning equipment has been designed to reduce shrinkage through the control of mechanical action, cleaning time, heat and chemistry. Wentz (1996) noted that consumers can detect shrinkage of more than 2 to 3%. The Center for Neighborhood Technology (Chicago, Illinois) evaluated the dimensional stability of a random sample of 460 garments that were wet cleaned. Among the woven garments, 27% showed shrinkage or stretching of 2–4%, and 11% had over 4% shrinkage. Shrinkage of knitted garments was greater: 21% shrank more than 6% (Patton, 1996). In spite of the shrinkage noted during wet cleaning of some garments, it is less than would occur during home laundering. A round robin trial organized by European Wet Cleaning Committee revealed that the shrinkage occurring during a 'gentle' wet cleaning process was approximately half of that which resulted from home washing; for a 'very gentle' process, it was about one quarter.

A case study (EPA, 1998b, p. 2) lists some of the benefits and challenges of wet cleaning relative to dry cleaning:

- Effects on clothes: no chemical smell; whites are whiter, water-based stains are more easily removed, better soil removal from some items. Challenges: shrinkage, possible dye bleeding, wrinkling, more care needed to finish garment.
- Types of clothes: cotton, linen, wool, silk, leather/suede, wedding gowns, garments decorated with beads and sequins. Challenges: wrinkling of acetate linings, antique satin, gabardine, some complex structured garments.
- Environmental effects: hazardous chemicals are not used nor waste produced; no air pollution results and there is reduced potential for contamination of water and soil. Challenge: more water is used.

Wet cleaning of cold weather clothing

There are a number of reasons why professional wet cleaning is appropriate for many cold weather garments. Manufacturers will in some cases specify professional wet cleaning on care labels and such instructions should be followed. The soils often found on cold weather clothing are more readily removed by aqueous cleaning with water and surfactants. Outerwear is typically soiled by air pollutants (acidic gases, carbon black), body excretions, food residues and direct contact with dirt (dust, pollen, metal oxides). Approximately 40% of the soil is water soluble and 10% is solvent soluble (Hasenclever, 1996). Professional dry cleaners in Europe tend to wet clean garments that are not so well handled by organic solvent cleaning: garments with microporous membranes, sports wear, protective rain wear, very dirty garments, and speciality items. Hasenclever further stated that heavy work wear tends to hold a considerable amount of organic solvent and solvent losses are high when cleaning these items because some solvent remains after a normal drying period. Wentz (1996) suggested preferred methods for cleaning various textiles by grouping them in a scale with two extremes: non-aqueous cleaning and aqueous cleaning. Cold weather items (overcoats, parkas, windbreakers, raincoats, sweaters, blankets, and sleeping bags) were grouped nearest the aqueous end of the scale. In order for garments to retain their original properties after wet cleaning, they should be thoroughly rinsed of detergent residues. The water resistance of rain gear should be checked and water repellent finishes reapplied if necessary.

13.4 Problem areas for maintenance of cold weather clothing

There are several aspects of much cold weather clothing that create special problems for their maintenance. Some of these are discussed below; some are highlighted further in Section 13.5.

13.4.1 Highly contaminated cold weather work clothing

There are some occupations where, at least in cold climates, working outdoors in frigid temperatures and the likelihood of contaminating one's winter work clothing are almost inevitable. Such occupations include, for example, workers in many parts of the oil and gas sectors and firefighters or other emergency personnel (see Fig. 13.2). In these cases there is concern for both the health and hygiene of the wearer and the fact that dirt and grease fill up air spaces and lower insulation. In addition, many contaminants may reduce or eliminate the effectiveness of such properties as flame resistance and/or the high heat-flux thermal protection provided by the clothing system (Crown *et al.*, 2002, 2004; Makinen, 1992). Contaminants may build up on such clothing because, at least in the oil and gas sector, outer garments such as parkas tend not to be cleaned very often relative to the more regular cleaning of coveralls, shirts or pants (Chandler and Crown, 2002). Thus, it is important in such circumstances to remove contaminants when they occur before they are ground into the fabric or otherwise become more difficult to remove.

The difficulty of removing grease or oily contaminants such as motor oil is highlighted in advice to members of the US military about cleaning their cold weather parkas: 'Grease spots are not removable because you can't clean your parka at a high enough heat to get out the stain without damaging the parka's tape' (Anon., 2006). There are often other components such as membranes that cannot withstand high laundering temperatures. However, laundering at high

13.2 Firefighters in dirty, frozen gear.

temperatures may not be the most appropriate procedure in any case. For many of today's synthetic outer fabrics, washing at very high temperatures may open up the fibre structure and allow the contaminant to be embedded even more deeply in the material.

In an extensive study of the effectiveness of several cleaning methods, Stull *et al.* (1996) found that dry cleaning, and even aeration, were most effective in removing many common contaminants found in structural firefighting turn-out gear. Recent experience suggests that appropriate wet cleaning procedures may also be effective. While aeration is suitable for some contaminants and is usually available unless temperatures are extreme, dry cleaning or professional wet cleaning is not always accessible to workers in remote locations. Crown *et al.* (2004) conducted research aimed at developing practical care procedures to help maintain the protective quality of flame resistant workwear laundered by workers at remote field sites, simulating domestic laundry conditions. Especially in the case of aramid fabrics, the use of laundry pre-treatments (either an industrial degreaser or one of two domestic pre-laundry sprays) was necessary to remove motor oil from the fabric to the point where at least some of its flame resistance was maintained. Neither industrial nor domestic detergent alone was as effective. If not thoroughly removed through rinsing, however, pre-treatment products may also increase the clothing's flammability. Current follow-up research is demonstrating that as the number of contamination/laundry cycles increases, so does the difficulty in removing built-up contamination (Mettananda and Crown, in press). Very careful control of the laundry process (water/garment/detergent ratio; water hardness and wash pH; water temperature; sufficient rinsing), is necessary in order to ensure that pre-treatment products and contaminants are removed and that redeposition of the oil and particulates does not occur.

13.4.2 Garments with waterproof breathable components

Waterproof breathable fabrics are widely used for outerwear including fashionable rain garments, hiking and skiing wear, military garments, and protective work wear. These fabrics and their properties are described in detail in Chapter 4. The outer layer normally comprises a tightly woven fabric treated with a durable water repellent (DWR) and either a waterproof, non-porous hydrophilic coating or a breathable porous membrane layer. With different fabric assemblies, garments can be made that keep the wearer warm and dry. The fabric assembly breathes by transmitting water vapour away from the wearer but it does not permit water droplets to pass from the outside to the wearer. Moist air from insensible perspiration and sweat can be trapped in the clothing layers and in cold climates can exist as vapour, liquid or ice (Lomax, 2007). If ice forms, it can damage the coating or membrane. Successfully cleaning these garments so that their function is not compromised depends on factors such as the structure and composition of each layer in the fabric assembly, the method of cleaning, how often it is cleaned and whether the garment has been physically damaged in wear.

Most breathable porous membranes are hydrophobic and breathe very effectively as long as the pores are not enlarged by stretching at elbow and knees, or contaminated. Contamination may occur from contact with body oils, particulate dirt, salt, lotions, and detergent/surfactant residues remaining after washing or dry cleaning. To prevent contamination and reduce stretching, a thin hydrophilic sealing layer is sometimes added to the base of microporous membranes (Mukhopadhyay and Midha, 2008). Since there are no pores or voids in the structure of non-porous hydrophilic films, contamination by soils and residues is not normally considered a problem. A third method of making a breathable fabric is to combine a microporous membrane with a hydrophilic layer on top.

Recommended cleaning procedures

W. L. Gore and Associates give instructions (www.gore-tex.com) for cleaning outerwear containing their breathable water repellent membrane and restoring water repellency. They always recommend following the manufacturer's directions for cleaning, but provide details for safe cleaning. A pre-wash spray may be used to pre-treat oil-based stains and soiled areas before washing. Garments may be machine washed in warm water (40 °C) using a powdered or liquid detergent without fabric softener or bleach. An extra rinse in clean water is advisable to ensure that no detergent residues remain. They recommend tumble drying at a warm temperature because the dryer heat is said to 'reactivate' the DWR finish on the garment. If ironing is necessary, a steam iron (warm setting) may be used with press cloth between the iron and garment. Professional dry cleaning is also an appropriate way to clean garments with a Gore-Tex membrane, but the dry cleaner must be willing to provide special care for such a garment. Because dry cleaning solvents contain a surfactant and perhaps other additives, a garment with a microporous membrane should be given an extra rinse with freshly distilled solvent containing no additives. If water does not bead up on the garment after washing and drying, the outer fabric should be sprayed with a DWR finish. DWRs are not permanent and need to be replaced regularly. They are available from sporting goods stores or retailers of outdoor wear.

13.4.3 Underwear and socks

Thermal underwear, usually a single layer, poses less difficulty in cleaning than cold weather outerwear with multiple layers and components. However, due to the close proximity of under garments to the wearer's skin, maintaining an acceptable level of cleanliness is important for health and good hygiene. Commonly, cold weather underwear comprises long-sleeved tops and long-johns, made from natural fibres such as wool, cotton or silk, or synthetic fibres, often micro-fibres, such as polypropylene or polyester. Synthetic fibre undergarments are generally easier to launder, although wash and dry temperatures should be kept low.

A frequent aesthetic problem relating to the care of underwear concerns build-up and retention of human body odours in the fabric. Synthetic fibres tend to retain and emit stronger odours generated from the body than do fabrics of natural cellulosic or protein fibres (McQueen *et al.*, 2007). This may be due to oils such as sebum and apocrine sweat from the body being adsorbed onto the fibres and yarns of the synthetic garments, and bacteria transferred from the skin continuing to produce odour even after the garment has been removed from the body (McQueen *et al.*, 2008). Therefore frequent laundering, possibly after each wear, is recommended. As underwear is unlikely to be shared among users there is little risk of cross-contamination from resistant strains of bacteria.

Socks should be kept clean and dry in order to maximize the thermal insulation they provide and to avoid bacterial infection. In cold weather climates, socks are often layered to increase thermal insulation. Two or more layers of lightweight socks can insulate feet more effectively than one pair of heavy socks because of the dead air trapped between layers (Kuklane *et al.*, 2000). Thermal insulation may also be affected by washing and repeated wearing of socks. After washing, wool blend socks containing at least 65% wool provided slightly higher insulation than before washing (Kuklane *et al.*, 1999). This slight increase in insulation was related to an increase in thickness of the socks and trapping of more air in a structure described as 'looser' after washing. The insulation decreased over time because of loss of fibre and thinning of the socks. It is important that socks are thoroughly dried after washing and between wearings as moisture in socks reduces the insulation provided by air trapped within the sock and between layers and footwear (Kuklane *et al.*, 1999).

13.5 Care of cold weather clothing – case studies

Most jackets, coats, and trousers intended for cold weather comprise a series of layers, with each layer intended to perform a specific function. There may be an outer shell with properties to protect the wearer from the wind and rain (see Section 13.4.2), an inner filling to provide insulation against the cold and a lining to provide a soft comfortable inner surface which may also wick moisture away from the wearer and protect the insulation materials from perspiration and dirt. In selecting an appropriate method of care, all components that comprise the garment need to be taken into consideration.

13.5.1 Outer layer

The DWR finish applied to the outer fabric is usually a silicone or a fluoropolymer product that is not permanent but will age and require replacement with

290 Textiles for cold weather apparel

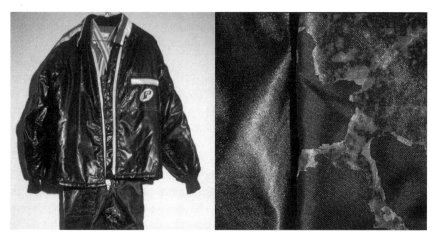

13.3 Jacket with a polyurethane coating and a detail of one partially delaminated panel following cleaning.

age, use and multiple cleanings. As well, these finishes may be removed by dry cleaning solvent and/or rendered ineffectual by dirt, detergent residues or fabric softeners. Whatever method of cleaning is selected, thorough rinsing and complete removal of the detergents used are critical. Reapplication of the repellent finish may be necessary. The application of heat in tumble drying may help to redistribute the finish over the fibres of the fabric. Applying heat through ironing/pressing is also suggested by manufacturers of some spray-on finishes, but pressing is not recommended when a heat-sensitive insulation layer is present in the garment.

Polyurethane membranes or coatings may be damaged by dry cleaning. In the example in Fig. 13.3, the coating on certain panels of the jacket was adversely affected by the cleaning solvent. The polyurethane swelled, adhered to itself and delaminated from the shell fabric during the dry cleaning process. Not all panels of the garment were damaged, indicating differences in the solvent swelling properties of the polyurethane coatings. Polyurethane coatings may also be degraded by light, even when they are found on the inner surface of a fabric. Light reaches the coating and initiates the deterioration, so that proper storage away from light is important.

13.5.2 Insulation layer

Synthetic fibre insulation, usually polyester or polyolefin fibres, are sensitive to heat and such fibres can shrink and eventually melt if exposed to temperatures that exceed their thermal transition temperatures or melting temperatures. Such temperatures may be encountered in tumble drying, pressing or ironing; therefore, precautions must be taken to avoid high temperatures when drying and finishing fibre-filled garments. Heat and pressure can also flatten a fibre bat and

Care and maintenance of cold weather protective clothing 291

13.4 Shrinkage of the heat-sensitive insulation layer in the garment has puckered the outer shell and lining fabric.

reduce its insulation properties. This is a problem for heat-sensitive polyolefin fibres (e.g., Thinsulate^{TM}), especially if incorrectly finished, and possibly for very fine polyester fibres. The heat and pressure of commercial steam pressing, for example, will damage this type of insulation. In the example in Fig. 13.4, a fibre bat shrank after exposure to a drying temperature that was too hot for the insulation. The outer fabric of polyester/cotton and the rayon lining were not affected but the lining is puckered by the shrunken dimensions of the heat-sensitive insulation layer.

Heat-sensitive, polyolefin materials also find their way into winter jackets as non-woven scrims or as ticking, baffles or tapes used to secure insulation materials. They may shrink, melt and harden, causing puckering and distortions to the garment, as in Fig. 13.5. In Fig. 13.6, the olefin ticking inside the down-filled jacket shrank. The other garment components were unaffected by the temperatures encountered in cleaning. In many jurisdictions these interior components, comprising less than 5% of the garment by weight, may not be identified on the content label. Problems with heat-sensitive materials can occur in tumble drying or steam finishing; cool drying temperatures are recommended after any cleaning process, as is avoiding finishing with pressure.

Non-woven fibre insulations of any fibre content can also be physically damaged (i.e., torn, shifted, compacted) by the mechanical actions of the cleaning process, so a gentle cycle with reduced agitation is advisable in any cleaning process. Wool fibre bats are also used as the insulation layer of cold weather garments. The fibres are unaffected by heat, but if they have not been given a shrink resistant treatment, they may felt, shrink and stiffen in washing, thus losing their insulation value and distorting the garment shape. Garments containing wool require dry cleaning or reduced-agitation wet cleaning, unless they are labelled as washable.

Down insulation is sensitive to the mechanical actions of cleaning processes.

292 Textiles for cold weather apparel

13.5 Polyolefin seam tapes used in the interior of the garment have contracted and hardened with exposure to heat during tumble drying.

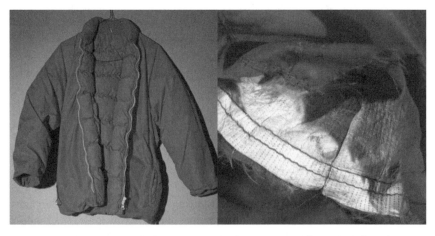

13.6 Down-filled jacket and a detail of the heat-sensitive polyolefin baffles that shrank during cleaning.

Care and maintenance of cold weather protective clothing 293

13.7 Down-filled coat with oily residues of detergent and soil on the outer shell fabric.

Loss of loft or permanent shifting of the down results in a reduction of the insulation value. Flattening and shifting can be caused by mechanical, physical actions, breaking up the down, especially while it is wet or saturated with dry cleaning solvent. In Fig. 13.7 the coat has oily looking stains known as swales (dark irregular residues deposited on the shell fabric). Uneven drying of down insulation can result in such swales. The shell fabric dries more rapidly than the down filling. The hollow shafts of feathers in the down mixture retain solvent and dry slowly. The filling also retains soils and detergent making effective rinsing difficult. Following washing, complete drying of the down insulation is important as well to prevent micro-organism (mildew) growth. Slow drying at a moderate temperature until the down is completely dry will ensure the down separates and regains its original loft and insulating properties after washing or dry cleaning.

13.5.3 Other components

Lining materials normally do not encounter problems in cleaning except when associated with insulation as noted above, but fleece materials used for lining

294　Textiles for cold weather apparel

13.8 Jacket with a distorted shape and a detail of the heat-sensitive piping that shrank during cleaning.

may become pilled with fibres tangled if exposed to excess heat and mechanical action. If the fleece contains a heat-sensitive fibre, shrinkage is also a possible problem. Attraction of lint to fleece occurs if garments are not cleaned separately or with similarly coloured items.

Piping trim may be sensitive to heat and may pucker and/or shrink during drying or finishing. In Fig. 13.8, the decorative piping used along the seams of the jacket contained an inner cord of polypropylene. In dry cleaning, temperatures greater than the shrinkage temperature of olefin (> 75 °C) are routinely encountered. The cord inside this piping was heated above its shrinkage temperature, causing it to contract and stiffen and to permanently distort the garment shape. The care label in the jacket did not indicate any heat restrictions, none were taken, and the trim was damaged.

Fake fur trims, often comprising acrylic or modacrylic fibres, are sensitive to heat and steam and may be matted or flattened after cleaning and tumble drying. Real fur trim on hoods may not be washable, but may be dry-cleanable or removable for cleaning.

13.5.4 Problems related to use and storage

Ideally garments should be stored on a padded hanger in a cloth (not plastic) garment bag to protect from light and dust, and with adequate room so that the insulation is not overly compressed. Degradation of materials through exposure to UV light, or photo-oxidation, will occur if closet doors are left open and garments are not moved for long periods of time (e.g., between winter seasons). Many dyes, fibres and finishes used in speciality cold weather clothing are sensitive to UV radiation from sunlight. The outer fabric may fade or weaken, and the coating and finishing materials may degrade and no longer provide the water repellency or wind resistance expected. In Fig. 13.9, the white nylon fabric

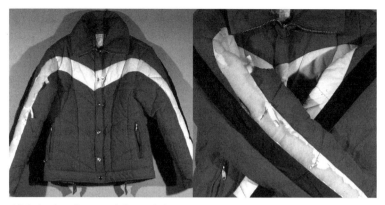

13.9 Jacket damaged through improper storage and a detail of the photo-degradation to the white nylon fabric of one sleeve.

on one sleeve of the jacket has shredded while the remainder of the garment is in good condition. The problem occurred while the jacket was stored in a closet with one sleeve exposed to light and the remainder of the garment protected within the closet. The white nylon has a thick coating of titanium dioxide to give it a matt white appearance and the pigment acted as a photo-sensitizing agent, causing more rapid degradation of the white stripe of fabric than of the other nylon fabric in the jacket.

13.6 New developments

13.6.1 New developments in home washing machines

In 2005, some appliance manufacturers introduced HE washers that incorporate steam as part of the cleaning cycle. The steam may be introduced in three ways, depending on the wash cycle chosen: steam during the wash cycle, steam as a sanitation step with the steam being injected at the end of the wash cycle, and steam as a 'refresh' cycle where a few dry but lightly soiled garments are tumbled in steam over a 20 minute period to relax wrinkles and diminish odours. Steam is generated from a separate water supply to the machine. With the first option, where steam is used during the wash cycle, the wash water temperature is automatically set to 'warm'. As the wash cycle progresses, the water is slowly heated and steam is introduced into the top of the tub (Anon., 2008). The steam enters at 100 °C and energy is transferred to the garments as the steam cools and condenses. The small volume of wash water pools in the bottom of the drum. The clothes rotate through the water and the steam-filled space above the water so that, according to manufacturers, there is optimal removal of organic stains (food and blood) and temperature-sensitive stains (oil). The steam function is designed for very dirty loads that need improved stain treatment, and loads with multiple types of stains (Anon., 2008). The HE machines with steam function offer a number of cycles and a selection of temperatures for different types of garments and cleaning needs. If cleaning cold weather clothing, the temperature and cycle should be carefully chosen, especially if there are heat-sensitive components in the garments. Pre-treatment of oily stains may still be necessary depending on the fibre content of the garments and severity and nature of the stains. The ability of the steam function to provide enhanced soil removal has not been confirmed by published research.

13.6.2 Solvents to replace perchloroethylene

Liquid carbon dioxide (CO_2) is a solvent that interests many dry cleaners as they search for a replacement for PCE. Unlike PCE, it is non-toxic, biodegradable and requires no hazardous waste removal. At room temperature carbon dioxide can be liquified under high pressure in a closed system to produce an effective solvent for a number of oily soils (EPA, 1998c). Equipment for cleaning with liquid CO_2 is expensive, however, the wash liquor (solvent plus detergent) can be filtered, distilled and the CO_2 recycled. To achieve good cleaning results, pre-spotters and detergents must be used. Liquid CO_2 does not remove particulate soils as well as PCE unless more mechanical action and appropriate detergents are used. Researchers found that large particulate soils like sand can be removed with increased mechanical action. Small particles such as carbon black and clays, however, require surfactants that effectively reduce the adhesive forces between the particles and fibres (van Roosmalen *et al.*, 2003). Colour loss from dyed garments after repeated cleaning in liquid CO_2 was less than that observed after cleaning with PCE (EPA, 1998c).

A number of other new solvents have been developed as possible replacements for PCE. Some of these solvents can be used directly in PCE equipment or hydrocarbon machines with little or minor retrofitting, but others require purpose-built equipment. Table 13.2 lists the physical properties of some of these solvents.

13.7 Sources of further information and advice

The following websites have valuable information related to subjects covered in this chapter.

- Sierra Trading Post – has definitions on its website of many trade names used in outdoor clothing.
  http://www.sierratradingpost.com/Glossary.aspx
- DWR Durable Water Repellent finishes – Sources of Information on Storm Waterproofing sprays and detergents for DWR fabrics
  http://www.stormwaterproofing.com/article%2001.htm
  http://www.stormwaterproofing.com/article%2002.htm

Table 13.2 Properties[a] of perchloroethylene and possible alternative solvents

Solvent name	Manufacturer	Chemical formula	Boiling point, °C	Flash point, °C	Vapour pressure, mmHg/25°C	Water in solvent solubility, ppm	Kauri-butanol value
Perchloroethylene	Dow, PPG, Vulcan	C_2Cl_4	121	none	18.2	105	90
Stoddard solvent	various	mixed hydrocarbons	150–212	38–60	0.5–0.27	negligible	29–45
DF 2000™	Exxon Mobil	synthetic hydrocarbon	190–205	64	0.49	negligible	27
Green Earth®	Green Earth Dow Corning	decamethyl pentacyclo-silane	211	77	0.3	250	13
Rynex	Rynex Holdings	Dipropylene glycol t-butyl ether	67	93	32	>50,000	70
Drylene™800	Chemica-Sud SNC	Petroleum distillate	195–209	>93	not available	<100	27
EcoSolv™	Chevron Phillips	C_{10}–C_{13} iso paraffins	83	61	not available	not available	26–27
DrySolv™	Enviro Tech	n-propyl bromide	not available	none	not available	not available	not available

[a] Flick, 1990; State Coalition for Remediation (no date); EnviroTech, 2008

- Cleaning and restoring water repellency of Gore-Tex® products
  http://www.gore-tex.com/remote/Satellite/content/care-center
- Gelanots
  http://www.gelanots.com/pages/microfiber.html
- Information by Roger Caffin FAQ – Rainwear – parkas and trousers, etc.
  http://www.bushwalking.org.au/FAQ/FAQ_Rainwear.htm
- Cleaning advice for DWR fabrics:
  http://www.lifeinsuranceforseniors.net/node/9637
- Insulation/fleece:
  Meida® is a polypropylene fibre material to be washed between 30 and 40 °C and can be ironed between 120 and 130 °C
  http://www.meida.citymaker.com/page/page/1436945.htm
- W L Gore and Associates:
  www.gore-tex.com

13.8 References

Ankeny M, Ruoth B, Keyes N, Quddus M, and Higbee L (2006), 'Evaluation of various laundry platforms on the color retention and surface appearance of 100% cotton knit fabric', *Book of Papers, AATCC International Conference & Exhibition*, 240–249.

Anon. (2005), 'Univar's Caftlon NPE-substitute surfactants now available across Europe', *Focus on Surfactants*, Vol. 2005(6), 3–4. Retrieved on 27 October 2008 from http://www.sciencedirect.com.

Anon. (2006), 'Grease, acid destroy clothing', *Preventive Maintenance Monthly*, PS 638, 46–47, USA Superintendent of Documents.

Anon. (2008), 'Washers and dryers: Green buying guide, *Consumer Reports*, 73(2), Retrieved from www.greenerchoices.org/ratings.cfm?product=washer.

Barrett T W and Moran G J (2004), 'Update on emerging infections: News from the Centers for Disease Control and Prevention'. *Annals of Emergency Medicine*, 43(1), 43–47.

Boliek J E and Jensen A W (1994), *Kirk–Othmer encyclopedia of chemical technology*. New York: John Wiley.

Chandler K and Crown E M (2002), 'How clean is clean? Management and worker perspectives on maintaining thermal protective workwear', *Proceedings, International Textile and Apparel Association Conference*, August, New York.

Crown E M, Feng A and Bitner E (2002), 'How clean is clean? Controlled wear/decontamination study of thermal protective workwear', *Proceedings, International Textile and Apparel Association Conference*, August, New York.

Crown E M, Feng A, and Xu X (2004), 'How clean is clean enough? Maintaining thermal protective clothing under field conditions in the oil and gas sector', Special Issue, *International J. Occup. Safety & Ergonomics*, 10, 247–254.

den Otter W A (1996), 'Report on the European wet cleaning Committee', in Reppert L and Speare L, *Apparel Care and the Environment: Alternative Technologies and Labeling* (EPA 744-R-96-002), 107–114, Retrieved on 30 October 2008 from http://www.epa.gov/dfe/pubs/garment/apparel/doc.pdf.

Easter E P and Ankenman, B E (2006), 'Evaluation of the effect of low water/low energy on clothes care', *Book of Papers, AATCC International Conference & Exhibition*. Atlanta GA, 223–239.

Enviro Tech (2008), 'Drysolv™ Material Safety Data Sheet', Melrose Park, Il: Enviro Tech International, Retrieved on 24 October 2008 from http://www.dctco.com/drysolv_msds.pdf.

EPA (1998a), *Cleaner Technologies Substitutes Assessment for Professional Fabricare Processors* (EPA 744-B-98-001), Washington, DC: Environmental Protection Agency. Retrieved on 22 May 2008 from *http://www.epa.gov/oppt/dfe/pubs/garment/ctsa/sumfctsa.htm*.

EPA (1998b), *Case study: Wetcleaning systems for garment care* (EPA 744-F-98-016), Washington, DC: Environmental Protection Agency. Retrieved on 9 April 2008 from http://www.epa.gov/dfe/pubs/garment/wsgc/wetclean.htm.

EPA (1998c), *Case study:Liquid carbon dioxide (CO_2) surfactant system for garment care* (EPA 744-F-98-018), Retrieved on 2 November 2008 from http://www.epa.gov/dfe/pubs/garments/lcds/micell.htm.

Flick E W (1990), *Industrial Solvents Handbook*, 4th edn, Norwich, NY: William Andrew.

Hasenclever KD (1996), 'Report of professional wet cleaning in Europe', in Reppert L and Speare L, *Apparel Care and the Environment: Alternative Technologies and Labeling* (EPA 744-R-96-002), 101–105, Retrieved on 30 October 2008 from http://www.epa.gov/dfe/pubs/garment/apparel/doc.pdf.

ISO 3758:2005 *Textiles – Care labeling code using symbols*. Geneva: International Organization for Standardization.

Kruessmann H (1996), 'Status of the European (international) care labeling', in Reppert L and Speare L, *Apparel Care and the Environment: Alternative Technologies and Labeling* (EPA 744-R-96-002), 115–128, Retrieved on 30 October 2008 from http://www.epa.gov/dfe/pubs/garment/apparel/doc.pdf.

Kuklane K, Holmér I, and Giesbrecht G (1999), 'Change of footwear insulation at various sweating rates', *Applied Human Science* 18(5), 161–168.

Kuklane K, Gavhed D, and Holmér I. (2000). 'Effect of the number, thickness and washing of socks on the thermal insulation of feet', in Kuklane K & Holmér I, *Ergonomics of Protective Clothing*, Stockholm: National Institute for Working Life, 171–178.

Kurz J (1996), Textile care research programs in Germany, in Reppert L and Speare L, *Apparel care and the environment: Alternative technologies and labeling* (EPA 744-R-96-002), 63–81, retrieved on 30 October 2008 from http://www.epa.gov/dfe/pubs/garment/apparel/doc.pdf.

Lomax R G (2007), 'Breathable polyurethane membranes for textile and related industries', *Journal of Materials Chemistry*, 17, 2775–2784.

Makinen H (1992), 'The effect of wear and laundering on flame-retardant fabrics', in McBriarty J P and Henry N W, *Performance of protective clothing, fourth volume* (ASTM STP 1133), Philadelphia, PA: American Society for Testing and Materials, 754–765.

McCoy M (2001), 'Additives: Where all the magic is', *Chemical and Engineering News*, 79 (3), 26–31.

McCoy M (2005), 'Dry cleaning dreams', *Chemical and Engineering News*, 83(46), 19–22.

McCoy M (2007a), 'Going green', *Chemical and Engineering News*, 85 (5), 13–19.

McCoy M (2007b), 'A new kind of clean', *Chemical and Engineering News*, 85 (12), 29–31.

McCoy M (2008), 'Greener cleaners', *Chemical and Engineering News*, 86 (3), 15–23.

McQueen R H, Laing R M, Brooks H J L and Niven, B E (2007), 'Odor intensity in apparel fabrics and the link with bacterial populations', *Textile Research Journal*, 77 (7), 449–456.

McQueen R H, Laing R M, Delahunty C M, Brooks H J L and Niven B E (2008), 'Retention of axillary odour on apparel fabrics', *Journal of the Textile Institute*, published online DOI: 0.1080/00405000701659774.

Mettananda C V R and Crown E M (in press), 'Quantity and distribution of oily contaminants present in flame resistant thermal protective textiles', accepted for publication in *Textile Research Journal*.

Mukhopadhyay A and Midha VK (2008), 'A review on designing the waterproof breathable fabrics Part I: Fundamental principles and designing aspects of breathable fabrics, *Journal of Industrial Textiles*, 37(3), 225–262.

National Institute of Drycleaning (1959), *NID Correspondence School: Synthetic Drycleaning*, Silver Spring, MD: NID.

Obendorf S K (2004), 'Microscopy to define soil, fabric and detergent formulation, characteristics that affect detergency: A review', *AATCC Review*, 4 (1), 17–23.

Patton J (1996), 'Results and conclusions from wet cleaning demonstration projects, in Reppert L and Speare L, *Apparel Care and the Environment: Alternative Technologies and Labeling* (EPA 744-R-96-002), 129–136, Retrieved on 30 October 2008 from http://www.epa.gov/dfe/pubs/garment/apparel/doc.pdf.

Soap and Detergent Association (2005), 'High efficiency washers and detergents', Retrieved on 4 November 2008 from http://www.cleaning101.com/laundry/HE.pdf.

Star A and Vasquez C (1999), 'Wetcleaning equipment report', Chicago, Il: Center for Neighborhood Technology, Retrieved on 25 April 2008 from http://www.cnt.org/wetcleaning.

State Coalition for Remediation (no date), *Physical properties of chemicals used in drycleaning operations*, Chicago: SCRD. Retrieved on 2 November 2008 from http://www.drycleancoalition.org/download/solvent_table.xls.

Stull J, Dodgen C R, Connor M B and McCarthy R T (1996), 'Evaluating the effectiveness of different laundering approaches for decontaminating structural fire fighting protective clothing', in Johnson J S and Mansdorf S Z, *Performance of protective clothing, fifth volume* (ASTM STP 1237), West Conshohocken, PA: American Society for Testing and Materials, 447–468.

Thiry M C (2008), 'It's a new laundry day: Environmental concerns have changed the landscape of the laundry room', *AATCC Review*, 8 (3), 22–31.

van Roosmalen M J E, van Diggelen M, Woerlee G F and Witkamp G J (2003), 'Dry cleaning with high pressure carbon dioxide – the influence of mechanical action on washing results', *Journal of Supercritical Fluids*, 27(1), 97–108, Retrieved on 31 October 2008 from http://www.ingentaconnect.com/content/els/08968446/2003/00000027/00000001/art00212.

Wentz M (1996), Textile care technology spectra and care labeling issues, in Reppert L and Speare L, *Apparel Care and the Environment: Alternative Technologies and Labeling* (EPA 744-R-96-002), 83–100, Retrieved on 30 October 2008 from http://www.epa.gov/dfe/pubs/garment/apparel/doc.pdf.

Appendix: Examples[a] of home laundry detergents tailored for special purposes

Detergent type/examples	Comments

Cold water detergents:
  Tide Coldwater (North America)
  Dash Cool Clean (Italy)
  Ariel Cool Clean (UK)
  Persil (Germany)
  Care Cold Wash (Denmark)
  Purex for Cold Water (USA; UK)
  Quick Dissolving Tide (North America)
	Liquids or powders designed to work effectively in water at 20–30 °C. All components are chosen for their ability to work at low temperatures; eg. catalytic bleach activators for percarbonate and perborate bleaches.

High efficiency detergents:
  Ultra Tide HE, Tide Free HE
  Gain HE
	Low sudsing detergents are intended for high efficiency washing machines that use less water than regular machines.

Compact liquid detergents:
  2X Ultra Tide
  All Small and Mighty
	Super concentrated liquids are about three times more concentrated than regular liquid detergents. Containers are small, use less plastic and are less costly to ship.

Two-in-one detergent/softener:
  Tide with a Touch of Downy
  All Cleans and Softens
  Arm & Hammer Plus a Touch of Softener
  Purex Plus, Dash 2-in-1
  Persil with a Touch of Silan
	Detergents with softener additives do not soften clothes as effectively as a softener added during the rinse or drying cycle. These detergents are not recommended for clothing that has been treated with a flame retardant finish or for clothing with mircroporous membranes

Eco-friendly detergents:
  Purex 100% naturally derived ingredients, Persil, SA8 Bioquest
  Tide Pure Essentials, Dow Ecosurf
	Baking soda and citrus extracts are used as builders in Tide Pure Essentials.

Colour safe detergents:
  Tide – Bleach Alternative Color Safe
  Tide Wear Care
  Cheer with Liquifiber-color protection
  Black Magic
	Black Magic is a detergent with a cationic fixing agent to bind loose black dye to fibres and keep black clothes black.

Detergents and softeners with anti-odour additives:
  Purex with Renuzit, Fresh Magic
  Tide with Febreeze Freshness
  Downy softener with Febreeze Fresh Scent,
  Febreeze Clean Wash
	Febreeze Clean Wash is a liquid auxiliary that is added to the wash cycle to enhance odour removal.

[a]Specific products are named for information only, and are not necessarily recommended by the authors.

Part III

Cold weather clothing applications

14
Cold weather clothing for military applications

R. A. SCOTT, Colchester, UK

Abstract: This chapter deals specifically with the technical and scientific requirements for military combat and protective clothing systems for cold weather operations. The importance of providing efficient cold weather clothing is highlighted by reference to historical problems faced by forces in past conflicts. The special requirements are discussed, together with the incompatibility problems which designers and scientists encounter with combat clothing systems. Performance details of fibres and materials are given and compared for the thermal insulation, waterproofness, and vapour permeability of UK forces clothing systems. The special requirements for electrically heated clothing are outlined. The chapter concludes by outlining future research on smart/reactive materials and techniques for military systems.

Key words: military combat clothing, incompatibilities in combat clothing systems, thermal insulation, water-vapour permeable textiles, electrically heated clothing, smart/reactive textiles, shape-memory materials, electrospinning, nanotextiles, smart pore fabrics, phase-change materials.

14.1 Introduction

Military forces throughout the world need to be equipped with protective combat clothing made from the best technical textiles available. Unlike the threats to civilians, who face involuntary accidents and hazards in the work-place, military forces in war face many complex hazards which are deliberately aimed at maiming or killing them.

Modern military operations demand flexibility, reliability, and a wide-spectrum performance from personal equipment. The most demanding situations are faced by dismounted Army, Navy and Air Force personnel that include infantry, marines, and Special Forces. Their individual roles mean that they need to wear, use and carry a wide range of clothing to cope with living in natural environments ranging from extreme cold, rain, snow, and winds through to baking heat, dust, and solar radiation. Rapid reaction forces may be required to move at short notice to operational theatres ranging from arctic and mountain, through tropical to hot desert environments.

The clothing requirements are less demanding for those forces who operate from the relative comfort of ships, land bases, and airfields, or where they have

306   Textiles for cold weather apparel

access to powered vehicles such as armoured fighting vehicles (tanks), artillery and transport. Developments over recent years have involved the integration of functional requirements in combat clothing to produce multi-role systems incorporating high-performance technical textiles in a minimum number of separate layers (Brown, 2004).

The role of fighting forces in NATO has changed over the last few decades. During the post World War II Cold War UK marines and infantry units were extensively trained in arctic and mountain warfare, with annual winter survival training in Norway and inside the Arctic Circle. The emphasis on clothing issues was to provide warm, light, durable thermal insulation, improved rain and moisture management, and better personal items such as sleeping bags, tents, footwear, and load carriage. The major strategy was to counter any hostile action by USSR Warsaw Pact forces in Northern Europe.

The demise of the USSR and the break up of Warsaw Pact forces led to a complete change in emphasis in terms of world threats. Following terrorist attacks in the USA and Europe in the early 21st century, the invasion of Iraq to unseat Saddam Hussein, and the military occupation of Afghanistan to oppose the Taliban terrorists meant that military operations were now taking place in hot, arid countries. The major clothing problems here are associated with keeping combatants cool whilst still providing the necessary protection against bombs, bullets, fire and chemicals. These opposing requirements cause many problems for scientists and technologists to solve.

14.2 History of military cold weather operations

There have been numerous examples of extreme cold climates affecting military success. Perhaps the best documented examples surround the attempts by armies to invade Russia during their extremely cold winters. Napoleon made this catastrophic mistake in 1812, and incurred heavy losses. It is also thought that during the Korean War in the 1950s that more casualties occurred through adverse weather conditions than through enemy action (Chappel, 1987).

Perhaps the best documented disasters occurred during World War II. Adolf Hitler and his armies made the same mistake – not just once in 1941–42, but again in 1942–43, showing that military leaders do not seem to learn from previous mistakes. During Operation Barbarossa in 1941 the advance on Moscow went very well in the summer and autumn, and a Russian observer at that time commented 'the Germans marched into Russia wearing smart uniforms and shiny boots' (Thompson, 1973). However, in October heavy rains set in, slowing the advance in a sea of mud. This was followed by freezing conditions and snow. The German Ostheer army was brought to a halt only 65 km (40 miles) from Moscow. The troops were exhausted and badly equipped for keeping warm in such conditions, as the German high command did not think that clothing was an important issue, despite urgent requests from field commanders.

In contrast, the Russian Siberian Army which they faced were well equipped with sheepskin coats, quilted pants, fur hats, felt-lined boots, and white camouflaged oversuits. They counter-attacked viciously in the snow, inflicting losses of about 743,000 casualties, plus many more (about 350,000) who were sick, often from frostbite. Hitler would not allow withdrawal from Moscow, despite warnings from his senior officers. This event was perhaps a turning point in the war, as the Russians had shown that it was possible to defeat the mighty Nazi war machine (Darman, 2007; Willmott *et al.*, 2004).

The next catastrophe occurred in the winter of 1942–43 when the German 4th and 6th Armies had advanced to the outskirts of Stalingrad. The Russians counter-attacked in late November in a pincer movement which encircled the Germans. By early January 1943 re-supply operations by air from the Luftwaffe failed and they were forced to surrender, despite Hitler's insistence that they carry on the fight. The Soviets captured 91,000 prisoners and recovered about 250,000 frozen axis corpses in the snow. The total losses were estimated at about 800,000 dead. Of the 91,000 captured only a handful survived the long trek back to Germany, as they succumbed to starvation, illness and the intense cold.

These failures were due to a range of problems including weapon and vehicle failures in the cold, but cold injuries accounted for a significant proportion of the casualties.

In more recent history the Argentinians invaded the south Atlantic Falklands (*Malvinas*) islands in 1982. United Kingdom forces set out to re-capture what was considered to be UK sovereign territory. The Infantry and Royal Marines units had to make 'wet' landings, i.e., wade ashore, and then traverse the inhospitable terrain in the wet autumnal weather. Many cases of non-freezing cold injury occurred – known as 'trench foot' from World War I. The problems were partly due to poor footwear, but the environment caused the majority of injuries.

14.3 General military clothing requirements

The physical and economic requirements for military protective combat clothing are listed in Table 14.1.

14.3.1 Textiles for environmental protection

Protection against the environment is considered to be the top priority, as the weather is omnipresent whether operations are carried out in peacetime, during training, in peace-keeping roles, or in full-scale war. The range of environmental conditions for military forces ranging from extreme cold through temperate to extreme hot are detailed in standard NATO publications (Defence Standard 00-35, 1996). Unlike civilians, military personnel cannot choose to operate only in fine weather, nor can they take control of their work-rates. Dismounted infantry

Table 14.1 General requirements for military combat clothing

	Comments
Physical property	
Low weight and bulk	Items have to be carried by dismounted troops
High durability	Must maintain functional performance in the field
Low maintenance	Must operate reliably in adverse conditions
Long storage life	Can be 10–20 years
Easy care	Smart, easily laundered, non-iron, low shrink
Anti-static	To avoid incendive or explosive sparks near weaponry
Economic property	
Readily available	From competitive tendering in global industry against a standard or specification
Minimal cost	Paid for with public funds through taxation
Repairable	In the field or HQ workshops
Decontaminable or disposable	If contaminated by nuclear, biological or chemical weapons

forces tend to operate in short bursts of high activity, running whilst carrying equipment and weapons. This activity is interspersed with long periods of inactivity, lying immobile in trenches or temporary cover. This is the most difficult operational situation for protective clothing. Excessive sweating followed by inactivity must be avoided in cold climates, as sweat wetted clothing cools the wearer rapidly, leading to hypothermia (see Fig. 14.4). Soldiers are taught the following *aide-mémoire*:

- Keep it Clean
- Do not Overheat
- Wear in Layers
- Keep it Dry

At the other extreme, high activity in a hot climate wearing layers of protective clothing such as body armour, can lead to hyperthermia (heat stress). Both situations are life and health threatening. The requirements for environmental protection are detailed in Table 14.2.

This chapter concentrates on textiles for cold weather clothing, but there are many other military requirements in an integrated clothing system:

- ballistic protection from bomb fragments, bullets, and high speed projectiles
- wide spectrum camouflage of outer layers, including:
  - UV reflection – from snow
  - visible range spectrum – to match the colours of the background
  - near infra-red (NIRR) – to prevent detection by image intensifiers and low light TV
  - far infra-red (FIRR) – to prevent detection by thermal imagers

Table 14.2 Protection against the natural environment

Property	Comments
Water vapour permeable	
Waterproof	For cold/wet climates
Water repellent	
Windproof	To prevent wind chill and dust
Snow shedding	External clothing
Thermally insulating	Against extreme cold
Air permeable	For hot climates
Rot resistant	When stored damp
UV light resistant	Prevent fading of colours
Biodegradable	If discarded or buried

- radar – detection by movement at long range
- acoustic – noise caused by movement
• protection from flames, heat and flash
• protection from nuclear, biological and chemical weapons (NBC)
• directed energy weapons – eye protection against laser rangefinders and target designators

Details of these requirements can be found in earlier publications (Scott, 2000, 2005).

14.4 Incompatibilities in combat clothing systems

Clothing systems are by necessity a compromise between a range of functional performance attributes. This leads to incompatibilities between conflicting requirements. Figure 14.1 illustrates the possible incompatibilities which scientists and technologists are attempting to overcome (Holmes, 1965). Although this chapter concentrates on cold weather clothing, many other requirements impinge on the performance of a complete clothing system.

Figure 14.1 indicates potential incompatibilities by a cross (X). Starting with a basic requirement – low weight and bulk – we can see problems associated with thermal insulation, ballistic protection, durability, flame retardance (because synthetic fibres such as nylon and polyester offer low weight and bulk, but are difficult to render flame retardant). Far infra-red camouflage (thermal) requires bulky insulated clothing. Synthetic continuous filament fibres provide good snow-shedding properties, but introduce an acoustic rustle or swish problem.

Another critical area surrounds that of water vapour permeability (for comfort) vs. water proofness, wind proofness vs. high air permeability vs. insect proofness and chemical warfare protection. Much R&D effort has been expended in recent years to minimise this set of problems.

A further set of incompatibilities concerns wide-spectrum camouflage and concealment. Visual camouflage must also incorporate Near IR reflectance to

310   Textiles for cold weather apparel

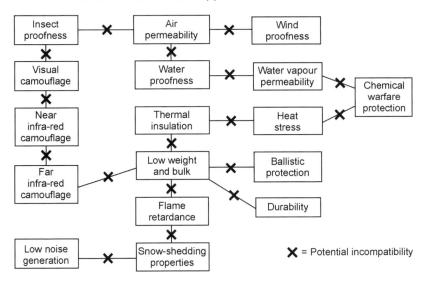

14.1 Incompatibilities in combat clothing system requirements.

match background vegetation when viewed through image intensifiers or low-light TV. Adding to this far IR to counter thermal imagers adds further complexity, as it requires thermal insulation or thermally reflective coatings. Arctic white camouflage needs to be worn as an additional layer over the standard woodland camouflage pattern. Since this chart was produced in the 1960s many significant improvements have been made by scientists and technologists to minimise these incompatibilities.

14.5 Biomedical aspects of protective combat clothing

Clothing comfort is dependent on many factors such as fit, flexibility, surface roughness, and dermatitic skin reactions, which are all associated with next-to-skin materials. Additionally, there are psychological aspects which are somewhat esoteric. The right military clothing can engender pride and a sense of identity in the wearer. Moreover, protective clothing can make the wearer feel safer and more willing to take risks (the macho image). A properly equipped force can instil a feeling of superiority over their adversaries.

Perhaps the most important factors are the thermo-physiological properties of textiles. The human body regulates its temperature by sweating during periods of high activity in hot climates. This sweat must leave the skin as a vapour in order to impart evaporative cooling through latent heat changes. Clothing interferes with vapour dissipation, leading to condensation of liquid on the skin and in the clothing layers. In cold climates this must be avoided at all costs, as sweat wetted clothing exhibits a loss of thermal insulation (see Fig. 14.4).

Humans regulate temperature in the cold by shivering and vaso-constriction of capillaries in the extremities (hands and feet). These effects are limited, and adding insulated clothing layers is necessary for comfort and survival. However, the utility range of insulated clothing is limited if activity levels vary significantly. Inactivity requires the addition of more layers, but walking, running, or carrying loads means that layers must be discarded.

It is imperative that thermo-physiological measurements can be made on materials and clothing assemblies. These tend to be carried out using standardised tests in laboratories. Wearer trials can then assess the subjective responses by the wearer to the clothing. Controlled human trials using instrumented subjects in stressful conditions are the most realistic measures of suitable performance, but these tend to be limited by high costs and ethical considerations these days. These limitations have accelerated the development of more realistic simulations of human interactions with their environment and activities. Instrumented manikins which are heated, which perspire and move are now used quite widely, although the costs of building them are prohibitive. There are thought to be about 100 worldwide (McCullough, 2005; Holmér, 2005; Rossi, 2005).

Laboratory measurements of water vapour permeability or resistance of textiles can be made using simple cup tests such as BS 7209, 1990. Laboratory measurements of thermal resistance can be made using a relatively simple Guarded Hotplate apparatus (Togmeter) to ISO 5085, 1996. The Tog is the ISO unit of thermal resistance, such that: 1 Tog = m^2 K/10 Watts. A more complex method, ISO 11092, 1993 utilises a sweating guarded hotplate apparatus (a skin model). This can measure water vapour resistance, the buffering capacity of underwear to sweat evolution, and the thermal resistance of textile materials, or complete clothing assemblies (see details of the use of these tests in later sections of this chapter).

14.6 Underwear materials

Next-to-skin clothing is primarily a hygiene layer. The thermal properties tend to be subsidiary to the tactile properties, and the way that the material handles moisture (mainly perspiration) in order to remove it from the skin. The tactile properties are associated with fit, flexibility, drape, handle, roughness, and dermatitic skin reactions (Goldman, 1988).

A significant proportion of the population has a true allergic reaction to certain fibres or treatments, particularly scaly wool fibres. Some military applications such as aircrew, naval action clothing, and armoured fighting vehicle (AFV) crews need underwear which is made from non-thermoplastic synthetic fibres to minimise contact melt/burn injuries.

The ability of underwear to remove or dissipate perspiration from the skin is critical for highly active dismounted land forces such as infantry, marines and

Table 14.3 Sweat buffering capacity of a range of knitted underwear materials

Underwear fabric description	Buffering index Kf	Ranking
100% cotton 1 × 1 rib (olive)	0.644	5=
100% hollow polyester 1 × 1 rib (olive)	0.641	5=
100% quadrilobal polyester 1 × 1 rib (olive)	0.720	4
70% hollow polyester/30% cotton 2-sided rib	0.731	3
67% hollow polyester/33% cotton double jersey	0.765	1=
64% hollow polyester/36% cotton double jersey	0.764	1=
72% quadrilobal polyester/28% cotton 2-sided rib	0.645	5=
63% quadrilobal polyester/37% cotton double jersey	0.635	8

special forces, as mentioned earlier. Their activities range widely from rapid movement on foot or skis carrying heavy loads, to immobilisation for long periods when lying in defensive positions, or on covert reconnaissance duties.

The capacity of underwear materials to handle pulses of sweat from the body (the buffering capacity) can be measured using the sweating guarded hotplate apparatus to ISO 11092, mentioned above. This test simulates the condition where the undergarment is lying on the wearer's wet skin. The passage of water through the material is measured at intervals. This gives an indication of the amount of water that has been evaporated from the skin to the environment, and also that which has been absorbed by the material. The buffering capacity (Kf) thus calculated has values between 0 (no water transported) to 1.0 (all water transported). Values above 0.7 are indicative of 'good' wicking performance. Table 14.3 shows the results for a wide range of experimental knitted underwear materials (Hobart and Harrow, 1994).

Overall, the results spread over a small range of Kf values between 0.635 and 0.765. The 100% cotton rib fabric was formerly the UK in-service cold weather underwear which ranked 5th along with specialist polyester and polyester/cotton wicking materials. The specialist quadrilobal polyester fabric (Coolmax®) was ranked 4th, 5th and 8th. The best performers were based upon hollow polyesters blended with cotton in a double jersey construction. It can be surmised that the fabric structure has a large influence on the wicking efficiency as well as the nature of the fibres.

14.7 Thermal insulation materials

Thermal insulation is an intrinsic property of combat clothing layers which provides an advantage in cold climates, but is a distinct disadvantage in hot climates during high activity operations. Textile fibres and structures form very light, resilient, easy-care, durable, and efficient insulators, although the fibres themselves merely act as a medium with a large surface area for trapping still air. Air is trapped on the surfaces of fibres, and in the interstices between them.

Table 14.4 Special synthetic fibres for thermal insulation

Polymer type	Trade names	Description
Polyester	Numerous	Solid fibres used in knitted and brushed fabrics for fleece garments, often made from recycled bottles.
Polyester	Coolmax® (Du Pont)	Quadrilobal cross-section to wick moisture. Used in sports underwear.
Polyester	Thermastat® (Du Pont)	Hollow tubular fibre, used initially for cold weather underwear.
Polyester	Hollofil® (Du Pont)	Single hole hollow tubular fibre. 6 dtex version used in sleeping bags and duvets.
Polyester	Quallofil® (Du Pont)	Hollow fibres with 4- or 7- hole versions. For sleeping bags and duvets.
Polyolefin (PE)	Thinsulate 'M'® (3M)	Microfibre, melt blown spun, dense felts for lining gloves, boots, etc.
Polyolefin (PE)	Thinsulate 'CS'®	Mixtures of microfibre and conventional solid fibres for clothing.
Polyacrylate	Inidex®	Special flame retardant filling.
Acrylic	Numerous	Solid fibre, soft handle, for pile fabrics.

The thermal resistance of still air is about 25–30 times greater than that of fibrous polymers. An efficient insulation medium typically comprises about 5–20% of fibres and 80–95% air (Cooper, 1979).

There are secondary effects on insulation efficiency that are influenced by the diameter of the fibres. Large numbers of fine fibres have a high specific surface area, thus trapping more still air. These fine fibres tend to form dense, thin, felt-like battings that resist compression well, so are ideal for insulating footwear and handwear. Thinsulate® is such a material, based upon melt blown polyolefin fibres, including microfibres. Conversely, battings made from large diameter hollow tubular fibres have excellent 'loft' or resilience, enabling them to recover from compression and maintain thickness. This makes them ideal for sleeping bags, duvets and outer wear. Table 14.4 lists special synthetic fibres used for modern thermal insulation.

Synthetic fibres have certain practical advantages over natural fillings such as down, feathers and kapok. They are rot resistant, hygienic, easy care, dry rapidly after wetting or laundering, resistant to compression damage, and are cheaper than down and feathers. They have in many cases superseded natural fillings for arduous heavy duty military use.

14.7.1 Measurements of thermal efficiency of textiles

Thermal resistance measurements using standard equipment such as the Togmeter to ISO 5085 give limited information. Military forces need insulation which is lightweight and compact, so the efficiency of insulative properties is much more important. Thus, we tend to measure warmth/thickness and warmth/

314   Textiles for cold weather apparel

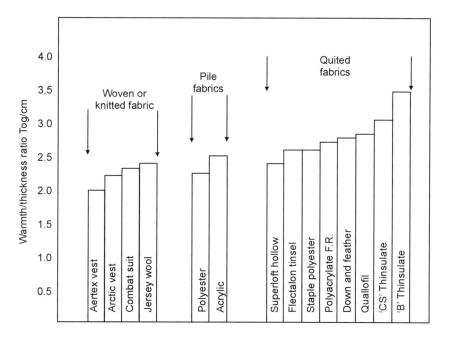

14.2 Warmth/thickness ratios for textiles.

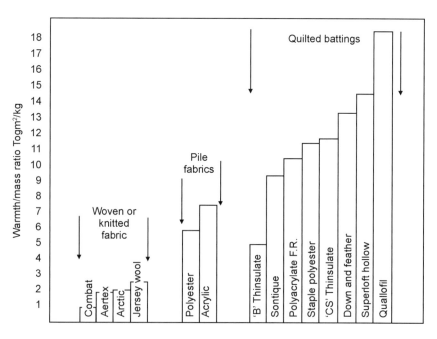

14.3 Warmth/mass ratios for textiles.

mass ratios. Figures 14.2 and 14.3 show measurements for a wide range of textiles for clothing, from woven and knitted fabrics through pile fabrics to fibrous battings and quilts (Scott, 2000).

It is clear from Fig. 14.2 that the warmth/thickness ratios only vary between about 2.0 and 3.5 Togs/cm. The two Thinsulate® microfibre types exhibit the highest values, as expected from the previous discussion. We can conclude that a certain thickness of any type of textile – whether woven, knitted, pile or quilt – will exhibit a narrow range of thermal insulation values.

By contrast, Fig. 14.3 compares the same materials on a warmth to mass basis. Here the differences are significant, ranging from about 1–3 Tog m^2/kg for woven and knitted fabrics up to a maximum of 18 Tog m^2/kg for quilted battings. Pile fabrics exhibit intermediate values between 6 and 7 Tog m^2/kg . The conclusion is that quilts and non-woven battings are the most efficient insulators. Performance specifications for insulation media should specify both a maximum mass and a minimum thickness in order to achieve optimum efficiency of the fibre filling. Table 14.5 lists thermal insulation materials for cold weather clothing and sleeping bags in UK military service.

14.7.2 The effects of moisture on thermal insulation

Protecting the individual against cold/wet climates is the most challenging problem, as it is essential to keep all thermal insulation layers dry. The thermal

Table 14.5 The range of thermal insulation materials in UK military service

Item	Description	Specification
Materials for sleeping bags	A range of polyester fillings based on hollow 4- and 7-hole fibres	UK/SC/5226
Warm weather sleeping bag	One × 100g/m^2 batting using 80% 4-hole and 20% 7-hole polyester	UK/SC 5492
Temperate sleeping bag	One × 150 g/m^2 batting using 100% 4-hole polyester	UK/SC/5609
Arctic sleeping bag	Two × 100 g/m^2 battings using 80% 4-hole and 20% 7-hole polyester	UK/SC/5610
Sleeping bag fleet (Royal Navy)	Zirpro® treated F.R. wool filling with Proban® F.R. cotton cover	UK/SC/4594
Boots, combat, cold/wet weather (lining)	Thinsulate® microfibre polyolefin	—
Shirt, cold weather (Norwegian)	Cloth, cotton, knitted, plush terry loop pile, 300 g/m^2	UK/SC/5283
Jacket, fleece pile	Cloth, knitted, polyester fleece, double faced, green	UK/SC/5412
Jacket and trousers thermal	Mixture of polyester hollow and micro-fibre battings, 200 g/m^2, 20 mm thick	UK/SC/5919

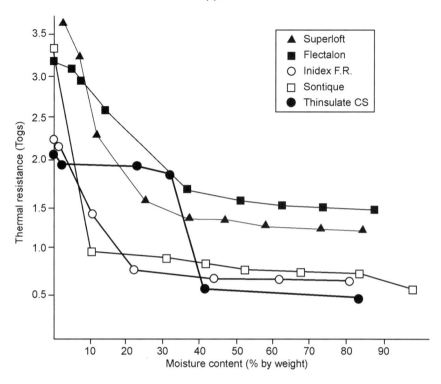

14.4 Loss of thermal resistance in wet fibrous insulation.

resistance of wet fibrous insulation media is significantly reduced, due to the replacement of low conductivity still air with high conductivity water in the structure. Fibrous battings also tend to soak up large amounts of water by capillary action. Some claim that synthetic fibre fillings are not deleteriously affected by the presence of moisture, i.e., do not lose thermal insulation when wet. However, Fig. 14.4 shows the effect of increasing amounts of water on a range of synthetic fibre fillings. In most cases these battings can lose between 40 and 75% of their dry insulation when they contain 50% by weight of moisture.

14.8 Waterproof/water vapour permeable materials

These are known as WWWW materials, as they offer the multiple functional properties of Waterproofness, Water vapour permeability, Water repellency, and Windproofness. They attempt to solve the perennial problems associated with keeping humans dry and comfortable, whilst allowing perspiration vapour to dissipate freely. Traditional waterproof barrier materials were based upon PVC, rubber, or polyurethane coatings or films on textile substrates, which were usually based upon woven continuous filament nylon or polyester fabrics. They were waterproof but did not allow sweat vapour to escape, leading to discomfort or severe illness in extreme cases. There are three main types of WWWW materials in current use.

14.8.1 Types of WWWW materials

High density woven fabrics

These are made from fine fibre yarns woven with a very tight sett, and treated with a water-repellent finish. Ventile® was the original example, developed by the Shirley Institute (now British Textile Technology Group) in Manchester, UK during World War II for RAF pilots who were forced to 'ditch' or crash-land in the cold North Sea. Ventile® is made from fine Sea Island cotton fibres in a low twist yarn which is woven tightly. This was then originally treated with a Velan® water-repellent stearamido derivative. It is still used by RAF aircrews, polar explorers and other military nations (Higginbottam, 1996).

Modern analogues are mainly of Japanese origin, based upon woven microfibre polyester and nylon multifilament yarns with fluorocarbon finishes. Trade names in recent years have included Teijin Ellettes® and Unitika Gymstar® (Scott, 1995).

Microporous coatings and membranes

These are now widely available in many variants. These membranes are typified by having microporous voids of effective sizes between 0.1 and 5 micron. The most well known example of these is Gore-Tex® which utilises a shock-expanded, biaxially oriented polytetrafluoroethylene (PTFE) film. The membranes look fibrillar under the microscope, possessing an extremely high void volume, and hence high vapour permeability. PTFE is also very inert, chemically resistant, and extremely hydrophobic, so that it repels liquid water strongly. Gore-Tex hybrid membranes for apparel usually incorporate an inner continuous coating of a hydrophilic polymer to resist contamination of the pores by sweat residues, and to resist penetration by low surface tension liquids such as organic solvents. The PTFE membranes are laminated by discontinuous spots of adhesive to a range of polyester or nylon fabrics for rainwear.

Other widely available products are based upon polyurethane coatings or films. The pores can be produced by different methods: (i) by incorporating salt crystals in the liquid polymer film which are subsequently leached out by aqueous treatments, leaving micro voids or (ii) preparing polyurethane films in an organic solvent which is miscible with water. Whilst still wet water vapour or steam is passed through the film, causing phase separation of the polymer. This widely used process produces microporous voids or blow-holes. The size of these pores can vary between 0.1 and 10 micron (Scott, 1995).

Examples of commercial products in recent years have included Cyclone®, Triple Point®, Drilite®, Aquatex®, and Porelle®. These polymers can be coated directly or laminated to filament woven fabrics. Three layer laminates are manufactured by bonding lightweight (say 50 g/m²) knitted or non-woven fabric to the back (the inside surface of a garment) to protect the film and improve durability.

Hydrophilic solid coatings and films

These are pore-free films. As such they possess the potential for a high resistance to liquid penetration. However, water vapour permeability rates through these films are inversely proportional to their thickness (Lomax, 1990).

Diffusion of water vapour through a solid film is achieved by the incorporation of hydrophilic functional groups along the molecular chains of the polymer. Such groups as –O–, –CO–, –OH, or –NH$_2$– are used which form reversible hydrogen bonds with water. Water vapour molecules diffusing through the film pass stepwise along the chains, a process which is facilitated by introducing pendant side groups to prevent close-packing of adjacent chains (Lomax, 1990). Block copolymers of polyurethane and polyethylene glycol (PEG) have been widely used. The original research on this type of polymer was funded by the UK Ministry of Defence, SCRDE, at Colchester, Essex in the 1970s. It has been made widely available under a number of different trade names, including the original Witcoflex Staycool®, Ventflex®, Keelatex®, and Isotex®.

Another type of hydrophilic copolymer is based upon a modified polyester, onto which polyether groups have been incorporated to impart hydrophilic properties with a limited degree of swelling. Commercial products such as Sympatex® are available as pre-prepared 10–25 micron films, which can be laminated to suitable filament fabrics as two-layer, three-layer, or lightweight drop liner forms (Drinkmann, 1992).

14.8.2 Relative performances of vapour permeable barrier fabrics

This class of materials are now widely available worldwide, and their performance is continuously improving (Weder, 1997). It is not possible to give detailed specific performance data for all these products, but they can be compared in general representative groups using the star rating system shown in Table 14.6.

14.8.3 WWWW materials in current UK military usage

These are listed in Table 14.7. Performance specifications annotated with 'PS' are those where the type of membrane or coating is not detailed as a commercial product. Any material which meets the strict performance requirements will be accepted.

Table 14.6 Relative performance of WWWW materials

Type of barrier fabric	Water vapour permeability	Water-proofness	Cost	Comments
Ventile type woven fabrics	*****	*	Medium to high	Original material for RAF immersion suits. Modern synthetic versions available
PTFE hybrid laminates	*****	*****	Very high	Market leader for highest performance
PTFE laminates	*****	*** to *****	Med/high	
Microporous polyurethanes	*** to *****	*** to ****	Medium	Widely available, reasonable durability
Hydrophilic polyurethanes	** to ***	*** to *****	Relatively low to high	Widely available in many forms
Hydrophilic polyesters				
Traditional impermeable	Not significant	*** to *****	Relatively cheap	Waterproof but uncomfortable

Key: * = poor ***** = excellent

Table 14.7 WWWW materials in current UK military usage

Material description	Specification number
Cloth, laminated, moisture vapour permeable, olive, No. 1 for arctic mittens	DC/S&TD/PS04-95 UK/SC/4778
Cloth, laminated, nylon/PTFE/nylon, waterproof, moisture vapour permeable, No. 2 for gaiters, snow, GS	UK/SC/5535
Cloth, laminated, nylon/PTFE/nylon, waterproof, moisture vapour permeable, No. 3 for foul weather clothing	UK/SC/4978
Cloth, laminated, waterproof, moisture vapour permeable, navy, olive, disruptive pattern, No. 4 for foul weather clothing	DC/S&TD/PS 1063
Cloth, laminated, waterproof, moisture vapour permeable, No. 5	DC/S&TD/PS 1016
Cloth, laminated, waterproof, moisture vapour permeable, temperate DP, Near IRR, for foul weather suits Army, Royal Marines, Royal Navy, RAF, Special Forces	UK/SC/5444
Cloth, coated, microporous polyurethane on texturised polyester, cedar green, for aerial erector's suit	UK/SC/5070
Cloth, laminated, waterproof, moisture vapour permeable for sock liner, extreme cold weather	S&TD/PS 04-96
Cloth, laminated, nylon PTFE/polyester, for one-man tent	UK/SC/4960
Cloth, ventile cotton, water-repellent, olive, for coverall immersion, aircrew, and smock, swimmer canoeist	UK/SC/4940

14.9 Materials for current UK combat clothing systems

At the time of writing the UK Combat Clothing System is based on Combat Soldier 95. There have been modifications to some of the materials and clothing items since its introduction in 1995, particularly the combat suit and thermal suit. The main items are listed in Table 14.8.

14.9.1 Thermal and water vapour resistance data for UK combat clothing systems

The thermal resistance (R_{ct}) and water vapour resistance (R_{et}) values for the current Combat Soldier system were measured using a guarded sweating hotplate apparatus to ISO 11092 (The Hohenstein Skin Model). The values for each layer are additive, although the true total is somewhat higher, due to air-gaps between the layers, (Congalton, 1997). From these figures the water vapour permeability index (imt) can be calculated, such that: $i_{mt} = S \times R_{ct}/R_{et}$, where $S = 60\,\text{PaW}^{-1}$. The i_{mt} has values from 0 (totally impermeable) up to 1.0 (totally permeable). Table 14.9 shows the results for combat clothing layers. Note that even a fleece fabric which is not wind proof, possesses an appreciable vapour resistance (R_{et}),

Table 14.8 UK combat clothing system materials

Clothing item	Material description	Specification number
1. Underwear	100% knitted cotton 1 × 1 rib, olive or 100% quadrilobal polyester (Coolmax®)	UK/SC/4919 UK/SC/—
2. Norwegian shirt	Knitted cotton, plush terry loop pile, olive	UK/SC/5282
3. Combat trousers and shirt	67/33% polyester/cotton, 2 × 1 twill	UK/SC/5843
4. Field jacket, windproof	100% cotton gaberdine, 2 × 2 twill Woodland DP, near IRR	UK/SC/5878
5. Fleece pile jacket	100% polyester, knitted fleece double faced, olive	UK/SC/5412
6. Waterproof rainsuit	3 layer laminate, nylon/membrane/ nylon waterproof, water vapour permeable	UK/SC/5444
7. Jacket and trousers, thermal	Woodland DP, near IRR Mixture of polyester hollow and microfibre	UK/SC/5919
8. Oversuit, white, snow	Batting, 200 g/m², 20 mm thick. Plain woven filament nylon with UV camouflage finish	UK/SC/3811

Table 14.9 Thermal and water vapour resistance of UK combat clothing materials

Textile layer	R_{ct} (m²K/W)	R_{et} (m²Pa/W)	i_{mt}
Cotton underwear	0.03	5.1	0.3
Norwegian shirt	0.05	8.6	0.3
Polyester fleece	0.13	13.4	0.6
Lightweight combat suit	0.01	4.3	0.2
Windproof field jacket	0.005	4.8	0.1
Waterproof/vapour permeable rainsuit	0.003	11.2	0.01

but a high i_{mt}. Also note that the rain suit exhibits a relatively high R_{et}, although this is classed as having a good performance in terms of vapour permeability. Compact woven fabrics and laminates exhibit very low intrinsic thermal resistance.

14.9.2 Water vapour resistance of footwear

Leather military footwear for cold climates can be fitted with a breathable liner or 'sock'. Its main purpose is to improve the waterproofness of the leather boots. Measurements of the R_{et} of boots and liners have indicated that the determining factor is the boot, which has a high vapour resistance compared with the liner (Scott, 2000). Leather boots tend to absorb moisture during a typical day's wear rather than dissipating it. Removing the boots overnight allows them to dry out.

14.10 Military hand- and footwear for cold climates

Keeping the body extremities warm is a significant problem, as vasoconstriction in cold climates tends to divert blood away from the hands and feet, to keep the vital organs warm. This can lead to frost nip or frost bite, or to loss of fingers and toes, and complete limbs in the worst cases. Building up thin layers of insulation on hands and feet is the recognised solution.

14.10.1 Cold weather handwear

Keeping the hands warm without encumbrance is a problem. In mild winter conditions a cape leather five-fingered glove is adequate. For mechanics and electricians fingerless knitted gloves are used, so that a high level of manual dexterity is maintained.

At much lower arctic temperatures and during inactive periods it is necessary to wear mittens – inner ones made from thin insulators, and an outer made from a thin WWWW material to keep the insulation dry.

At very low temperatures bare skin can stick to cold metal objects. If manual dexterity is not to be impaired, a thin aramid heat resistant contact glove is used.

Table 14.10 UK military cold weather handwear and footwear

Description	Specification
Handwear	
Gloves, Combat Mark 2 (leather)	UK/SC/5405
Gloves Fingerless, Technicians and Mechanics (leather palm)	UK/SC/5067
Gloves, Contact, Combat (Aramid FR)	UK/SC/5406
Gloves, Extreme Cold Weather (Task Gloves)	UK/SC/5547
Mittens, Outer, Extreme Cold Weather	UK/SC/5483
Mittens, Inner, Extreme Cold Weather	UK/SC/5484
Gloves, Electrically-Heated, for Seacat Missile Operators	UK/SC/4118
Gloves, Electrically-Heated, for Rapier Missile Operators	UK/SC/4751
Footwear	
Boots, Combat, Assault (leather)	UK/SC/5259
Boots, Ski March (leather)	—
Boots, Tent, Extreme Cold Weather	UK/SC/4643
Gaiters Snow Mk 2 (WWWW)	UK/SC/4736

Heat resistance is required if handling hot utensils on a gas stove or fire (see Table 14.10).

14.10.2 Electrically heated gloves

Encumbering the hands with layers of insulation leads to loss of manual dexterity, which means that delicate manual operations such as firing weapons, operating keyboards, or writing become impossible. Some ground-to-air (Rapier) or sea-to-air (Sea-Cat) missiles require skilled operators to acquire targets, launch the missile, and then guide it to its target by joystick. The operators may have to sit out in the open in all weathers. To cope with this, electrically heated gloves have been developed by UK MOD. These are of five-fingered knitted construction. Fine stainless steel element wires (300 strands of 12 micron wire) which are electrically insulated with PTFE are knitted in with the nylon yarns. These thin, flexible gloves run from a low voltage (24–28 volt) power supply to prevent contact short-circuiting (see Table 14.10).

14.10.3 Cold weather footwear

Thermal insulation is built up in layers, starting with a wool/nylon sock, a WWWW breathable type liner or 'sock', and an outer leather boot. The outer boot sole is fitted with cleats to fit ski bindings, and has a tread pattern to minimise slipping on ice and snow. To keep the feet warm during the night tent boots are provided, which are made from polyester battings enclosed by staple filament nylon fabric.

To protect the lower limbs from cold wet snow a gaiter is worn over the trousers. In the arctic this is made from a WWWW material, with a heavy texturised nylon outer fabric. There also exists a wet weather gaiter which is made from heavy duty 'thornproof' filament nylon (see Table 14.10).

14.11 Research and development of future materials

14.11.1 Improved thermal insulation

Temperature control in active humans in a cold climate currently involves carrying or wearing several layers of insulated clothing. If the activity level or the temperature changes the individual has to stop to add or remove layers to maintain comfort and safety. This is a particular problem for infantry soldiers, marines and special forces, as they have to carry everything they need to survive and fight. There is a great tactical advantage in having one garment whose thermal properties can be adjusted at will.

Variable pile fabrics

This tactical requirement led to the invention of variable insulation media, mimicking the action of fur and feather-bearing animals, who can adjust the thickness of their coats and plumage according to climatic changes. The author took out a patent on 'Variable Insulation Pile Fabric' (GB 2,234,705B) in February 1993 on behalf of the Secretary of State for Defence, UK. The basic principle uses 3-D textile structures or pile fabrics which can be erected or collapsed to effect a change in thermal resistance. At the time of writing several structures have been examined.

The first of these utilised modified weaving machines is normally used to produce velvet plush pile fabrics. The finished material consisted of two parallel cotton fabrics linked by long threads, the latter in a mixture of 100% cotton or monofilament polyester yarns. The monofilaments provide resilience and support to keep the two cotton fabrics apart in the relaxed state. This structure can be collapsed manually by the wearer, using zips and separators in a parallelogram action. The thermal performances of some of these Resilitex® structures are shown in Table 14.11. Measurements were made using a Tog-meter to ISO 5085 on loom-state fabrics. The samples were compressed using the top plate of the guarded hotplate apparatus (Scott, 2003).

The results show appreciable changes in thermal resistance between the erect and compressed state. This change is augmented in a practical gilet type of garment, since the action of pulling the fabrics down to collapse the pile also removes any air-gaps between other inner layers of clothing. These structures are experimental, and further work to optimise the performance is required. At the time of writing the French firm, Tissavel, can produce such 3-D spacer drop-stitch fabrics in thicknesses from 24 to 45 mm (Tissavel SA Ltd, 2001).

324   Textiles for cold weather apparel

Table 14.11 Thermal resistance data for experimental woven variable pile fabrics

Sample	Mass g/m²	Thickness (mm)	Thermal resistance (Togs)	Change in thermal resistance*	Warmth/ weight (Tog m²/kg)
Resilitex S 25025					
Erect	842	14.57	1.97	×3.90	2.34
Compressed	842	4.4	0.505		0.60
Resilitex S 25027					
Erect	895.5	16.73	2.10	×4.77	2.33
Compressed	895.5	4.0	0.44		0.48

* = Ratio of erect/collapsed resistance values.

Another set of experiments utilised 3-D warp-knitted fabric structures on a modified Karl Meyer double-needle-bar RD 7 machine. The connecting pile yarns were a mixture of polyester monofilaments to provide compression resistance, and polyester/cotton yarns to provide air-trapping insulation. These fabrics were collapsed in their natural resting state, and required to be pulled up to the erect state. Test results showed that this form was less efficient than the Resilitex® open 3-D structures. Liba also produce rib raschel machines for the production of spacer fabrics, although the range of thicknesses is limited (Bohm, 2002).

Inflatable insulated garments

Inflatable clothing is an obvious approach to providing adjustable thickness in a clothing system – the thicker the space the higher the insulation value, up to a limit where the air is in convective motion. The main problem is associated with the water vapour impermeability of most air-holding structures, which results in the trapping of perspiration in the inner clothing.

An early approach to solving this problem was invented by UK MOD scientists (Elton, 2001). The idea was patented by (Forshaw, 1999) in GB 2,242,609B in 1999. This involved the use of inflatable rubber tube lattices or grids, with air spaces between the fine tubes. The tubes could vary in diameter from 3–5 mm up to 10–15 mm.

Another approach by W L Gore & Associates was patented in about 1996, although no details are available (Gore Enterprises, 1999). This used the air-tight properties of second-generation Gore-Tex® laminates to form a simple waistcoat, sealed around the edges and inflated by mouth.

Shape-memory alloys and polymers

Shape-memory materials return to a pre-determined form above a given transition temperature. They can, therefore, be used to change shape and produce air gaps automatically when the ambient temperature changes. Research work has been carried out both on Nitinol, a nickel-titanium alloy in wire form, and polyurethane block copolymers in thin film form (Russell, 2000). The work concentrated on the rapid production of an insulating air-gap in the high temperatures associated with a fire, in order to protect humans from skin burns. However, work on suitable disc-shaped films of block copolymer urethanes could be used to form air gaps when the temperature dropped markedly, providing increased thermal insulation automatically (Russell and Congalton, 2001).

Phase-change materials

Material additives to buffer temperature changes in clothing materials have been developed since the 1970s. For textile applications they consist of micro-spheres containing a range of homologous series paraffin waxes with differing crystalline temperatures. When heated they take in latent heat of melting, and conversely, release latent heat on cooling and re-solidification. The micro-spheres can be applied as a coating to textiles, or dispersed within the fabric. The most well-known commercial products for clothing are made by Gateway Technologies of Boulder, Colorado, under the trade name Outlast®. An account of their construction and performance has been published by Pause (1995). These have been examined by UK MOD, US Department of Defense, and other researchers. Overall, the results indicate that the temperature buffering effects were quite small and transient (Leitch and Tassinari, 2000; Shim and McCullough, 2000).

14.11.2 Smart pore fabrics

There are cold/dry climatic conditions where outer clothing needs to be wind resistant when the wearer is inactive, but allows sweat vapour to escape when the wearer is active. Adaptive ventilation has been investigated by the Bio-mimetics Dept at Reading University in association with UK MOD (Jeronimidis, 2000). Coated fabrics were produced to mimic the action of pine cones, which open and close in response to humidity changes. The coating was a hydrophilic PU on a light knitted fabric, with many small U-shaped holes or pores punched in the fabric. The 'valves' or pores obtained in this way open up when the internal humidity is high, due to hygral expansion of the coating. They close when the humidity decreases, providing a windproof cover. The principle was demonstrated successfully, but there were practical problems in producing a high enough density of small pores in the material.

14.11.3 Electro-spinning technologies

Electro-spinning is a process by which a suspended droplet of polymer solution or melt is charged to a voltage high enough to overcome surface tension forces. Fine jets of liquid are fired towards a grounded target, forming continuous multifilaments of the polymer. The fibres are extremely fine and may form ultra thin layers of interconnected webs of fibres. Electro-spun nonwoven mats may be thought of as microporous membranes. It is possible to spin fibres directly onto 3-D screen forms, or even thin protective skins for the human body. Selection of suitable polymers opens up diverse possibilities for future WWWW materials or very thin insulation media (Gibson et al., 1999).

14.11.4 Electrotextiles, sensors, actuators, and nanotechnology

An appreciable amount of effort is being expended in the field of smart interactive textiles for military and civilian use. The aims are widespread, including electro-optical materials for sensing life functions of soldiers remotely, sensing climatic or battlefield conditions in order to actuate adjustable protection, wearable electronics, transmission of power and signals in clothing, self-heated clothing, and many other possibilities, (Tassinari and Leitch, 2004).

When fullerene carbon nanotubes are spun in a polymer/nanotube composite onto a thin fabric support, the nanocomposites exhibit piezo-electric properties for sensing strain. These carbon nanotubes have very high electrical conductivity – up to 35 times that of conventional piezo-electric devices. This direct and reverse conversion of mechanical to electrical energy can create a platform for developing the next generation of smart textiles for membrane structures, distributed shape modulation, and energy harvesting (Laxminarayana and Jalili, 2005).

14.12 References

Bohm C (2002) 'Rib Raschel for Spacer Fabrics'. *Knitting International*, February, pp. 1–37.

Brown M (2004) 'Integrated Soldier Technology'. *Proc. International Soldier Systems Conference* (ISSC 2004). Boston, MA, 13–16 December.

BS 7209 (1990) 'British Standard Specification for Water Vapour Permeable Fabrics'. *Appendix B. Determination of water vapour permeability index*. BSI, London.

Chappel M (1987) 'The British Soldier in the 20th Century'. *Part 5, Battledress 1939–1960*. Wessex Publishing, Devon, pp. 8–16.

Congalton D (1997) 'Thermal and Water Vapour Resistance of Combat Clothing'. *Proc. International Soldier Systems Conference* (ISSC 97). Ed. R A Scott, Colchester, Essex, October, pp. 389–404.

Cooper C (1979) 'Textiles as Protection against Extreme Winter Weather'. *Textiles*, Vol. 8, No. 3. Shirley Institute, Manchester, pp. 72–83.

Darman P (2007) *World War II – A day-by-day History*. Sandcastle Books, Worcs., UK, pp. 143–144.

Defence Standard 00-35 issue 2 (1996) *Environmental Handbook for Defence Materials*. Chapter 1-01, Table 2, p. 5. Def. Stan. Ops. Glasgow, UK.

Drinkmann M (1992) 'Structure and Processing of Sympatex® Laminates'. *J. Coated Fabrics*, Vol. 21, pp. 199–211.

Elton S F (2001) 'Adaptable Materials for Smart Clothing'. *Conference on Textiles for the Future*. Tampere University, Finland, 16–17 August 2001.

Forshaw P (1999) 'Inflatable Grids for Adjustable Insulated Clothing'. Patent GB 2,242,609B.

Gibson P, Shreuder-Gibson H, Pentheny C (1999) 'Electro-spinning Technology. Direct Application of Tailorable Ultra-thin Membranes'. *J. Coated Fabrics*, Vol. 28, pp. 63–72.

Goldman R F (1988) 'Biomedical aspects of Underwear'. *Handbook on Clothing*, ed. L Vangaard. Chapter 10. NATO Research Study Group 7. Natick, Mass. USA.

Gore Enterprises (1999) 'Puffed up Protection'. *New Scientist*, 14 August, p. 7.

Higginbottam C (1996) *Water Vapour Permeability of Aircrew Clothing*. MOD, DERA. Report PLSD/CHS 5/TR 96/089, December.

Hobart A-M, Harrow S (1994) 'Comparison of Wicking Underwear Materials'. Defence Clothing & Textiles Agency, unpublished internal report, Colchester, Essex.

Holmér I (2005) 'Textiles for Protection against the Cold'. Chapter 14 in *Textiles for Protection*, ed. R A Scott. Woodhead Publishing, Cambridge, pp. 384–386.

Holmes G T (1965) *Combat Clothing and Textiles*. 8th Commonwealth Conference on Clothing & General Stores. Ottawa, Canada.

ISO 11092 (1993) *Textiles – Determination of Physiological Properties. Measurement of Thermal and Water Vapour Resistance under Steady State Conditions*. ISO, Geneva.

ISO 5085 (1996) *Determination of Thermal Resistance using a Guarded Hotplate Apparatus*. ISO, Geneva.

Jeronimidis G (2000) *Biomimetics: Lessons from Nature for Engineering*. 35th John Player Memorial Lecture. Inst. Mechanical Engineers, London.

Laxminarayana K, Jalili N (2005) 'Functional Nanotube based Textiles – the Next Generation with enhanced Sensing capabilities'. *Text. Res. J.*, Vol. 75, No. 9, pp. 670–680.

Leitch P, Tassinari T (2000) Interactive Textiles – New Materials for the New Millennium Part 1. *J. Industrial Textiles*, Vol. 29, No. 3, pp. 173–190.

Lomax G R (1990) 'Hydrophilic Polyurethane Coatings'. *J. Coated Fabrics*, Vol. 20, pp. 88–107.

McCullough E A (2005) 'Evaluation of Protective Clothing Systems using Manikins. Chapter 9 in *Textiles for Protection*, ed. R A Scott. Woodhead Publishing, Cambridge.

Pause B (1995) 'Development of Heat and Cold Insulating Membrane Structures with Phase Change Materials'. *J. Coated Fabrics*, Vol. 25, July, pp. 59–68.

Rossi R (2005) 'Interactions between Protection and Thermal Comfort'. Chapter 10 in *Textiles for Protection*, ed. R A Scott. Woodhead Publishing, Cambridge, pp. 245–246.

Russell D A (2000) 'First experiences with Shape Memory Materials in Functional Clothing'. *Int. Avantex Symposium*, Frankfurt, Germany, 27–29th November 2000.

Russell D A, Congalton D (2001) 'Shape Memory Materials for Variable Insulation'. *Int. Soldier Systems Conference* (ISSC 2001). Bath, Avon, UK, September.

Scott R A (1995) 'Coated and Laminated Fabrics'. Chapter 7 in *Chemistry of the Textile Industry*, ed. C M Carr. Blackie Academic & Professional, London, pp. 234–243.

Scott R A (2000) 'Textiles in Defence'. Chapter 16 in *The Technical Textiles Handbook*. eds. Horrocks A R, Anand S A. Woodhead Publishing, Cambridge, pp. 425–460.

Scott R A (2003) 'Variable Pile Insulation Fabrics'. Unpublished Internal Research Review. MOD S&TD Caversfield, UK, March.

Scott R A (2005) 'Textiles for Military Protection'. Chapter 21 in *Textiles for Protection*, ed. R A Scott. Woodhead Publishing, Cambridge, pp. 597–621.

Shim H, McCullough E A (2000) 'The effectiveness of PCMs in Outdoor Clothing'. *Proc. NOKOBETEF 6 Conference on Protective Clothing*. Stockholm, Sweden, 7–10 May, pp. 90–93.

Tassinari T, Leitch P (2004) 'Interactive Textiles for Soldier Systems'. *International Soldier Systems Conference*, Boston, MA, December.

Thompson L (1973) *The World at War*. Television Documentary Series. Producer J Isaacs. Thames Television and Imperial War Museum, London.

Tissavel SA Ltd (2001) 'Three Dimensional Fabrics'. *Technical Textiles International. International Newsletters*, p. 19.

Weder M (1997) 'Performance of Breathable Rainwear Materials with respect to Protection, Physiology, Durability and Ecology'. *J. Coated Fabrics*, Vol. 27, pp. 146–168.

Willmott H P, Cross R, Messenger C (2004) *World War II*. Dorling Kindersley, London, pp. 133–134.

15
Protective clothing for cold workplace environments

I. HOLMÉR, Lund University, Sweden

Abstract: Human survival and function in cold climates are strongly dependent on optimal and appropriate use and application of technical means to maintain the human body at or close to a normal heat balance. Clothing is a suitable means of controlling heat exchange with the environment. Clothing must be flexible and adjustable to care for the great variations in activity and climate. International standards are available that allow the assessment of cold stress in terms of a required clothing insulation (I_{REQ} – ISO 11079). Thermal insulation is the single, most important property of the cold weather clothing assembly. Methods are available for its determination using either thermal manikins (ISO 15831) or literature data bases (ISO 9920). Principles for design of cold protective clothing are based on the multi-layer principle. New fabrics, smart textiles and intelligent functions may provide added value.

Key words: clothing insulation, cold stress, wind protection, design, methods.

15.1 Introduction

In many parts of the world, in particular in Northern countries and in countries with a continental climate type, the air temperature during several months of the year is low and often well below zero. People working outdoors are then subject to cold stress during large parts of the work shift. Such work includes, for example, farming, forestry, fishing, building and construction work, but also security work, firefighting and search and rescue work. Outdoor cold work is difficult to control and much determined by weather conditions. Indoor cold work is found all over the world in cold stores and cold rooms, where frozen and fresh food is stored and produced. Indoor cold exposure can be well controlled in terms of constant air temperatures and low air velocities. The most obvious and important protective measure against cold exposure, outdoors as well as indoors is the appropriate choice of clothing (Hassi *et al.*, 2002; Holmér *et al.*, 1997; Holmér, 2002, 2005a,b).

15.2 Directives and standards

Personal protective equipment for workplaces are regulated in Europe by two Directives. Directive 89/656 (1989) contains minimum health and safety requirements for the use by workers of personal protective equipment at the workplace. Directive 89/686 (1989) contains regulations related to the manufacturing, testing and marketing of personal protective equipment. Several international standards have been developed for testing of clothing against cold and foul weather and classification of their protective performance.

- EN 342 (2004) Protective clothing – Ensembles for protection against cold.
- EN 343 (2004) Protective clothing – Protection against rain.
- EN 511 (2005) Protective gloves against cold.
- EN 14058 (2005) Protective clothing – Garments for protection against cool environments.
- EN 14360 (2005) Protective clothing against rain – Test method for ready made garments – Impact from above with high energy droplets.
- ISO EN 15831 (2003) Thermal manikin for measuring the resultant basic thermal insulation.
- ISO EN 9237 Textiles – Determination of permeability of fabrics to air.

The most important physical properties for determining a clothing ensemble is its protective performance in cold environments are

- thermal insulation
- water vapour resistance
- permeability to air
- permeability to water.

The methods for measurement of these properties are described elsewhere in this book. Thermal insulation is measured on a complete ensemble with a standing or walking thermal manikin (EN 342, ISO 15831). The insulation value is presented in the marking of the clothing (EN 342). This insulation value in EN 342 is a resultant insulation value ($I_{cl,r}$), i.e., it is determined under the influence of wind and walking movements. The standard, basic insulation value (I_{cl}) is measured with a standing manikin in calm air. This is the value found in most literature (see Table 15.1). Normally, sweating is quite limited under cold conditions. At high activity levels, however, sweating and evaporative heat exchange may become important. The water vapour resistance value of the ensemble is then important for heat exchange. For an overall assessment of the cold protection, the permeability of the outer layer to air and water is important as well. Wind and water penetrating the outer layer into deeper layers may significantly reduce the insulation value. Table 15.1 provides basic insulation values for selected clothing ensembles.

Table 15.1 Basic insulation values (I_{cl}) of selected garment ensembles measured with a thermal manikin (modified from ISO-11079, 2007)

Clothing ensemble	I_{cl} m²°C/W	clo
1. Briefs, shirt, fitted trousers, socks, shoes	0.10	0.6
2. Briefs, shirt, coverall, socks, shoes	0.13	0.8
3. Briefs, shirt, trousers, smock, socks, shoes	0.14	0.9
4. Briefs, undershirt, shirt, overall, calf length socks, shoes	0.16	1.0
5. Briefs, shirt, trousers, vest, jacket, socks, shoes	0.17	1.1
6. Briefs, shirt, trousers, vest, coverall, socks, shoes	0.19	1.3
7. Underpants, undershirt, insulated trousers, insulated jacket, socks, shoes	0.22	1.4
8. Briefs, T-shirt, shirt, fitted trousers, insulated coverall, calf length socks, shoes	0.23	1.5
9. Underpants, undershirt, shirt, trousers, cardigan, coat, hat, gloves, socks, shoes	0.25	1.6
10. Underpants, undershirt, shirt, trousers, cardigan, coat, overtrousers, socks, shoes	0.29	1.9
11. Underpants, undershirt, shirt, trousers, cardigan, coat, overtrousers, socks, shoes, hat, gloves	0.31	2.0
12. Undershirt, underpants, insulated trousers, insulated jacket, overtrousers, overjacket, socks, shoes	0.34	2.2
13. Undershirt, underpants, insulated trousers, insulated jacket, overtrousers, coat, socks, shoes, hat, gloves	0.40	2.6
14. Undershirt, underpants, insulated trousers, insulated jacket, overtrousers, parka, socks, shoes, hat, gloves	0.46	3.0
15. Undershirt, underpants, insulated trousers, insulated jacket, down overtrousers, down parka, socks, shoes, hat, gloves	0.46	3.8
16. Polar clothing systems	0.46–0.70	3–4.5
17. Sleeping bags	0.46–1.4	3–9

15.3 Protection requirements

The protection offered by a clothing ensemble with a given insulation value is dependent on the climatic conditions and the activity level. Climatic conditions are defined by

- air temperature
- mean radiant temperature
- wind velocity
- humidity.

Air temperature and wind are the two most important factors. Mean radiant temperature in many cases acts to relieve cold stress as it is similar to or higher than air temperature. The latter occurs, for example, in the presence of solar radiation. At very low temperatures the air is dry and water vapour pressure close to zero. As evaporative heat exchange is mostly small (most of it is

Table 15.2 Examples of metabolic energy production associated with different types of work (modified from ISO-8996, 2004)

Class	Average metabolic rate W/m²	Examples
0 Resting	65	Resting
1 Low	100	Light manual work; hand and arm work; driving vehicle in normal conditions; casual walking (speed up to 3.5 km/h).
2 Moderate	165	Sustained hand and arm work; arm and leg work; arm and trunk work; walking at a speed of 3.5 km/h to 5.5 km/h.
3 High	230	Intense arm and trunk work; carrying heavy material; walking at a speed of 5.5 km/h to 7 km/h.
4 Very high	290	Very intense activity at fast pace; intense shovelling or digging; climbing stairs, ramp or ladder; running or walking at a speed greater than 7 km/h.
Very, very high (2 hr)	400	Sustained rescue work, wild land fire fighting. Endurance time less than 2 hours.
Intensive work (15 min)	475	Structural fire fighting and rescue work. Endurance time less than 15 minutes.
Exhaustive work (5 min)	600	Fire fighting and rescue work, climbing stairs, carrying persons. Endurance time less than 5 minutes.

actually humidified exhaled air), the humidity plays little role in heat exchange. From about 0 °C and up it becomes increasingly important.

The metabolic energy production is closely linked to the activity level (Table 15.2). In most activities all of this energy converts into heat. The large amounts of heat production during heavy work facilitates the maintenance of heat balance also under very cold conditions.

The clothing insulation required (I_{REQ}) for adequate protection in the cold can be determined with the I_{REQ}-index (ISO-11079, 2007). I_{REQ} is shown in Fig. 15.1 for combinations of air temperatures and activity levels. I_{REQ} is a resultant insulation value that clothing must provide under the actual conditions. If compared with I_{cl} values (Table 15.1), these values must be corrected for wind and body motion. Such correction factors can be found in ISO 11079. The

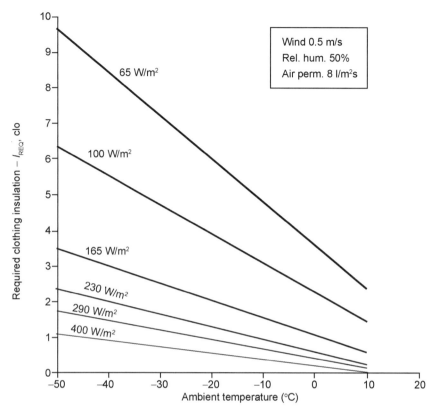

15.1 Required resultant clothing insulation, I_{REQ}, for combinations of activity and ambient temperature calculated according to ISO 11079.

standard also provides a link to a computer programme that makes the necessary calculations and corrections (URL). It is readily seen that I_{REQ} is high at low activity and increases sharply when it gets cold. During high activity very little clothing insulation is required.

If the user knows the insulation value, either as I_{cl} value from Table 15.1 (ISO-9920, 2007), or as I_{clr} value from EN 342, the conditions for heat balance can be assessed with ISO 11079. If I_{clr} is equal to or higher than I_{REQ}, clothing protects sufficiently. If I_{clr} is less than I_{REQ} the body will cool down and lose heat from tissues. Depending on the rate of change in tissue heat loss times can be calculated after which certain, defined body temperatures are reached. These times may represent recommended, longest exposure times. Figure 15.2(a) shows the recommended exposure time for light work with different insulation values of the clothing ensemble (I_{cl} value). With more insulation exposure times become longer for a given ambient temperature. Figure 15.2(b) shows the effect of wind (up to 12 m/s) on exposure times for light work. An I_{cl} value of 3.0 clo provides less protection in wind and exposure times become shorter for a given

334 Textiles for cold weather apparel

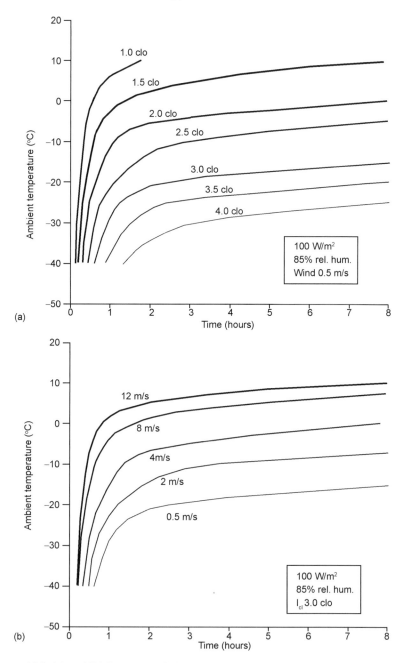

15.2 (a) and (b) Recommended exposure time for light strain during moderate work (100 W/m²) and for clothing ensembles with different basic insulation values (I_{cl}). Outer layer is assumed to be wind proof. Examples of clo values for clothing ensembles are given in Table 15.1. (b) shows the effect of wind on stay times for a 3 clo ensemble.

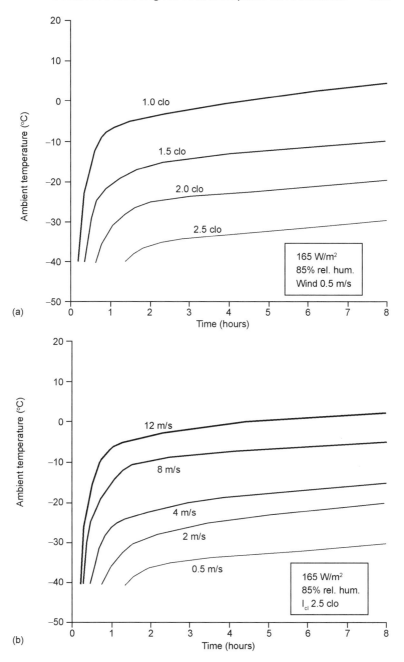

15.3 (a) and (b) Recommended exposure time for light strain during moderate work (165 W/m²) and for clothing ensembles with different basic insulation values (I_{cl}). Outer layer is assumed to be wind proof. Examples of clo values for clothing ensembles are given in Table 15.1. (b) shows the effect of wind on stay times for a 2.5 clo ensemble.

temperature. Figure 15.3(a)–(b) shows similar information for moderate work rate. The effect of wind is shown for a 2.5 clo ensemble. The air permeability of the outer layer for all examples is $8 \, l/s \, m^2$.

15.4 Clothing for cold protection

A few key words can summarize the principles for cold weather clothing

- still air
- dry clothing
- layering.

Protecting the body against heat losses in the cold is best done by creating layers of still air on top of the skin. These layers are formed by fibres, fabrics and garments and by spaces between them. Typically, a cold protective ensemble is built up by several layers of garments on top of each other. The layers, apart from contributing to the total insulation, serve specific purposes. Fabrics must be kept as dry as possible. Moisture in clothing impairs its insulative capacity; also clothing is difficult to dry, in particular in outdoor conditions.

15.4.1 The multi-layer principle

At least three distinct layers with different functions (apart from contributing to insulation) can be defined. The inner layer is worn directly on top of the skin and controls the microclimate temperature and humidity. With low activity the layer must reduce air movements. With high activity, heat and moisture should be transported from the layer to cool the skin. Moisture control can be performed by absorption, by transportation to next layers or by ventilation. Absorption reduces skin humidity and retains relative comfort, but moisture remains in the clothing system and may be detrimental to heat balance at a later stage. Using hydrophobic textiles next to skin quickly increases humidity and moisture is transferred to outer layers. The advantage is an increased awareness of discomfort, but moisture remains in the clothing system. The ventilation principle requires a vapour barrier worn next to skin. Microclimate humidity quickly rises with sweating, but water vapour cannot escape to outer layers. The humid microclimate forces the wearer to open up the clothing and ventilate the microclimate. This principle is preferably used in resting and low activity conditions.

The absorbing technique may be useful for low to moderate activities with limited sweating. The transporting principle may be applied to all kind of activities, but in particular to high activity with sweating. They are most suitable for sports events and long term exposures. Another important factor for cold protection is moisture control. This means that wetting of layers must be avoided at all time (from inside by sweating or from outside by rain or snow). If this is not possible the consequences of moisture accumulation must be controlled.

The middle layer provides most of the insulation. It comprises one or several garments of thicker material depending on the requirements. Choice of textiles is not critical as long as good insulation is provided. Non-absorbing materials should be selected for long-term exposures with limited heating and drying opportunities (expeditions, etc.).

The outer layer must provide protection against environmental factors such as wind, rain, fire, tear and abrasion. In occupational contexts, protection may also be required against, for example, chemical and physical agents. This layer can also add to the total insulation by the provision of insulation liners and battings.

15.4.2 Protection against wind and precipitation

As already mentioned in a previous section, protection against wind and precipitation is important, in particular in outdoor work. Protection must be provided by the outer layer. The air permeability of the fabric of this layer is critical and is measured according to ISO 9237. Garment design must be such that openings can be closed. Sleeve cuffs facilitate the control of convective heat exchange of the arm. The aim is to prevent ambient cold air penetrating the surface layer and reaching inner, warm layers. The ensemble fabrics must not be too soft, as strong wind compresses material and layers and enhances heat losses. The following formula is used in ISO 11079 for prediction of wind and body motion effects and computes a correction factor for the basic I_{cl} value.

$$I_{cl} = \frac{I_{cl,r} + (0.092 \cdot e^{(0.15 \cdot v_a - 0.22 \cdot v_w)} - 0.0045)/f_{cl}}{(0.54 \cdot e^{(0.075 \cdot \ln(ap) - 0.15 \cdot v_a - 0.22 \cdot v_w)}) - 0.06 \cdot \ln(ap) + 0.5} - 0.085/f_{cl} \qquad 15.1$$

where ap is the air permeability in l/m² s, v_a is the air velocity in m/s and v_w is the walking speed in m/s. The effect of wind on recommended exposure times for light and moderate work in 3.0 and 2.5 clo (I_{cl}), respectively is shown in Figs 15.2(b) and 15.3(b).

It should also be mentioned here that protection of the head is important for the maintenance of a good heat balance. The human head stays warm and may lose large amounts of heat, in particular in windy conditions, if not sufficiently protected by thick hair or head gear. An insulated hood is extremely useful in minimizing heat losses from neck and head regions.

Wet clothing loses some of its insulative power. In conditions with rain or melting snow, water must be repelled by the outermost layers in order to keep the inner layers dry. Layers of clothing that have become wet due to sweating or leakage of water, should be replaced by dry garments.

15.4.3 Indoor cold work

Indoor cold work does not expose the worker to wind and precipitation of any great magnitude. This gives more freedom for composition of the ensemble and for selection of garments. The inner layer should serve the same function as previously described. The middle and the outer layers may both serve as insulative layers. The middle layer may be a standard shirt or sweater and trousers and the outer layer a heavy insulated overall, parka or jacket and trousers. Type and nature of work also affects the selection of clothing. Frequent shifts in workrate require flexible clothing that can be easily adjusted, opened and closed. During breaks and rest pauses at normal room temperature the insulative clothing should be taken off, ventilated and warmed. Preferably, the worker should have a second set of protective clothing stored at room temperature for change. This set should be donned after the break, while the used ensemble is warmed up and dried. This procedure is particularly valuable for handwear and footwear.

15.4.4 New materials and fabrics

Technological development has brought a number of new fibres and fabrics with interesting properties related to their thermal function.

Moisture control

As long as one can stay dry, the choice of material (natural or synthetic) for the various layers is a matter of taste or other preferences and requirements. Moisture control is critical to staying warm in the cold. The body loses constantly some moisture by skin diffusion, but when sweating large amounts of moisture may be produced in a short time. As previously mentioned, sweating must at all costs be avoided. Having said that, appropriate adjustment of clothing and selection of fabrics allow some control of moisture transport and accumulation. Cotton is not suitable next to the skin as it absorbs a lot of moisture and clings to the skin when getting wet. Wool can be used as a next-to-skin fabric and may keep the skin relatively dry. When the fabric becomes saturated, however, the moisture control is reduced. Many synthetic textiles are hydrophobic and the moist air moves from skin through the fabric to the next layer.

Breathable fabrics

In foul weather good protection against rain and snow is required. However, waterproof fabrics may interfere with evaporative heat exchange. During activity the person gets wet from inside due to sweating instead of from outside. Microporous materials help to solve this problem to some extent. The small pores allow water vapour to pass but stop liquid water. This works reasonably well in temperate and warm climates. Due to the 'cold wall' principle it becomes less valuable in colder climates. During intermittent cold and warm exposures (in and out) it may allow absorbed moisture to escape in warm conditions. Also textiles of this kind are often highly windproof, which is beneficial in cold environments.

Improved insulation

Wool and down fabrics are highly insulative due to the very nature of the fibre materials. Modern synthetic textiles such as battings made of polyester hollow fibres or polyolefin microfibres resemble in a way the natural materials and provide good insulation per unit thickness.

As a spin-off from space technology, reflective materials (mostly aluminized fabrics or fibres) are used in garments and survival kits. The idea is that much of the heat the body loses through radiation will be reflected back to the skin. Such an ensemble will transmit less overall heat and the net insulation is higher compared to a similar one without a reflective layer. Practical tests, however, have shown that the net effect is small – in certain conditions negligible. The reasons are several. Radiation heat loss is only a minor part of the overall heat loss in the cold, in particular in the presence of wind and/or body movements (10–15%). Reflection of radiation requires spacing of layers, which is difficult to achieve and maintain. Most reflective fabrics are impermeable and interfere with moisture transfer. Aluminized insoles for shoes are common but provide no additional insulation compared to soles of similar thickness without aluminium. Gloves and socks with aluminium threads in the fabric do not give higher insulation compared to those without.

Intelligent textiles

In recent years several new types of materials and fabrics have been put on the market that contains some active component. Examples are phase-change materials (PCM), inflated tubings, shape-memory materials and electrical heating. Fabrics containing PCM respond to cooling by releasing heat from a range of waxes in the fibre or fabric (fibre content changes from liquid to solid phase). When the person gets warm, for example, during high activity, the solid PCM melts by body heat and helps to remove excessive heat. By choosing a certain temperature for the phase change the fabric (in a garment) could assist the wearer's thermoregulatory adjustment to hot and cold environments. Although the principle is physically sound, the PCM fabrics on the market contain insufficient amounts of the phase-change material. The heat transfer involved is small, difficult to measure and almost impossible to perceive (Shim *et al.*, 2001; Ying *et al.*, 2003; Weder and Hering, 2000). Special vests with 1–2 kg of PCM have been shown to have a higher capacity to exchange heat with the body (Gao *et al.*, 2007).

Inflatable fabrics, in principle, should allow the expansion of thickness of the ensemble, thereby increasing the effective insulation. Fabrics with a system of thin tubings can be inflated by blowing air into them. The effect is a thicker layer for that particular garment that should add insulation. Shape-memory materials change their geometry with lowered temperature and make the fabric a little bit thicker. The effect may be significant for the fabric itself, but the final effect on thermal insulation in a multi-layer ensemble is probably small.

Electrically heated elements incorporated into fabrics have been available for many years. The disadvantages, so far, have been the low capacity of portable batteries, and the durability of the wiring system. Today more powerful, long-lasting, portable batteries are available. Fabrics have been introduced with built-in conducting properties that are more durable and easy to design (Andersson and Seyam, 2004; Wang, 2007). This concept is likely to be most beneficial to the heating of hands and feet, but garments are already available on the market that have built-in batteries and are charged from the mains supply.

15.5 Sources of further information and advice

- International Society for Environmental Ergonomics:
  www.environmental-ergonomics.org
- European Society for Protective Clothing:
  www.es-pc.tno.ne

15.6 References

Andersson, K. and Seyam, A. M. (2004) The road to true wearable electronics. *Textile Magazine*, 1, 17–22.

Gao, C., Kuklane, K. and Holmér, I. (2007) Cooling effect of a PCM vest on a thermal manikin and on humans exposed to heat. In Mekjavic, I. B., Kounalakis, S. N. and Taylor, N. A. S. (eds) *Envirnonmental Ergonomics XII*. Portoroz, Slovenia, BIOMED.

Hassi, J., Mäkinen, T., Holmér, I., Påsche, A. and Al., E. (2002) *Handbook for Cold Work*, Oulu Regional Institute of Occupational Health.

Holmér, I. (2002) Risk assessment for cold work. *Croner's management of health risks*, Special report, 1–8.

Holmér, I. (2005a) Protection against cold. In Shishoo, R. (ed.) *Textiles in sport*. Cambridge, Woodhead.

Holmér, I. (2005b) Texiles for cold protection. In Scott, R. (ed.) *Textiles for protection*. Cambridge, Woodhead.

Holmér, I., Granberg, P. O. and Dahlström, G. (1997) Cold. In Stellman, J. (ed.) *Encyclopedia of Occupational Health*. Geneva, ILO.

ISO-8996 (2004) Ergonomics – Determination of metabolic heat production. International Standards Organisation.

ISO-9920 (2007) Ergonomics of the thermal environment – Estimation of the thermal insulation and evaporative resistance of a clothing ensemble. International Standards Organisation.

ISO-11079 (2007) Ergonomics of the thermal environment. Determination and interpretation of cold stress when using required clothing insulation (IREQ) and

local cooling effects. International Standards Organisation.
Shim, H., McCullough, E. and Jones, B. (2001) Using phase change materials in clothing. Textile Res. J., 71, 495–502.
Wang, S. (2007) *Intelligent thermal protective clothing*. Hong Kong, Hong Kong Polytechnique University.
Weder, M. and Hering, A. (2000) *How effective are PCM-materials? Experience from laboratory measurments and controlled human subject tests*, St. Gallen, Publisher.
Ying, B., Kwok, Y., Li, Y., Yeung, C. and Song, Q. (2003) Thermal regulating functional performance of PCM garments. *International Journal of Clothing Science and Technology*, 16, 84–96.

16
Footwear for cold weather conditions

K. KUKLANE, Lund University, Sweden

Abstract: In cold climates all user requirements for footwear (insulation, waterproof, vapour permeability, drying, etc.) cannot be met easily due to their conflicting nature. Defining user conditions is important. Cold may be roughly divided into the three ranges (>+5; +5 to −10; <−10 °C). The choice of proper footwear may be based on this approach. It also reduces some of the conflicting user requirements. Under any defined user condition it is important to take proper care of the feet and the footwear. The best performance can be achieved when considering already at the design stage that not only the materials in footwear but also the whole foot-sock-footwear system should work together.

Keywords: cold protective footwear, foot performance in cold, thermal insulation, moisture management, user and design requirements.

16.1 Introduction

Cold can be defined as a temperature below a certain limit value or as a performance loss due to the low physiological temperatures (e.g. cold fingers), cold sensation (distraction phenomenon), the extra protective clothing worn (added energy cost and restricted movement) or a combination of these factors. Even ordinary clothes, e.g., a business suit, can be considered as protection against the cold. It would be uncomfortable to sit in an office in just your underwear even at the ordinary room temperature (around 20 °C). Although an individual would not, for example, run a race in a suit as effectively as in sportswear it would not be considered a performance inhibitor at the office. Thus, the decrease in performance caused from normal work-wear and protective clothes could not be considered as an effect of cold. In spite of that these garments allow us to work at lower environmental temperatures without cold related discomfort.

In a report Dyck (1992) drew a long list of factors that must be considered depending on user requirements. According to him, in cold climates footwear must be insulated and protect insulation in case of puncture; be waterproof; be able to absorb and transmit sweat vapour (liquid sweat accumulation is to be avoided in cold weather); be permeable or adequately ventilated; and be able to dry and/or drain rapidly. This chapter covers the factors mentioned above and also raises issues related particularly to cold climate footwear. It has to be kept in mind that not all of the above requirements can be fully met, due to their conflicting nature.

16.2 Criteria for cold protective footwear

In order to design footwear for cold weather it is important to define what is meant by cold for the various user conditions. Ordinary street shoes may keep the feet warm and comfortable within a temperature range of −5 to +25 °C but this must be in conjunction with adequate clothing on the upper body combined with the body's own thermal reaction and heat redistribution. However, when moisture is present between +15 and +20 °C, and this is combined with minimal activity, discomfort from the cold may occur. Using this information and combining it with what is known about thermal properties of footwear, cold can be divided into three ranges. The choice of suitable footwear may be made based upon these criteria. It also reduces some of the conflicting user requirements within a particular range.

1. For temperatures above +5 °C no specific footwear insulation is required. Most ordinary shoes and occupational footwear have insulation around and above $0.20 \, m^2 \, °C/W$. This would be enough to keep feet warm at temperatures down to +5 °C if the user is active. The use of thicker socks, e.g., sports socks in terry cloth, adds insulation and helps to keep the feet warm. The main issue at this temperature range is whether moisture is present. Footwear should be chosen to avoid external moisture from entering it and to allow internal moisture to leave it. If the likelihood of exposure to external moisture is known in advance then the choice would be more simple.
2. The temperature range between +5 and −10 °C is the most complicated to deal with due to changing weather around the freezing point of water (0 °C). In this range, the footwear should be chosen by considering two factors: protection from external moisture and the need for high insulation. Snow can stick to and melt on footwear as a result of the surface temperature of the footwear staying above 0 °C (Fig. 16.1). In this temperature range the water vapour permeable but water resistant membrane properties may still be effective as the surface temperature of footwear generally stays above 0 °C.
3. At temperatures below −10 °C there is less external moisture available to penetrate the footwear. However, moisture from sweating does not easily leave footwear due to low temperatures in the outer material of the footwear. In these conditions it is important to have a high level of insulation in the footwear and to have internal moisture management properties. There is also increased need to dry the footwear between periods of usage.

In any situation regardless of the environmental and user conditions, it is important to take proper care of the feet and the footwear. It is especially

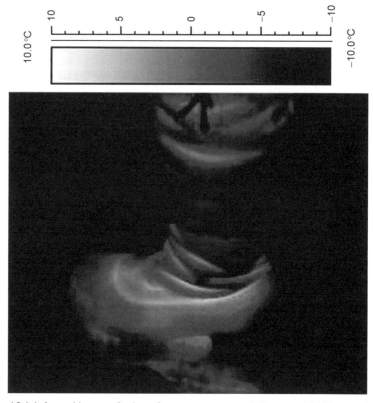

16.1 Infra-red image of winter footwear on a metal plate at −10 °C in a stable state after 90 minutes of the measurements. The mean surface temperature of the thermal foot model was kept at 34 °C.

necessary to consider the different ways of drying the footwear. It is also important to realise that certain usage environments will change the requirements for the user, e.g., footwear for work in food industries with specific temperatures and work with water, or hiking boots for extreme cold that may need to consider the risks of water under a weak ice layer.

16.3 Feet in cold

16.3.1 Foot heat sources

The extremities of the human body are more affected by exposure to the cold than other body parts. The hands and feet have a surface area which is large in relation to their volume. Extremities have little local metabolic heat production because of their small muscle mass and this falls with the tissue temperature. For example, each foot may generate up to 2 W, but at tissue temperatures below about 10 °C this may be reduced to about 0.2 W (Oakley, 1984).

The heat balance of extremities relies greatly upon the heat input by blood circulation from the body core. Blood flow to the extremities is under thermoregulatory control and is often reduced in the cold when heat production is moderate or low. The heat supplied by normal blood flow to the foot reaches over 30 W in warm conditions or during exercise, but it is reduced by the cold and can fall below 3 W (Oakley, 1984). A foot at a mean temperature of 35 °C has about 160 kJ of heat above an ambient temperature of 0 °C. If the mean tissue temperature falls to 5 °C then the heat content of the feet is still 23 kJ (Oakley, 1984).

16.3.2 Factors affecting foot cooling

The temperature of the foot is related to a number of different factors, e.g., activity, insulation and cleanliness. The feet are comfortable when the skin temperature is about 33 °C and the relative humidity next to the skin is about 60% (Oakley, 1984). Feet start to feel cold at toe skin temperatures at around 25 °C, while discomfort from cold is noted at temperatures under 20–21 °C (Enander et al., 1979). Further decrease of the foot skin temperature below 20 °C is associated with a strong perception of cold (Luczak, 1991; Fig. 16.2).

The feet are often the only body parts that are in contact with the ground. During standing the contact between the soles and the ground can cause a great amount of heat loss. This is a further explanation to why the feet are so affected by exposure to the cold. The toe skin temperatures do fall rapidly, especially, when the person is inactive in cold environments. Work in the cold that involves long periods of standing requires high insulation clothing as well as very warm footwear and good insulation in soles is especially important. Considering the weight that is applied to the soles of footwear the insulation material used should not be easily compressed.

Although walking can increase heat loss from the feet by convection, it also reduces the contact time with the ground. Walking also results in increased heat production and better blood circulation to feet and in this way promotes higher foot temperatures. Cold toes may warm up during exercise, but cool rapidly again when the exercise stops.

In jobs that involve various activity levels in cold environments an individual has to deal with intensive sweating in the feet during high activity, as well as rapid cooling when the person is less active. This can also result in discomfort due to the high humidity concentration in footwear. Standing jobs, e.g., meat cutting in food industry, signalling at harbour, etc., involve cooling of the feet through intensive heat loss by conduction, and lower heat input from blood flow, especially if there is not much possibility for feet motion. With exercise it is possible to warm up the feet, but the exercise duration must be at least ten minutes in order to warm up the cold toes. An 8-hour long study was conducted (Rissanen and Rintamäki, 1998) which showed that at −10 °C foot and toe

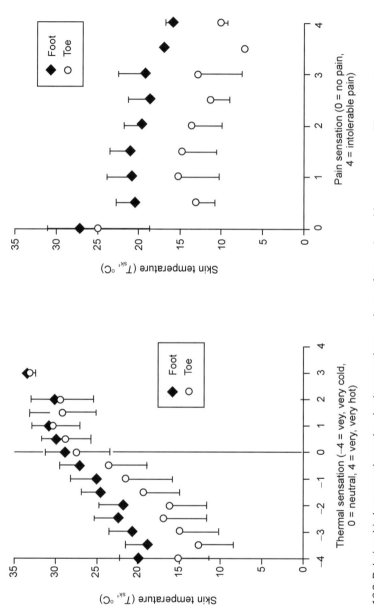

16.2 Relationship between thermal and pain sensations and mean foot and toe skin temperatures. The values include ratings during cold exposure, intermittent activity and warm up. Modified from Kuklane (2004).

temperatures increased during exercise (240 W/m²). The increased temperature of the feet during this exercise was found to be partly related to the increased blood flow as well as the redistribution of heat produced by the calf muscles by air circulation in the shoe. This study also found that temperature in the toes started to raise only after 15–20 minutes of the exercise. This delayed recovery of the toe skin temperature has also been observed in other studies (Ozaki *et al.*, 1998). Therefore, it is essential to pay special attention to protecting the toes from the cold.

The whole body thermal insulation has been found to affect the thermal condition of individual body parts and insulation in these individual parts affects the overall comfort of a person (Afanasieva, 1972). If the insulation of the body is inadequate and a person generally feels cold, one will often notice it in the feet. This is due to the skin temperature of the feet being normally the lowest because of the vasoconstriction. Cold feet may actually be a symptom of general cold discomfort (Fanger, 1972). On the other hand, if the feet are inadequately protected the feeling of cold discomfort will be dominating in spite of proper clothing on the rest of the body. Williamson *et al.* (1984) showed that dropping the temperature of the toes by 4 °C (from 28 to 24 °C) corresponded to a 14% increase in discomfort, while in hands the skin temperature drop of 7.0 °C (from 31 °C) increased the sensation of discomfort by 10%.

16.3.3 Cold and pain sensation in feet

The cold and pain sensations in feet are significantly related to foot skin temperature while there is present considerable individual variation. However, the temperature in toes is commonly lower than in whole foot when someone experiences the feeling of cold and pain. Thermally neutral sensation in both toes and in a foot as a whole is usually maintained if the skin temperatures stay above 25 °C, while during strong cold sensation the temperature of the toes are about 5 °C lower than mean foot temperature (Fig. 16.2). This is also the case in relation to the pain that can occur in a cold environment: there is no pain while temperatures stay above 25 °C, however, first signs of pain will appear when the toe temperature drops to about 15 °C. Increases in the sensation of pain can occur without considerable changes in the skin temperature of the toes and can become intolerable at around 10 °C. At 15 °C the cold receptors seem to be overridden by the pain receptors activity. Cold sensation is experienced due to the higher temperatures in other foot areas. Pain sensations are an important physiological alarm signal – something has to be done at once. If the skin temperature drops below 7 °C then numbness develops and risk of a cold related injury is greatly increased (Holmér *et al.*, 2003).

The temperature of the feet decreases rapidly during inactivity and the drop is quicker if footwear with low insulation is worn. Footwear with high insulation helps to prevent heat loss and slows the cooling process, and therefore provides better thermal comfort than footwear without insulation. The level of insulation in footwear has been found to correlate well with the foot temperatures that were measured on human subjects. Also, the cold and pain sensations in the feet correlate well with the level of measured footwear insulation (Kuklane et al., 1998). The level of insulation in certain footwear areas such as the toes and the heels became clearly noticeable in measured skin temperatures.

Tanaka et al. (1985) showed that when feet were submerged in cold water, the cold and pain sensations were strongest during the second minute of exposure when the constant skin temperature change was the quickest. Later the temperature drop slowed down, and pain and cold sensations reduced. In another study (Tochihara et al., 1995) a similar trend was noted where it was found it took a longer time with many short exposures to reach the same skin temperature than with fewer long exposures. There was lower pain and cold sensation during short exposures although final skin temperatures were approximately as low. The cold or pain sensations are often connected with a particular foot part such as the heel or more often toes. Cold sensation depends mostly on the temperature of the coldest part of the leg and it has been found that the toes are a key area that must be protected for comfort. From this knowledge it can be assumed that the temperature of localised areas can be the criterion to limit exposure to cold.

16.4 Foot and footwear related injuries in cold

Cold feet are one of the biggest sources of complaints for outdoor workers regarding the temperature of their working environment. The next biggest problems are connected with sweaty or wet feet, slipping, the fit of the footwear and protection from work hazards (Bergquist and Abeysekera, 1994; Gao et al., 2007a; Kuklane et al., 2000d, 2001). According to the statistics from Swedish National Board of Occupational Safety and Health on cold injuries at work, over 70% of the cold related injuries were on the hands and feet (Bergquist, 1995). Work performance of people depends to a large extent on their thermal status. For many occupations, e.g., foresters, farmers, industrial and construction workers, military personnel, etc., personal mobility is of great importance. Personal mobility largely depends on the legs and the feet, and the condition of these parts is largely dependent on the footwear that is worn. It is also important to consider that working conditions often dictate the specific footwear that is worn by the worker.

Soldiers in combat conditions often have to operate in both active and stationary situations. An additional problem in these conditions is that they often do not have opportunity to remove their boots to dry them out or to take the time to generally warm themselves up (Dyck, 1992; McCaig and Gooderson, 1986; Oakley, 1984). The role of an officer in setting adequate hygienic routines in field conditions is important.

Military history is one of the best documented and has the most information in relation to the effects of cold in an occupational environment. The effects of the cold in this environment can be seen looking back to many years ago, for example, to when Hannibal lost half of his troops crossing the Alps. From much closer eras there are records available on Napoleon's campaign to Russia; World War I; Hitler's campaign to Russia, etc. In Nordic history there were several cases where troops were not prepared for the cold, e.g., the retreat of the Swedish troops from Norway after the death of Carl XII and the Russians expecting quick victory in the Winter War against Finland, etc. (Paton, 2001).

Even in more recent years, e.g., Falkland conflict, both sides experienced many injuries in relation to the cold. The most common problems during the Falkland conflict were related to the poor performance of the boots and the condition of the feet (McCaig and Gooderson, 1986). A lot of records are available about the injuries that were experienced in relation to the cold such as frostbite on feet and trench-foot. Frostbite occurs when the skin temperature falls below its freezing point (0.6 °C), and tissue begins to freeze. The recovery period is accompanied by easily visible changes, such as blistering and gangrene (Oakley, 1984; Hamlet, 1997).

The incidence of trench-foot has been noted in environments with temperatures from just below to well above freezing. Some of the factors that can cause this condition are: the cold, moisture, immobility, the tightness of the boots and other restrictions to normal circulation. Typically, the first sign is loss of sensation in the toes (Oakley, 1984; Hamlet, 1997; O'Sullivan *et al.*, 1995).

In order to reduce the cold related injuries and make superior officers aware of the situation a method for detecting the risk level of an individual was developed in Northern Sweden (Linné, 2005). By a controlled exposure to cold the cooling of the feet was observed and occurrence of white spots recorded. White spots were found to be a warning sign for the risk of a cold-related injury. The occurrence of the spots may have appeared due to a previous cold-related injury or just sensitivity to the cold. It was also known they could be linked to a vibration related injury, e.g., work-related injury (not common in the younger soldiers) or from some types of sports activity that may involve vibrations, e.g., roller-skating daily on rough surfaces, eventually. Thermal cameras have also been found to be effective in observing these sorts of problems.

Endrusick (1992) studied various types of boots for the US Navy. He found that if the boots lacked an integrated steel safety toe, the individual was at a higher risk of severe foot injuries. In a questionnaire conducted by Bergquist and Abeysekera (1994) the highest reported problem was the thermal comfort of the footwear (57%). Of this, 43% related to discomfort and cold sensation associated with the steel toe cap and its alleged cooling effect. However, according to many studies (Bergquist and Holmér, 1997; Elnäs *et al.*, 1985; Påsche *et al.*, 1990) work shoes with steel caps and steel soles were not thermally different from the same models without steel reinforcements. In a study by Kuklane *et al.*

(1999b) it was found that steel toe caps actually added insulation to footwear with otherwise low insulation. However, the insulation differences that were found were not significant from the practical viewpoint when considering the different foot temperatures that were seen in the subjects.

The differences were observed in 'after effect' of steel toe-capped footwear. This effect could be related to a slower warming of toes in footwear with a steel toe-cap that has higher thermal inertia (ca. 100 g extra mass). After cold exposure the toe temperatures begin to warm up after 5–15 minutes of being in a warm environment or exercising (Kuklane *et al.*, 1998, 1999b; Ozaki *et al.*, 1998; Rintamäki *et al.*, 1992; Rissanen and Rintamäki, 1998; Tochihara *et al.*, 1995) and this is often the length of time for a break. If footwear is not taken off in this period the steel toe caps slow the warming of the foot and therefore keep the temperature of the foot and toes at a lower level. Because of this it is recommended to remove the footwear during breaks to let the feet breathe and let the socks and boots dry.

However, cold-related injuries in feet are not the most common problem of unsuitable boots in a cold environment as the most frequent problem is injuries caused by slipping and falling (Gao and Abeysekera, 2004b; Påsche *et al.*, 1990). The sole must be designed according to the intended use of footwear, to avoid slipping and stumbling. Chiou *et al.* (1996), Gao *et al.* (2004) and Rowland (1997) describe some methods for testing slipping and show interesting results. Often materials that have good friction on a lubricated metal plate may not be good for ice and snow as the materials tend to become hard in cold and, thus, lose their 'grip'. The materials may change their properties at temperatures below 0 °C and with any abrasion that occurs. Also, the friction properties of the ice and snow change with different temperatures (Gao *et al.*, 2003, 2004).

At temperatures above 0 °C ice and compacted snow may be covered with water film that acts as a lubricant. At low temperatures ice and snow turn less slippery. Even though the snow and ice covered walking surface stay at subzero temperatures, a warm sole of footwear may melt a thin water film between the sole and the walking surface, especially straight after leaving a warm indoor environment. It has also been found that the reactions and the experience of an individual may relate to the likelihood of slipping and falling accidents (Gao and Abeysekera, 2004a). It is common knowledge that on the first few days of a snowy and cold period the emergency units deal with a higher level of related injuries.

16.5 Footwear insulation

The insulation becomes a more important factor at lower environmental temperature and with activity: the generated heat is better trapped in boots with higher insulation. The present European (CEN) and international (ISO) standards for safety, protective and occupational footwear EN ISO 20344–

20347 (2004) do not measure the insulation of footwear. Instead they classify footwear as cold protective by an allowed 10 °C temperature drop inside the footwear during 30 minutes at an initial suggested temperature gradient of about 40 °C (+23 °C for conditioning and −17 °C for testing, respectively).

One study on this topic (Kuklane et al., 1999c) expressed strong doubts if the simple pass/fail test is correct for thermal testing. For example, the same footwear that helps to keep a good thermal comfort at −10 °C when walking may be too cold for standing at the same temperature or too warm to be used at +10 °C. A recent study in this area (Kuklane et al., 2008) has shown that footwear such as sandals and mesh-shoes can pass the simple cold protective footwear test. This shows that the test is not an effective way of testing footwear for protection against the cold. The major problem is that the protective footwear that has as low insulation as the footwear mentioned above may be classified as cold protective and in this way giving the user a deceiving safety feeling and exposing him/her to higher risks.

16.5.1 Measuring insulation

The thermal foot model (Fig. 16.3) is a physical model of a human foot that can be heated to and controlled at a given surface temperature (T_s), e.g., 34 °C. At a constant ambient temperature (T_a) the power stabilises at a level that depends on the temperature gradient between the model surface and environment, and insulation of the footwear. The power to the foot model in a stable environment is equal to the heat losses through the footwear. Knowing the area (A) of the foot model, its different zones (toes, sole, heel, etc.), and measuring power (P) to them, the temperature gradient allows the calculation of the total insulation (I_T) of the whole footwear or for separate zones by using this equation:

$$I_T = (T_s - T_a) \times A/P \qquad 16.1$$

The insulation can be calculated by this equation but it is also important to consider the condition the footwear is in and because of this further analysis maybe necessary. Insulation that is measured without the presence of wind (air velocity < 0.15 m/s) and without motion is called total insulation (I_T). This includes air layer insulation (I_a). Air layer insulation measured on a bare foot model is about 0.11 m²°C/W and if air layer insulation is subtracted from I_T then effective insulation is achieved (I_{clu}). Finally, if the surface area exposed to the air on a bare foot compared to wearing a boot is considered (f_{cl}, clothing area factor), basic insulation is acquired (I_{cl}). This value may be considered as the correct insulation of a sock or an item of footwear. If the measurements were carried out with the presence of wind and/or with motion simulation then a resultant total insulation is found ($I_{T,r}$). It may be corrected also as specified above, but resultant air layer insulation ($I_{a,r}$) has to be measured in the same wind and motion conditions, too. More information regarding these terms and

352   Textiles for cold weather apparel

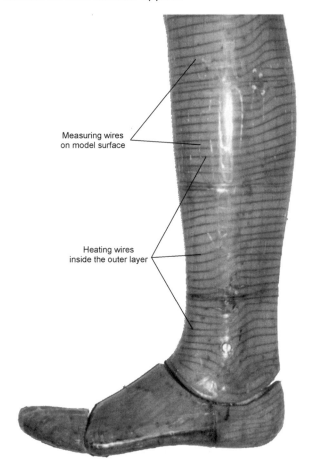

16.3 A thermal foot model in plastic material. Heating wires can be seen in plastic and measuring wires on the surface. Modified from Kuklane and Holmér (1998).

related calculations can be acquired in different standards definition sections (ISO 9920, 2007). The present chapter commonly refers to I_T and in few cases $I_{T,r}$ in text and figures if not pointed out differently.

If insulation, environmental temperature and activity level are known then it is possible to choose correct footwear for particular conditions. It is also possible to estimate the change in foot skin temperatures by considering dynamic changes and finding out recommended exposure times according to threshold limit values for cold sensation, strong cold sensation or pain sensation (2.2.8 212 7-06, 2006; Kuklane, 2004).

The thermal foot method (Bergquist and Holmér, 1997; Kuklane, 2004; Santee and Endrusick, 1988) allows an evaluation of the footwear and allows feedback to the manufacturers regarding the footwear as a whole and on individual parts areas. For example, the results from some footwear tests show insufficient insulation around the toe area, with this information the manufacturers would know that this could be improved. This method also provides useful information to customers and also allows the use of the results to predict the insulation needed for specific cold conditions, the exposure time that is recommended, what may occur in certain conditions and the recommendation for the use of the footwear (Lotens, 1989; Kuklane et al., 2000f).

The insulation of footwear can also be measured during wear-trials on the human subjects (4.3.1901-04, 2006). The insulation values from thermal foot model measurements correlate well with the level of insulation acquired on the human subjects (Kuklane et al., 1999a). The results correlate better if the subject is at thermal comfort. If the demand for total and local thermal comfort is not satisfied, then uncertainty of the human results grows showing higher insulation measured on human subjects than on a thermal model (Ducharme and Brooks, 1998; Ducharme et al., 1998; Kuklane et al., 1999a) and the extremities are more affected.

16.5.2 Use of insulation values

Figure 16.4 gives an example for choosing footwear based on the foot temperature for two activity levels (the footwear given as examples in this and several other figures are described in Table 16.1). The model assumes relatively even temperature and insulation over the whole foot surface. Based on the series of studies, certain footwear insulation could be suggested for certain temperature ranges (Table 16.2). Another approach would be to define critical foot skin temperatures and calculate required footwear insulation based on exposure time and ambient conditions (Afanasieva, 2006). This method considers the estimated cooling rate of the feet and also allows the use of footwear with lower insulation but only for a limited time.

Field studies have confirmed the relevance of using the thermal foot method for testing the level of thermal protection in footwear (Kuklane et al., 2000b, 2000c,

Table 16.1 Footwear that are referred to in the figures (modified from Kuklane et al., 1999c)

No.	Boot description
1	A rubber boot without lining
2	A leather boot without lining
3	A leather boot with warm lining
4	A winter boot with impregnated leather, Thinsulate and nylon fur lining
5	A three-layer boot for extreme cold consisting of two felt inner boots and nylon outer layer

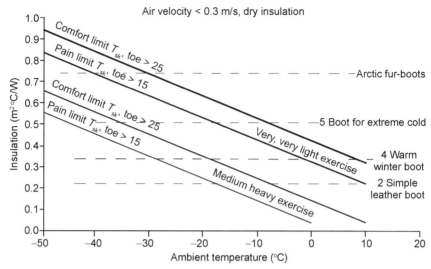

16.4 Required footwear insulation in relation to activity and ambient air temperature. Toe skin temperature above 25 °C corresponds to thermal comfort without strong sweating response, and above 15 °C corresponds to strong cold sensation but no pain. The figure includes some example footwear: a leather boot without warm lining (2); a winter boot with impregnated leather, Thinsulate® and nylon fur lining (4); a three layer boot for extreme cold consisting of two felt innerboots and nylon outer layer (5); arctic fur boot was measured on human subjects in Russia (Afanasieva, personal communication). Modified from Kuklane (2004).

2001). However, when choosing footwear for a certain activity level, it is also important to consider the effects of wind, walking and the moisture level (Fig. 16.5). If the individual is involved in a stationary activity in a cold environment, a high level of insulation in the sole of the footwear is essential. It may be recommended to have at least 0.05 to 0.10 m² °C/W higher insulation in the sole area than in the whole footwear depending on user conditions (Table 16.2).

Table 16.2 Recommended insulation of the footwear ($I_{T,r}$) and sole area (I_{sole}) to minimise discomfort from cold in some temperature ranges for relatively low activity (~100 W/m²) (modified from Kuklane, 2004)

Air temperature (°C)	Footwear (m²°C/W)	Sole (m²°C/W)
+15 to +5	$0.20 \leq I_{T,r} < 0.25$	$0.25 \leq I_{sole} < 0.30$
+5 to −5	$0.25 \leq I_{T,r} < 0.30$	$0.30 \leq I_{sole} < 0.35$
−5 to −15	$0.30 \leq I_{T,r} < 0.37$	$0.35 \leq I_{sole} < 0.42$
−15 to −25	$0.37 \leq I_{T,r} < 0.45$	$0.42 \leq I_{sole} < 0.55$
< −25	$0.45 \leq I_{T,r}$	$0.55 \leq I_{sole}$

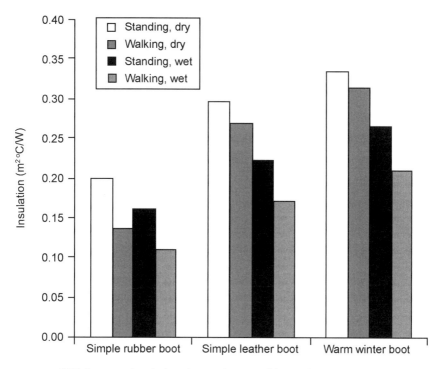

16.5 Footwear insulation change due to walking and sweating. Modified from Kuklane and Holmér (1997).

16.6 The effect of moisture in the footwear

Initial footwear insulation is an important factor to keep feet warm. However, the level of activity of the person and moisture level in footwear are also factors that influence the temperature of the feet. Cold sensation in the feet is often connected with low skin temperatures due to sweating and moisture on the feet. The footwear may be well insulated, but if the feet are wet, due to an outside or an inside source, the feet can easily start to feel cold. Dry fibres with air between them are good insulators. Water conducts heat about 23 times better than air while the insulation capacity of air does not considerably change through different humidity levels (the thermal conductivity of water vapour is within the same magnitude as air). If the condensed water gradually replaces air in and between the fibres it will cause further and faster cooling. After prolonged soaking in a wet environment the footwear may lose up to 35% of its insulation even in footwear that is intended for conditions where contact with water is expected (Endrusick, 1992).

16.6.1 Sweating in feet

The latest studies have shown that in a warm environment a foot may sweat about 30 g/h and in some cases even up to 50 g/h (Fogarty *et al.*, 2007; Taylor *et al.*, 2006). In these studies the subjects were exposed to high levels of heat stress. In the cold the whole body and especially the foot temperature will stay much lower and therefore sweat production is diminished. During relatively heavy exercise in the cold the average sweat rate stays around 10 g/h per foot (Rintamäki and Hassi, 1989; Hagberg and Holmér, 1985). During common occupational exposures the sweat rates are expected to lie between 3 and 6 g/h (Rintamäki and Hassi, 1989). Gran (1957) found that the sweat rate in feet changes in average from 3 g/h during a resting period to 15 g/h during high levels of activity. During very high levels of exercise the sweat rates may reach as much as 30 g/h per foot even in the cold (Kuklane *et al.*, 2001) or even up to the levels reported in heat, e.g., during sport activities at an elite level.

Gran (1957) found that foot temperatures are related to the amount of sweat that is absorbed in various shoe parts, and observed that the biggest humidity concentration can be found in the areas of the sole, heel and toes. This may be partly related to the lack of evaporation that can occur from these areas. Taylor *et al.* (2006) found that the soles had lower sweat rates than other parts of the foot. However, sweating in the sole area has also been shown to be related to stress factors, and in cold, when the sweating in other foot areas is suppressed the soles may have one of the highest sweat rates as the result of a stressful environment.

16.6.2 Moisture in footwear

Cold weather footwear, especially protective footwear for occupational use, is often made of impermeable or semi-permeable materials. Impermeable materials do not allow the water from the outside to make the insulation wet. At the same time, in such boots almost all the moisture from sweating remains inside. In some cases the sweat production could potentially cause a higher moisture content in the footwear than external water could. Footwear made from material such as leather have the ability to breathe dependent on the type of leather, treatment that may have been applied and the type of shoe-care that is used. The polish and leather treatment protect the material and prevent water from permeating into the footwear. However, work in a wet environment for a long period of time or the presence of snow can quickly wear off the protective layer. Considering this information, changing weather environments with wet and melting snow are the worst environments for the feet to remain protected (Martini, 1995).

In some cases an extra covering layer, i.e., outer sock made from watertight and/or vapour permeable material may help. However, different outer sock materials may not work effectively in conjunction with all footwear materials in various weather conditions. In some specific cases, for example, outer socks made from rubber or membrane material used over a leather boot did work effectively and did keep feet warm and dry at temperatures between −6 and −10 °C. In similar conditions, however, a nylon outer sock caused the leather to become wet and therefore caused discomfort for the wearer (Johansson, 2000).

This effect might be related to the thermal resistance of the material from which the outer sock is made. Most probably the nylon had a somewhat higher insulation level than the rubber and membrane materials, and therefore it kept a microclimate and the sock inner surface temperature at above 0 °C. This caused condensation to form on the inner surface of the nylon and when this came into contact with the boots the leather became wet. This did not occur in the rubber and membrane outer socks as the condensation froze and in this way moisture disappeared from the circulation. Although the nylon did not work effectively in this case it would have worked more effectively at lower temperatures. This highlights the importance of selecting the correct combination of materials to suit the environmental conditions.

During cold weather (below −10 °C) the water in the external environment is generally not a problem, except in certain jobs or activities where water is involved, e.g., fire-fighting, farm work, etc. In cold conditions the condensation caused from sweating could be a major problem. The moisture in footwear moves towards the colder surfaces where there is lower water vapour pressure and thus away from the feet. However, at a certain distance from a heat source, i.e., the feet, the humidity condenses and at the layer where the temperature drops under 0 °C, it may freeze (Johansson, 2000; Kuklane *et al.*, 2000e). Ice conducts heat around four times better than water. Because of this the insulation in the footwear is gradually reduced and the feet are exposed to rapid cooling. At lower temperatures the border, where humidity condenses and water freezes, becomes closer to the feet. This is not very common during occupational exposure but, for example, ice formation in footwear was reported during the bandy[1] world championship in Russia (Österberg, 1999). Severe cold (−40 °C) in combination with high sweating rates contributed to this. Also, under certain conditions the condensed water can move back towards the warmer areas by capillary action, creating a circulation for heat transport.

When reduction occurs in the insulation of footwear, the factors that need to be considered are the sweat rate of the feet, the evaporation and condensation rate, the absorption capacity of the footwear materials, the moisture transport within shoes and the environmental conditions (Kim, 1999; Kuklane and Holmér, 1998; Kuklane *et al.*, 1999d, 2000e). The various effects of moisture in protective clothing have been studied extensively during the last years (Havenith

---

1. Bandy is a game played on ice using skates, like ice-hockey, but played on a football field by teams of 11. Instead of a puck a little hard ball is used.

et al., 2008). It was found that moist layers (not fully saturated) may increase heat loss by about 5%, while the 'heat pipe' effect may increase the heat loss to around 40%. Evaporation can increase heat loss even further, but when thick highly insulated footwear is worn in the cold, this effect would most probably seldom reach 10%. The insulation in these cases may be called apparent insulation following the term used by Havenith *et al.* (2008). It is referred to as 'apparent' as apart from dry heat loss (convection, conduction and radiation) the insulation is calculated based on heat loss due to evaporation, wet conduction and evaporation-condensation, i.e., it includes also all changes that can cause heat loss due to the presence of moisture.

A sweat rate of 3 g/h may increase the heat losses from foot by 9–19% and a sweat rate of 10 g/h could increase the loss by 19 to 36% depending on the initial insulation of the footwear (Fig. 16.6). Footwear with high insulation does lose more in insulation capacity than footwear without a special liner at the same sweat rate. At the same time heat losses through moist footwear with low insulation may quickly grow equal to those from a bare foot in the same environment. In toe area the increase of heat losses due to moisture can be about 30% for boots without warm lining and about 45% in boots with higher insulation. This may be related to the internal air circulation in the footwear and the moisture that builds up in the colder and less insulated areas.

The evaporation that can occur in well insulated footwear is minimal at subzero temperatures (Kuklane and Holmér, 1998; Kuklane *et al.*, 1999d, 2000e;

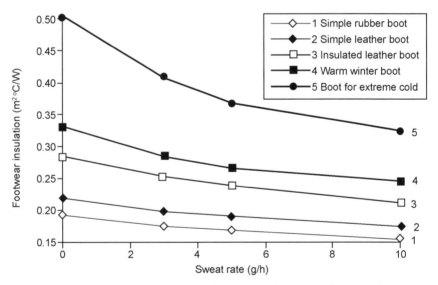

16.6 Change of footwear insulation at various sweating rates (apparent insulation). Length of each test was 90 minutes and the point values for every boot and sweat rate are based on last 10 minutes of measurements. Modified from Kuklane *et al.* (1999c).

Rintamäki and Hassi, 1989) and is usually less than 5%. During prolonged exposure to the cold without a chance to dry the footwear, e.g., military activities or hiking, the materials may become soaked and this can occur in all materials. For example, natural felt is a material that maintains a good level of insulation even if sweating occurs. It can absorb a large amount of moisture so that the feet stay relatively dry for long periods. Thick felt insoles (8–10 mm) can withstand compression well and effectively reduce heat loss from the feet. However, during long periods of exposure to the cold without being able to dry the footwear, even felt will become saturated. If the felt soling of the footwear becomes fully wet, the thermal qualities of the felt will disappear, and a cold-related injury may occur (Linné, 2000).

As previously mentioned, the different sweat rates affect the rate of heat loss. Strong sweating results in higher heat loss. However, the heat loss will stabilise when the balance is reached between the sweat rate and transportation, evaporation and condensation (Fig. 16.7). Further reduction of apparent insulation depends mostly on wetting of insulation layers that increases heat conductivity. This is related to the absorption capacity of the footwear material; a low capacity is connected to a continued loss of insulation depending on whether the wet material is next to the skin and appearance of free water in the sock and footwear.

If sweating of the feet reduces or stops, footwear has been found to regain some of the lost insulation (Kuklane et al., 1999d). This effect seems to depend

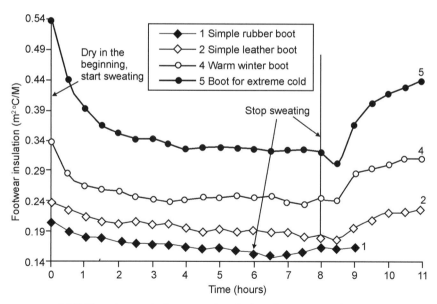

16.7 The change in footwear insulation due to sweating over one day (apparent insulation). The water supply was switched on during 8 hours (6 hours for boot 1) followed by 3 hours without water input. Modified from Kuklane *et al.* (1999c).

on the drying of the materials that are closer to the foot, i.e. sock, and if this does occur it will reduce conductive heat losses from the foot towards more distant and cooler footwear layers. Also, in this case, the temperature gradient dependent driving force of the 'heat pipe' effect is lower and takes place in the portion of the material package that is further away from the foot. The rate of the increase in apparent insulation does depend on the foot skin temperature. If the temperature of the foot is low the increase in insulation and drying process will be slower as the low temperature corresponds to the lower water vapour pressure difference between the foot and the distant layers of the footwear, and environment. There is always some perspiration from the feet, but it is important to avoid heavy sweating in a cold environment. The situation with sweating may be improved with the use of hygroscopic socks, e.g., woollen or wool blend, and changing the socks after heavy sweating in order to keep the feet warm and dry.

16.6.3 Drying of footwear

When special means for footwear drying are not available, they might not be able to get dry overnight or even over a few days (Kuklane *et al.*, 2000e; Fig. 16.8). Multi-layer footwear with removable insulation layers will dry out much quicker than standard footwear. When full drying of the footwear is not possible, other methods should be used. It is important to change socks after heavy

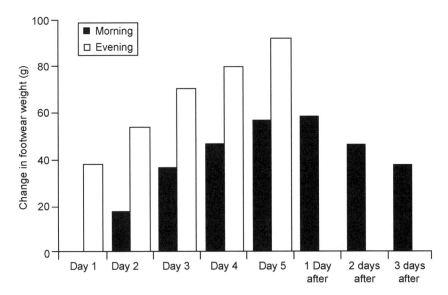

16.8 Moisture accumulation in footwear over one week. Tests were carried out at −10 °C with an up and down motion in order to simulate pumping effect. The tests under one day included water supply to the foot model and footwear system at the rate of 5 g/h for 8 hours, followed by storage at an ordinary room temperature until next morning. Modified from Kuklane *et al.* (2000e).

sweating. The absorbent materials, e.g., kitchen paper can be placed inside the footwear to help reduce the moisture. It is also beneficial to find a warm, dry and well ventilated area to leave the footwear to help with the drying process. It is important not to place footwear too close to extreme heat sources, such as flames or oven. This could damage and even shrink the footwear material or cause cracks to appear. In footwear without an absorbent lining the woollen socks can absorb moisture very effectively (Hassi et al., 2002). In this way the skin surface stays dry and comfort sensation could be maintained for a longer period. Disposable insoles made from an absorbent material similar to sanitary towels can also be an effective way of keeping the feet from becoming damp. In fact, sanitary towels have been used as such insoles in some situations.

16.7 Design of cold protective footwear

When designing footwear that is suitable for the cold weather, there are many different factors that need to be considered such as mobility, protection against the cold, insulation, waterproofing to keep the feet dry, permeability, durability, weight, and the fit (Oakley, 1984). Safety shoes that are worn in a cold climate have to protect the feet from work hazards and at the same time offer thermal comfort to the wearer (Bergquist and Abeysekera, 1994). The total foot comfort is determined by the interaction between socks, soles and shoes. The shoe should fit well on the foot. It should be large enough for socks and allow the toes to move (Nielsen, 1991).

16.7.1 Reducing heat losses

Cold feet are often experienced by many people, whether it is during working hours or leisure time. If an individual is correctly informed about properties that can offer protection from the cold, e.g., insulation, this can help them to select the most appropriate footwear. Work is currently being conducted at research groups to improve the methods of measuring insulation, water vapour resistance of footwear, and how to effectively disseminate this information to people. Thermal insulation is most important for preserving heat, as was discussed in Sections 16.5 and 16.6, and will be discussed in connection with several other topics below.

Auxiliary heating

For many years the possibility of using auxiliary heating, e.g., electrical heating, in footwear has been studied (Goldman, 1964). However, the heating methods that have been looked at would not be easily utilised in the footwear due to various reasons, e.g., the increased weight of the footwear, restricted space inside footwear, the size of the footwear and ease of movement and any factors that may affect balance during walking. The invention of new, small yet powerful batteries has made this idea more feasible, however, there are also several other aspects to consider such as the electrical heating equipment that would be used, activities and logistics (Haisman, 1988). Some questions related to using a heat source in footwear are also covered in the next section.

PCM in footwear

Various types of PCM (phase-change material) are now used in the apparel industry (Gao *et al.*, 2007b; Shim *et al.*, 2001). The problem with these materials is that it is not an easy task to fit them into footwear. When making footwear that utilises this type of material it is initially important to establish where it would be best placed within the footwear. The effectiveness of materials containing PCM is dependent on the amount that is present in the product. Thus, the second area to consider is what quantity of PCM must be added to the footwear in order to see a physiologically significant effect in the feet. If the temperature at which the phase change occurs (melting point) is not considered along with the location of the PCM within the footwear then the PCM might not have any effect at all under various ambient conditions.

Let us look at an example where the ambient air temperature is around $-10\,°C$, the indoor temperature is about $20\,°C$, and the foot temperatures of an individual are at an average of $32\,°C$. If simultaneously the PCM in footwear has melting point at $26\,°C$ (just above the threshold of cold sensation) and it is placed just below outer layer of the footwear then in this situation it would not start to melt before going outside, as the footwear would be put on just before leaving an indoor environment. The temperature gradient towards skin is $6\,°C$, whilst towards environment it is $36\,°C$, and most of the insulation will remain between the foot and the PCM. Because of this there is no chance for the PCM to melt and show any effect.

If the PCM was placed within the inner layer of the footwear, (still considering the short time putting on the footwear and entering the outdoor environment), the material may still have problems with melting effectively in times of low activity, whilst at a high activity it may melt. This would then allow the material to work as a heat source in later periods of low activity. Despite the fact that the materials can be seen to be effective, the actual melting process of the PCM may be a long process. A temperature difference of $6\,°C$ between the feet and PCM would not be enough to melt the PCM crystals within, for example, a short exertion of energy (10 minute run) and thus would not cause the positive effect of PCM during lower activity that follows. It should also be considered that not all of the heat from the feet is going into the melting process of the PCM and on the other hand, not all of the heat generated by the PCM is going into keeping the feet warm due to the high temperature gradient towards the environment. Also, the use of socks with various insulation levels would modify the outcome. To use this approach effectively, the footwear would need to be stored at temperatures above 26 °C or alternatively, the footwear would need to be worn continuously. However, wearing the footwear continuously is not recommended for several reasons and one reason is that moisture accumulation would occur within the footwear.

It would be reasonable to select a PCM for footwear inner layer that stays liquid at room temperature, e.g., with a melting point around 18 °C. This temperature would approximately correspond to the onset of a strong cold sensation and increased distraction from it. If this type of PCM was used, the feet would cool slightly when the footwear was first put on, but only by the amount needed for warming up liquid mass of PCM. The PCM would start to harden and release heat when the inner layer of the footwear reaches about 18 °C. The temperature of the feet would still be higher than this, due to heat input from the body and of insulation of the socks. Field studies (Enander *et al.*, 1979; Kuklane *et al.*, 2000b, 2000c, 2001; Williamson *et al.*, 1984) have shown that in the cold working environment, the skin temperature of the feet will often shift around 20 °C depending on the activity. Thus, PCM with a melting point around 18 °C would presumably be most effective when used within the inner layer of the footwear. However, the footwear would still need to be stored at temperatures above 20 °C, and should not be left in an environment where the temperatures are below 18 °C. Even at 20 °C the melting process may take a long time due to the low temperature gradient and overshoot of the crystallisation start point (Gök *et al.*, 2006). Footwear driers that use circulation of warm air would increase the speed of this process and could also improve the performance of the footwear.

PCM is most effective the closer it is to the skin, i.e., if it was used within socks. Ideally there should be no additional layers in between that could diminish the effect, this would ensure the insulation from outside environment is maximised. The socks could also be worn indoors and this would cause the PCM to melt long before going outside. Considering this situation a PCM with a melting point of about 26 °C may be recommended. PCM with a melting point below 24 °C may cause a prolonged cold sensation, and above 30 °C it may not melt due to low temperature gradient or if the feet were already colder than that.

In order to avoid discomfort from moisture, the PCM in socks should be combined with materials that can transport moisture away from the feet (polypropylene, polyester) and/or with materials with good moisture absorbing properties (wool). The sock materials that are chosen and the placement of them would depend mainly on the materials that are already present within actual footwear. For example, if a rubber boot without lining was used, moisture could easily create discomfort, so relatively thick or even double woollen socks may be recommended for use. When wearing a rubber boot with a felt lining, this would not be so important. A sock may integrate PCM with synthetic fibres on the inside of the sock, with a thicker wool blend layer in the middle for insulation, and with a thin synthetic layer such as polyester, on the outside for wear strength.

16.7.2 Layer by layer

The European footwear standards state that safety footwear should have insulation and insoles that cannot be removed without damaging the footwear (EN ISO 20345, 2004). This clause is meant to protect the user from increased injury risks caused by removal of protective layers. On the other hand, making it possible to remove the insulation layers and insoles would make it easier to dry the footwear (Kuklane et al., 2000e). It is known that it is highly important to keep the feet dry as this helps to keep the feet warm and comfortable. From the design viewpoint of safety boots the layers of material should be stiff enough and/or well fixed to the outer layer in order to avoid them moving around in the footwear. However, if the use of removable layers is prohibited, then proper socks and/or extra insoles should be used for moisture management (Hassi et al., 2002).

16.7.3 Size effects

The insulation properties of footwear, greatly depends on the amount of air that is present in the fabric and between the foot and the shoe and it has been found that the insulation could be increased with an extra pair of thick socks (Kuklane and Holmér, 1998; Kuklane et al., 2000a). However, it is important that the shoes are big enough to accommodate the thick socks. Attempting to increase the insulation of the feet by wearing thicker socks may not be efficient when wearing tight-fitting footwear, e.g., wearing two pairs of socks with a boot that is designed for one pair will push out the air and substitute it with conducting fibre. Compression of the foot may also occur and this can reduce the circulation in the feet and therefore cause the feet to become cold (Hassi et al., 2002; Kuklane et al., 1999a; Påsche et al., 1990). Figure 16.9 illustrates compression in relation to insulation and the effect it can have on the feet. Loose-fitting footwear (even though it has more area to accommodate air), does not provide a higher insulation if the space is not filled with a material that restricts the air motion (convection) inside the footwear. At the same time, loose-fitting footwear also affects performance and increases the risk of stumbling and falling.

16.7.4 Pumping effect and air permeability

During activities where feet are involved, air will move around in the inside of the footwear. The so-called 'pumping effect' is a good way to get rid of moisture. In ordinary shoes the pumping effect removes about 40% of humidity (Gran, 1957). In a cold environment the pumping effect may also remove a

Footwear for cold weather conditions

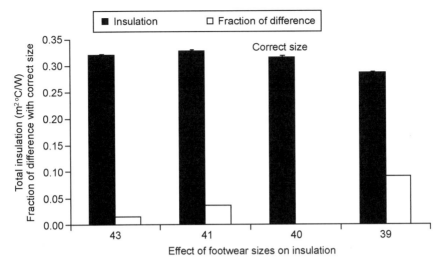

16.9 Effect of footwear size on cold protection of the feet. Fraction of difference shows difference from correct size.

considerable amount of heat, and due to this, in some cases it should be avoided. In the case of winter boots, however, the pumping effect during walking is minimal compared to other types of footwear (Kuklane *et al.*, 2000e; Rintamäki and Hassi, 1989). A reason for this may be that the insulation layer in winter boots sits relatively tight around the leg. If using mechanical solutions to enhance the pumping effect then the location of the inlet openings should consider the ground conditions (water, snow) and the risk of filling the air channels with moisture, particles and finally dirt, and there might be a need for self-cleaning ability. In the cold it is important that any air that is sucked into the foot area is pre-warmed, e.g., air has to pass via the calf area. The areas where the air is drawn in and expelled could be around the ankle or main part of the foot, or even the toe area. Whether this is a positive or negative occurrence in relation to the warmth of the foot, would depend entirely on the balance between the input of warm air, the input of moisture, and whether moisture is reduced in areas that are more prone to condensation, such as the toe area.

Footwear for protection against the cold should have an outer layer that is relatively airtight. High air permeability in cold wind allows quick cooling of the skin even in well insulated clothing. In winter boots the reduction in insulation due to walking is commonly less than 10% (Kuklane and Holmér, 1997; Kuklane *et al.*, 2000e). In footwear without a warm inner lining the effect is commonly bigger, i.e., about 30% (Bergquist and Holmér, 1997; Kuklane and Holmér, 1997). This reduction is partly due to the effect of increased external heat convection and partly due to the pumping effect. In the case of winter footwear with warm lining the air inside stays relatively still, while in more loose fitting footwear the air moves around more freely thus increasing the internal heat exchange. The combined effects of convection and moisture can reduce footwear insulation up to 45%.

16.7.5 Weight

Weight is another important factor to consider when choosing footwear. Several studies have shown that by increasing the footwear weight by 100 g the oxygen consumption of an individual will increase by 0.7–1.0% (Frederick, 1984; Jones *et al.*, 1984, 1986; Legg and Mahanty, 1986). The weight added to footwear is equivalent, in energy cost, to about five and more times the weight carried on the torso (Legg and Mahanty, 1986; Dorman *et al.*, 2005b; Soule and Goldman, 1969). In a cold environment the energy cost can be increased by up to 20% due to the protective clothing that is worn (Dorman *et al.*, 2005a, 2005b; Havenith and Dorman, 2007; Teitlebaum and Goldman, 1972). If the ground is covered by snow then walking in it, combined with the effect of the winter clothing can increase energy consumption by up to 65% depending on snow depth (Pandolf *et al.*, 1976; Smolander *et al.*, 1989). According to other sources (Headquarters, 2005) walking in soft, deep snow may even double the energy used, but this depends on the walking or running velocity of the individual.

It is always a trade-off between the protection level needed and the required performance or desired productivity when choosing any type of equipment, not only footwear. However, the weight of the insulation material that is needed to keep the feet warm is relatively small compared to the weight of other materials that are required to fulfil other protection requirements. Thus, the increase of footwear weight due to extra insulation should be acceptable in most industrial and leisure activities. However, more insulation material may increase the bulk of the footwear and this may become a limiting factor in respect to mobility.

16.7.6 Slip resistance

The choice of cold protective footwear sole material should, as well as other protection and physical requirements (durability for flexing, tear, etc.), consider the coefficient of friction (COF) on ice and snow at the intended exposure temperatures. Considering the exposure temperature ranges defined for cold protective footwear in Section 16.2 then the corresponding COF of sole material should be tested, for example, as follows. Footwear for temperatures above +5°C should have a sole material COF measured on moist or lubricated materials (tiles, metal). Footwear for temperatures between +5 and −10°C should have sole material COF measured on wet ice, and footwear for temperatures below −10°C should have sole material COF measured on hard ice and with footwear conditioned in cold. However, several tests have shown that on wet ice most shoe materials are very slippery (Gao *et al.*, 2003). Thus, these tests may not be very significant and shoes for the range of +5 to −10°C could be tested by both other methods instead, unless a new special sole material was developed for these conditions.

When footwear is used in an occupational environment it can be subjected to various combinations of temperatures and ground conditions, e.g., work in cold stores, on boats etc. Thus, any given friction value should be treated as additional information, and one has to have experience or acquire more information in order to choose footwear with correct soling.

16.8 Socks

An effective way to increase the insulation on the feet is to wear socks. The importance of socks for thermal comfort has been stressed throughout this chapter, as footwear and socks always work effectively together. This section will now discuss the advantages and effects of using different or more than one pair of socks.

The insulation value of socks is generally related to the thickness of the material and air trapped in and between the fibres (Fig. 16.10). There is a clear difference in insulation values between socks with various material thickness/ weight. If there is enough space within the footwear then it is recommended that thick socks or several pairs should be used in a cold environment. However, it has been found that two pairs of thinner socks provide a higher level of

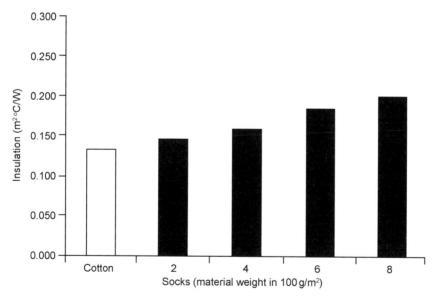

16.10 Total insulation of various single socks. Socks 2, 4, 6 and 8 are all made in wool and polyamide blend terry material. Cotton is an ordinary cotton-polyester sock with 70% cotton. It weighs as much as sock 2 (~20 g). Modified from Kuklane et al. (2000a).

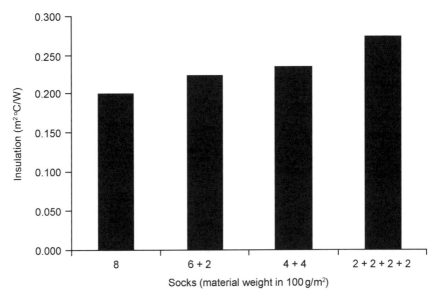

16.11 Insulation of sock combinations giving the total material weight 800 g/m². Modified from Kuklane *et al.* (2000a).

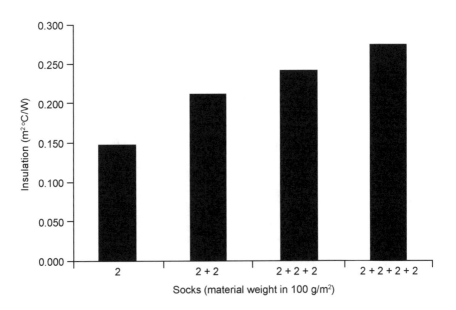

16.12 Effect of sock layers on the insulation on the example of the sock with material weight of 200 g/m². Modified from Kuklane *et al.* (2000a).

insulation than one pair of thick socks (Figs 16.11 and 16.12). Adding several layers of material increases the number of layers of air between the socks (Fig. 16.12). As the additional layers are added they start to compress the lower layers and then the effect of extra layers starts to diminish.

However, for a footwear–sock system the insulation increase may not be as clear as when looking at socks only. It depends on footwear insulation; boots with low insulation gain relatively more than well insulated footwear (Fig. 16.13). This can be explained easiest by following an example. If one looks, contact cooling of skin on a metal plate that has practically no thermal resistance then adding a thin plastic layer increases the resistance to heat losses enormously. However, if one has this metal plate already covered with, for example 1 mm thick felt layer, then adding the same plastic tape would practically have no measurable effect. However, when considering the prevention of moisture in the footwear, the quantity of the sock material is very important. Large amounts of moisture can stay within the socks, and this allows this moisture to be easily removed with just changing the socks. This is especially essential in well

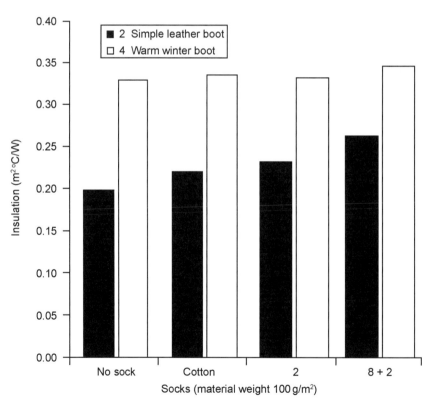

16.13 Insulation of footwear without socks and in combination with socks. Modified from Kuklane *et al.* (2000a).

insulated footwear during prolonged cold exposure and in footwear made of water vapour resistant and not absorbing materials, and for keeping the feet and footwear dry (Fig. 16.8).

There are several other points that show the importance of using several pairs of socks, beside additional insulation and the moisture management especially out in the field. If several pairs are available it can allow an individual to wash or repair a pair of socks whilst still having the other one on the feet. Wearing several pairs of socks could reduce the friction between the footwear and the feet by the friction occurring between the layers of socks instead. This could therefore reduce the risk of blisters in the feet as well as improve the wear length of the socks due to the presence of several layers (Kuimet, 2007).

16.9 References

2.2.8 212 7-06 (2006) Federal Service for Protection of User Rights and Human Wellbeing, Moscow.
4.3.1901-04 (2006) Federal Service for Protection of User Rights and Human Wellbeing, Moscow.
Afanasieva, R. F. (1972) *The Central Scientific Research Institute of Sewing Industry, Ministery of Light Industry of USSR*, Moscow.
Afanasieva, R. F. (2006) Personal communication.
Bergquist, K. (1995) *Dept. of Human Work Sciences, Div. of Industrial Ergonomics* Luleå University of Technology, Luleå, pp. 120.
Bergquist, K. and Abeysekera, J. (1994) *3rd Pan-Pacific Conference on Occupational Ergonomics* Ergonomic Society of Korea, Seoul, Korea, pp. 590–594.
Bergquist, K. and Holmér, I. (1997) *Applied Ergonomics*, **28**, 383–388.
Chiou, S., Bhattacharya, A. and Succop, P. A. (1996) *Am Ind Hyg Assoc J*, **57**, 825–831.
Dorman, L., Havenith, G. and THERMPROTECT_network (2005a) *The 11th International Conference on Environmental Ergonomics* (Eds, Holmér, I., Kuklane, K. and Gao, C.) Ystad, Sweden.
Dorman, L., Havenith, G. and THERMPROTECT_network (2005b) *The Third International Conference on Human-Environmental System ICHES' 05* Tokyo, Japan, pp. 439–443.
Ducharme, M. B. and Brooks, C. J. (1998) *Aviation, Space, and Environmental Medicine*, **69**, 957–964.
Ducharme, M. B., Potter, P. and Brooks, C. J. (1998) *The Eighth International Conference on Environmental Ergonomics* (Eds, Hodgdon, J. A. and Heaney, J. H.) The Eigth International Conference on Environmental Ergonomics, San Diego, California, USA, pp. 207–210.
Dyck, W. (1992) Defence Research Establishment, Ottawa, Canada.
Elnäs, S., Hagberg, D. and Holmér, I. (1985) Arbetarskyddsstyrelsen, 171 84 Solna.
EN ISO 20344 (2004) CEN, European Committee for Standardization, Brussels.
EN ISO 20345 (2004) CEN, European Committee for Standardization, Brussels.
EN ISO 20346 (2004) CEN, European Committee for Standardization, Brussels.
EN ISO 20347 (2004) CEN, European Committee for Standardization, Brussels.
Enander, A., Ljungberg, A.-S. and Holmér, I. (1979) *Scandinavian Journal of Work, Environment & Health.*, **5**, 195–204.

Endrusick, T. L. (1992) *Proceedings of the Fifth International Conference on Environmental Ergonomics* (Eds, Lotens, W. A. and Havenith, G.) Maastricht, the Netherlands, pp. 188–189.
Fanger, P. O. (1972) *Thermal comfort*, McGraw-Hill Book Company, USA.
Fogarty, A. L., Barlett, R., Ventenar, V. and Havenith, G. (2007) *The 12th International Conference on Environmental Ergonomics* (Eds, Mekjavic, I. B., Kounalakis, S. N. and Taylor, N. A. S.) Biomed d.o.o., Ljubljana, Piran, Slovenia, pp. 283–284.
Frederick, E. C. (1984) *Applied Ergonomics*, **15**, 281–287.
Gao, C. and Abeysekera, J. (2004a) *Safety Science*, **42**, 537–545.
Gao, C. and Abeysekera, J. (2004b) *Ergonomics*, **47**, 573–598.
Gao, C., Abeysekera, J., Hirvonen, M. and Aschan, C. (2003) *International Journal of Industrial Ergonomics*, **31**, 323–330.
Gao, C., Abeysekera, J., Hirvonen, M. and Grönqvist, R. (2004) *Ergonomics*, **47**, 710–716.
Gao, C., Holmér, I. and Abeysekera, J. (2007a) *Applied Ergonomics*, doi:10.1016/j.apergo.2007.08.001.
Gao, C., Kuklane, K. and Holmér, I. (2007b) *The 12th International Conference on Environmental Ergonomics* (Eds, Mekjavic, I. B., Kounalakis, S. N. and Taylor, N. A. S.) Biomed d.o.o., Ljubljana, Piran, Slovenia, pp. 146–149.
Gök, Ö., Yilmaz, M. Ö. and Paksoy, H. Ö. (2006) *Proceedings of the 10th International Conference on Thermal Energy Storage, ECOSTOCK*.
Goldman, R. F. (1964) *Fifteenth Alaskan Science Conference* (Ed, Dahlgren, G.) Alaska Division American Association for the Advancement of Science, Alaska, pp. 401–419.
Gran, G. (1957) *J Soc Leather Techn & Chem*, **43**, 182–197.
Hagberg, D. and Holmér, I. (1985) Arbetarskyddsstyrelsen, Solna.
Haisman, M. F. (1988) *Ergonomics*, **31**, 1049–1063.
Hamlet, M. (1997) *Problems with cold work*, Vol. Arbete och Hälsa 1998:18 (Eds, Holmér, I. and Kuklane, K.) National Institute for Working Life, Stockholm, pp. 127–131.
Hassi, J., Mäkinen, T., Holmér, I., Påsche, A., Risikko, T., Toivonen, L. and Hurme, M. (Eds) (2002) *Handbok för kallt arbete (Handbook for work in cold)*, available in Finnish, Norwegian and Swedish, Työterveyslaitos, Helsinki; Arbetslivsinstitutet, Stockholm; Thelma AS, Trondheim.
Havenith, G. and Dorman, L. (2007) *The 12th International Conference on Environmental Ergonomics* (Eds, Mekjavic, I. B., Kounalakis, S. N. and Taylor, N. A. S.) Biomed d.o.o., Ljubljana, Piran, Slovenia, pp. 150–152.
Havenith, G., Richards, M., Wang, X., Bröde, P., Candas, V., Hartog, E. d., Holmér, I., Kuklane, K., Meinander, H. and Nocker, W. (2008) *Journal of Applied Physiology*, **104**, 142–149. doi:10.1152/japplphysiol.00612.2007.
Headquarters, Department of the Army (2005) Technical Bulletin, TB MED 508, Washington, DC, USA.
Holmér, I., Geng, Q., Havenith, G., den Hartog, E., Rintamäki, H., Malchaire, J. and Piette, A. (2003) Arbete och Hälsa 2003:7, Arbetslivsinstitutet, Stockholm.
ISO 9920 (2007) International Standards Organisation, Geneva.
Johansson, H. (2000) Personal communication.
Jones, B. H., Toner, M. M., Daniels, W. L. and Knapik, J. J. (1984) *Ergonomics*, **27, 8**, 895–902.
Jones, B. H., Knapik, J. J., Daniels, W. L. and Toner, M. M. (1986) *Ergonomics.*, **29, 3**, 439–443.

Kim, J. O. (1999) *Textile Research Journal*, **69**, 193–202.
Kuimet, P. (2007) *Postimees.ee* Tallinn, Estonia, pp. http://www.postimees.ee/170807/esileht/siseuudised/277562.php.
Kuklane, K. (2004) *International Journal of Occupational Safety and Ergonomics*, **10**, 79–86.
Kuklane, K. and Holmér, I. (1997) *Problems with cold work*, Vol. Arbete och Hälsa 1998:18 (Eds, Holmér, I. and Kuklane, K.) National Institute for Working Life, Stockholm, Sweden, pp. 96–98.
Kuklane, K. and Holmér, I. (1998) *International Journal of Occupational Safety and Ergonomics*, **4**, 123–136.
Kuklane, K., Geng, Q. and Holmér, I. (1998) *International Journal of Occupational Safety and Ergonomics*, **4**, 137–152.
Kuklane, K., Afanasieva, R., Burmistrova, O., Bessonova, N. and Holmér, I. (1999a) *International Journal of Occupational Safety and Ergonomics*, **5**, 465–476.
Kuklane, K., Geng, Q. and Holmér, I. (1999b) *International Journal of Industrial Ergonomics*, **23**, 431–438.
Kuklane, K., Holmér, I. and Afanasieva, R. (1999c) *International Journal of Occupational Safety and Ergonomics*, **5**, 477–484.
Kuklane, K., Holmér, I. and Giesbrecht, G. (1999d) *Applied Human Science*, **18**, 161–168.
Kuklane, K., Gavhed, D. and Holmér, I. (2000a) *Ergonomics of protective clothing. NOKOBETEF 6 and 1st European Conference on Protective Clothing*, Vol. Arbete och Hälsa 2000:8 (Eds, Kuklane, K. and Holmér, I.) National Institute for Working Life, Stockholm, Sweden, pp. 175–178.
Kuklane, K., Gavhed, D., Karlsson, E. and Holmér, I. (2000b) *Ergonomics of protective clothing. NOKOBETEF 6 and 1st European Conference on Protective Clothing*, Vol. Arbete och Hälsa 2000:8 (Eds, Kuklane, K. and Holmér, I.) National Institute for Working Life, Stockholm, Sweden, pp. 75–78.
Kuklane, K., Gavhed, D., Karlsson, E. and Holmér, I. (2000c) *Ergonomics of protective clothing. NOKOBETEF 6 and 1st European Conference on Protective Clothing*, Vol. Arbete och Hälsa 2000:8 (Eds, Kuklane, K. and Holmér, I.) National Institute for Working Life, Stockholm, Sweden, pp. 71–74.
Kuklane, K., Gavhed, D., Karlsson, E., Holmér, I. and Abeysekera, J. (2000d) *Ergonomics of protective clothing. NOKOBETEF 6 and 1st European Conference on Protective Clothing*, Vol. Arbete och Hälsa 2000:8 (Eds, Kuklane, K. and Holmér, I.) National Institute for Working Life, Stockholm, Sweden, pp. 67–70.
Kuklane, K., Holmér, I. and Giesbrecht, G. (2000e) *The Third International Meeting on Thermal Manikin Testing*, Vol. Arbete och Hälsa 2000:4 (Eds, Nilsson, H. and Holmér, I.) National Institute for Working Life, Stockholm, Sweden, pp. 106–113.
Kuklane, K., Holmér, I. and Havenith, G. (2000f) *Applied Human Science*, **19**, 29–34.
Kuklane, K., Gavhed, D. and Fredriksson, K. (2001) *International Journal of Industrial Ergonomics*, **27**, 367–373.
Kuklane, K., Ueno, S., Sawada, S. and Holmér, I. (2009) *Annals of Occupational Hygiene*, **53**, **1**, doi:10.1093/annhyg/men074.
Legg, S. J. and Mahanty, A. (1986) *Ergonomics.*, **29**, **3**, 433–438.
Linné, A. (2000) Personal communication.
Linné, A. (2005) *The 11th International Conference on Environmental Ergonomics* (Eds, Holmér, I., Kuklane, K. and Gao, C.) Lund University, Ystad, Sweden, pp. 589.
Lotens, W. A. (1989) TNO Institute for Perception, Soesterberg.
Luczak, H. (1991) *Ergonomics*, **34**, 687–720.

Martini, S. (1995) Norwegian Defence Research Establishment, Kjeller.
McCaig, R. H. and Gooderson, C. Y. (1986) *Ergonomics*, **29**, 849–857.
Nielsen, R. (1991) *International Journal of Industrial Ergonomics*, **7**, 77–85.
Oakley, E. H. N. (1984) *Ergonomics*, **27**, 631–637.
Österberg, O. B. (1999) In *Svenska Dagbladet* Stockholm, 6 February, p. 47.
O'Sullivan, S. T., O'Shaughnessy, M. and O'Connor, T. P. F. (1995) *Annals of Plastic Surgery*, **34**, 446–449.
Ozaki, H., Enomoto-Koshimizu, H., Tochihara, Y. and Nakamura, K. (1998) *Applied Human Science*, **17**, 195–205.
Pandolf, K. B., Haisman, M. F. and Goldman, R. F. (1976) *Ergonomics*, **19**, 683–690.
Påsche, A., Holand, B. and Myseth, E. (1990) *Polartech '90. International Conference on Development and Commercial Utilisation of Technologies in Polar Regions*, pp. 325–334.
Paton, B. C. (2001) *Medical Aspects of Harsh Environments*, Volume 1. (Eds Pandolf, K. B. and Burr, R. E.), Textbooks of Military Medicine (Eds Lounsbury, D. E., Bellamy, R. F. and Zajtchuk, R.) Office of The Surgeon General at TMM Publications, Borden Institute, Walter Reed Army Medical Center, Washington, DC 20307-5001. pp. 313–349. (http://www.bordeninstitute.army.mil/published_ volumes/harshEnv1/harshEnv1.html).
Rintamäki, H. and Hassi, J. (1989) *Arctic Rubber, Scandinavian Rubber Conference* Tampere, Finland.
Rintamäki, H., Hassi, J., Oksa, J. and Mäkinen, T. (1992) *European Journal of Applied Physiology*, **65**, 427–432.
Rissanen, S. and Rintamäki, H. (1998) *European Journal of Applied Physiology*, **78**, 560–564.
Rowland, F. J. (1997) *Fifth Scandinavian Symposium on Protective Clothing* (Eds, Nielsen, R. and Borg, C.) Elsinore, Denmark, pp. 180–189.
Santee, W. R. and Endrusick, T. L. (1988) *Aviat Space Environ Med*, **59**, 178–182.
Shim, H., McCullough, E. A. and Jones, B. W. (2001) *Textile Research Journal*, **71**, 495–502.
Smolander, J., Louhevaara, V. and Hakola, T. (1989) *Ergonomics*, **32**, 3–13.
Soule, R. G. and Goldman, R. F. (1969) *J Appl Physiol.*, **27**, 687–690.
Tanaka, M., Yamazaki, S., Ohnaka, T., Harimura, Y., Tochihara, Y. and Matsui, J. (1985) *Journal of Sports Medicine and Physical Fitness*, **25**, 32–39.
Taylor, N. A. S., Galdwell, J. N. and Mekjvic, I. B. (2006) *Aviation, Space, and Environmental Medicine*, **77**, 1020–1027.
Teitlebaum, A. and Goldman, R. F. (1972) *J Appl Physiol*, **32**, 743–744.
Tochihara, Y., Ohnaka, T., Tuzuki, K. and Nagai, Y. (1995) *Ergonomics*, **38**, 987–995.
Williamson, D. K., Chrenko, F. A. and Hamley, E. J. (1984) *Applied Ergonomics*, **15**, 25–30.

17
Gloves for protection from cold weather

P. I. DOLEZ and T. VU-KHANH, École de technologie supérieure, Canada

Abstract: Gloves are an important part of personal protection in cold environments. Indeed, heat loss is most predominant at body extremities, especially in high wind conditions. This chapter presents the principal materials used in cold protection gloves as well as the effect of cold temperature on their properties. It also describes those among their requirements in terms of hazards and comfort/functionality characteristics that may be affected by cold temperatures. Finally, some examples of particular applications of cold protection gloves are provided as well as some indications about possible avenues of new developments.

Key words: gloves, cold environment, protection properties, hazards, comfort, functionality.

17.1 Introduction: key issues of gloves in cold environments

Gloves are an essential part of personal protection in cold environments. Indeed, heat loss is most predominant at body extremities (Abeysekera and Bergquist, 1996), especially in high wind conditions (Shitzer *et al.*, 1998). This is due to the fact that heat production in hands is mostly ensured by blood circulation (Imamura *et al.*, 1998), which is subjected to cold-induced vasoconstriction (Shitzer *et al.*, 1998; Ding *et al.*, 2007). As a result, fingers and hands are often affected by frostbite and other cold-related injuries like numbness, cold-induced Raynaud's syndrome and various skin sores (Spear, 2006; Long *et al.*, 2005; Merla *et al.*, 2001; Wilkerson, 1986). It can be noted that physical exercise has been shown to increase finger temperature and restore part of the manual functions (Imamura *et al.*, 1998). Indeed, hand and finger performances are also affected by cold temperature. An increase in accident rates when ambient temperature is lower than 19 °C has been reported (Nag and Nag, 2007). It was associated with a reduction in finger dexterity and sensitivity. While manual capabilities are mostly preserved down to a finger skin temperature of 15 °C, fine finger functions are almost entirely lost below 4.4 °C (Ding *et al.*, 2004).

The use of gloves for cold protection is not new. For example, Pliny the Younger reports how his uncle's shorthand writer was wearing gloves in winter in order not to slow down the Elder's work (Pliny the Younger, c. 100 CE). However, all issues are far from being solved. In a survey carried out in 1996 with workers performing outdoor work, 90% of them reported thermal discomfort related to gloves and 80% difficulty to work with gloves (Abeysekera and Bergquist, 1996). The major requirements involved in hand protection against cold include thermal insulation, water and wind resistance, breathability, permeability to water vapour as well as the preservation of functional and sensorial performance (Desai, 2008; Abeysekera and Bergquist, 1996). However, some of these requirements are contradictory. For example, the use of thick insulating materials needed for thermal protection implies a reduction in hand freedom of movement and finger dexterity, which are crucial for maintaining task performance. The standard strategy of superimposing layers of insulating material is therefore not always the best solution. Another difficulty is the need for selective and one-way transport properties for water and air; the entry of outside water, cold air and wind must be blocked while allowing sweat and eventual internal overheated air to escape (Nielsen, 1992). A delicate compromise is thus to be found between these seemingly contradictory requirements. Indeed, the achievement of both thermal protection and preservation of sufficient functionality and comfort is the key for glove acceptance and use.

One additional challenge with hand protection in cold environments is that cold is often not the sole hazard to be dealt with (Herman et al., 1992). For example, food processing, especially meat and fish preparation, involves a combination of cold temperature and high mechanical risks, in particular cutting and puncture (Vézina, 1990; Nag and Nag, 2007). Cold environments may also coexist with chemicals, for instance in the oil extraction industry (Spear, 2006). In the case of firefighters in cold climate areas, thermal protection includes both high and low temperatures, and must be combined with resistance to flame, water, hot vapours, industrial chemicals, cut, puncture, abrasion, as well as the various biologic and chemical substances associated to terrorist-type treats (Filteau and Shao, 1999; Stull et al., 2006). While some have resorted to wearing superimposed gloves corresponding to the different types of risks (Abeysekera and Bergquist, 1996), the loss in terms of dexterity and comfort in particular makes it a hardly acceptable solution.

In terms of regulations, European standards distinguish between cool and cold conditions, the limit being set to −5 °C (Heffels, 2006). Requirements include thermal resistance, air permeability, water and water vapour penetration resistance and resistance to tear of the outer layer. In Europe, gloves must be labelled according to the level of protection they offer against cold (see Fig. 17.1) as well as against the other types of risks they may be designed to protect from (The University of Edinburgh, 2005; Bobjer, 2004). In some cases, specific regulations apply to occupational activities concerned with particular risks, for example firefighters (NFPA, 2006) and cryogenic liquid handling (ASTM,

376 Textiles for cold weather apparel

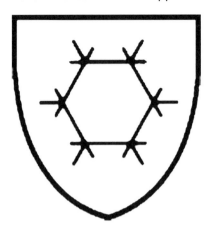

17.1 European label for protection against cold.

2006). It must be noted that cold environment conditions may be found both outdoors and indoors, with for example about 30% of cold workplaces in Germany being refrigerated rooms (Heffels, 2006).

17.2 Design, structure and materials used for hand protection in cold environments

Four major categories of designs can be identified in terms of cold protection for hands. Gloves cover each finger independently, offering the possibility of preserving some fine dexterity. An increased level of fingertip sensitivity can be provided by fingerless gloves, in which finger compartments are half-length and opened. They have been shown to provide an increase in skin temperature when tested down to 4 °C compared to un-gloved hands tested at the same temperature, however, not seemingly translating in an increased dexterity (Riley and Cochran, 1984a). On the other hand, mittens do not have separate finger compartments except for the thumb. The air circulation between the fingers as well as the reduced material surface area makes mittens a warmer design than gloves, however with a large loss in dexterity (Abeysekera and Bergquist, 1996). The last type of design is a hybrid between glove and mitten, in which an additional semi-detached pouch can encapsulate the four fingers as a mitten would do or be folded back and allow individual finger movement (Zuckerwar and Friedman, 2004). Usually, the four fingers are only half-covered as in the fingerless glove design. This design is of interest when fine dexterity is only occasionally needed.

In terms of structure, gloves (term used as a generic in the rest of the text independently of the question of design) for cold protection are often multi-layered constructions (Herman *et al.*, 1992). A thermal liner provides thermal insulation. It is often combined with an outer layer ensuring protection against the other risks, for example, mechanical or chemical. A membrane may be positioned between the thermal liner and the outer shell to provide waterproofness and breathability.

As air is one of the substances offering the lowest thermal conductivity (see Table 17.1), thermal liner materials are always designed to trap as much air as possible within their structure. Conventional materials include cotton fleece,

Table 17.1 Thermal conductivities of various insulating materials

	Thermal conductivity (W/m.K)
Air	0.026[a]
Feathers with air	0.023[b]
Silk	0.051[b]
Wool	0.059[a]
Cotton	0.071[b]
Polypropylene	0.12[b]
Polyethylene terephthalate (polyester)	0.15[b]
Leather	0.176[b]
Polymethylmethacrylate (plexiglass)	0.21[b]
Polyamide (nylon)	0.24[b]
Polytetrafluoroethylene (PTFE)	0.25[b]
Polyethylene low density	0.33[b]
Polystyrene	
Solid	0.1[b]
Foam ($d = 64$ kg/m^3)	0.033[b]
Poly(isobutylene-co-isoprene) (butyl rubber)	0.13[c]
Polyisoprene (natural rubber)	
Solid	0.14[b,c]
Foam ($d = 56$ kg/m^3)	0.036[b]
Polyvinylchloride (PVC)	
Solid	0.17[b]
Foam ($d = 56$ kg/m^3)	0.035[b]
Polychloroprene (neoprene)	
Solid	0.19[b,c]
Foam ($d = 112$ kg/m^3)	0.04[b]
Polybutadiene (BR)	0.22[c]
Polydimethylsiloxane (silicone rubber)	
Solid	0.25[b]
Foam ($d = 160$ kg/m^3)	0.086[b]
Poly(butadiene-co-acrylonitrile) (nitrile rubber) (NBR)	
Solid	0.28[c]
Foam ($d = 160$–400 kg/m^3)	0.036–0.043[b]
Polyurethane	
Solid	0.31[b]
Foam (air, $d = 20$–70 kg/m^3)	0.036[b]
Foam (CO_2, $d = 64$ kg/m^3) (90% closed cells)	0.016[b]

[a] from Shitzer *et al.* (1998), [b] from Speight (2005), [c] from Gent *et al.* (2001)

synthetic closed-cell foam and BOA fleece (Superior glove works, 2008a). With a velour construction and proprietary polyester fibres, Polartec® adds durability to lightweight warmth (Polartec®, 2007). However, the drawback of these materials is their bulkiness, which hampers dexterity (The University of Edinburgh, 2005). One solution has been to reduce the diameter of the fibres while increasing their density, for example with Thinsulate® developed by 3M (3M, 2008). It is a non-woven structure with 15 μm diameter fibres made of a blend of polypropylene and polyethylene terephthalate. A further improvement was made by DuPont with the production of polyester hollow fibres trapping air inside their core (Thor·lo, 2008). Thermolite® is based on that principle. Finally, even if they seem to have found more applications in heat protection, aramid fibres like Kevlar® do also present some interesting characteristics as insulation material against cold temperature (Berry *et al.*, 2008; The University of Edinburgh, 2005).

As the thermal insulation layer is often positioned close to the skin, it may also have to play a role in pulling humidity and sweat away from the surface of the skin. To that extent, wool offers some interesting characteristics as it is capable of absorbing up to 30% of its weight in water without the unpleasant feeling of dampness (Liya *et al.*, 2007). This moisture absorption process is also accompanied by an energy release which creates a warming effect. Raggwool is a blend of wool and nylon that offers increased wind and water resistance (Superior glove works, 2008b). Other materials like acrylic offer an even better alternative as they do not absorb humidity but carry it away along the surface of the fibres in a wicking effect (Thor·lo, 2008). DuPont developed some fibres with peculiar cross-section shapes, for example clover leaf or four-channel, to reinforce that wicking effect (see Fig. 17.2).

Recent developments in cold temperature gloves have considered the use of external heating. Indeed, numerical simulations have shown that a power input of 0.5 W per finger in leather and wool gloves leads to an increase in fingertip temperature of about 10 °C when placed in 0 °C still air conditions (Shitzer *et al.*, 1998). An even stronger effect may be expected since blood perfusion of heated fingers will increase and act as a heat source to further raise the finger temperature. With simulative spacesuit gloves designed for extravehicular activity, a heat input of 0.5 W per finger allowed maintaining the finger skin temperature above 15 °C in a −140 °C environment (Ding *et al.*, 2004). As an alternative to the use of solid metallic heaters, conductive textiles have been developed using various approaches. Metal-based yarns can be produced either as spun yarn with one metal thread, as pure metal thread or with metal coated filaments (Linz, 2007). Conducting polymers like polyaniline and polypyrrole can also be applied as an external coating on fabrics (Kaynak and Hakansson, 2005; Hakansson *et al.*, 2004), on yarns (Devaux *et al.*, 2007) or by fibre impregnation (Dall'Acqua *et al.*, 2004). Finally, conductive nanofillers like carbon nanotubes can be added to melt-spinnable polymers for the production of

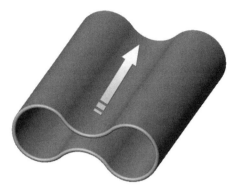

17.2 Principle of the channel moisture wicking effect.

nanocharged yarns (Devaux *et al.*, 2007). Electrically active structures can then be formed with these conductive yarns, for example, through specially patterned knitting (Dias, 2007). However, the use of external heating may be restricted to specific applications by battery weight constraint and power requirements.

As a more autonomous alternative, phase-change materials (PCM) can be used as a thermoregulating solution by taking advantage of the high latent heat absorbed and released by some chemicals at the solid-liquid phase transition (Meister *et al.*, 2005; Hayes *et al.*, 1993; Colvin and Bryant, 1998). Several PCM compounds, like polyethylene glycol (PEG) and paraffins, have melting and freezing temperatures situated within the physiologic range relevant to protective gloves (Meng and Hu, 2008). MicroPCMs consist of 3 to 100 micron diameter particles made of a microencapsulated PCM core in a polymer shell. MicroPCMs can be used as a coating at the surface of fibres and textiles (Shin *et al.*, 2005). However, this method is not optimal since heat capacity is limited and microPCMs tend to be washed away. MicroPCMs can also be wet spun within a polymer matrix and woven into a fabric (Hayes *et al.*, 1993). However, once again the thermal capacity of the fibres produced is limited.

Tests have also been conducted to embed microPCMs in polyurethane foams (Colvin and Bryant, 1998). The membranes obtained are less sensitive to compression-induced thermal loss. MicroPCMs have also been incorporated in adhesives to be used between fabrics (Carfagna and Persico, 2006) or in cellulose-based fibres (Meister *et al.*, 2005). To avoid the need for microencapsulation, a solid-solid PCM material was prepared by bulk polymerisation of PEG and polyurethane, the last one acting as the solid structure (Meng and Hu, 2008). These PEG-polyurethane fibres, produced by melt spinning, displayed a high heat storage capacity (about 100 J/g).

Compared to standard insulating materials, PCMs can provide a heating as well as a cooling effect, which is especially appropriate for example to prevent heat stress for cold climate areas firefighters (Colvin and Bryant, 1998). However, the effect is limited in time and can mostly be viewed as an increase in the transitory period before temperature equilibrium is reached. For example, numerical simulations of a 3 mm thick glove composed of a layer of insulating material and a layer of microPCM acrylic composite fibre knit placed in a 0 °C environment showed an increase from 30 s to 130 s in the time to reach steady state temperatures (Hayes et al., 1993).

Among the materials used as an outer shell in protective gloves for cold environments, leather appears as the preferred candidate. Indeed, it has a relatively low thermal conductivity, it is breathable and relatively waterproof, i.e., allowing water vapour to pass through but limiting liquid water penetration, it remains flexible at low temperature, and it offers a good resistance to puncture and abrasion and a good grip (Duncan, 1994; Betteni, 2004; The University of Edinburgh, 2005; Filteau and Shao, 1999; Thorstensen, 1993). However, the properties of leather are not consistent, depending on the location on the hide as well as varying from one animal to another (Betteni, 2004; Ward, 1974). They are also highly affected by water (Ward, 1974), except when treated with waterproofing products like neat's foot oil (Anon., 2003). Above freezing, leather tends to soften when wet and becomes more rigid below freezing (Filteau and Shao, 1999). In addition, it stiffens upon drying (Latourette et al., 2003). Finally, it only offers a limited resistance to cutting (Duncan, 1994).

As an alternative to natural leather, various types of synthetic leather have been proposed. Initially, synthetic leathers consisted in coated fabrics (Nagoshi, 1990; Civardi and Hutter, 1981). Polyvinyl chloride (PVC)-based synthetic leather does not absorb water and its properties are not affected by water (Anon., 2008). However, this material is not breathable and it is sensitive to environmental ageing. For its part, polyurethane offers an outstanding resistance to abrasion and a partial breathability, about one-tenth that of leather, while maintaining its flexibility at low temperature (Gee, 1975; Salm, 1999; Nagoshi, 1990). Poromerics were developed to provide a leather-like material with an improved breathability, about half that of leather (Nagoshi, 1990; Civardi and Hutter, 1981). The most interesting ones combine a microcellular polymer matrix with a non-woven reinforcement. For example, Amaretta^{TM} and Clarino^{TM} feature 0.1 μm diameter fibres in a microcellular polyurethane matrix, and offer breathability, flexibility, insensitivity to water as well as durability and abrasion resistance (Kuraray, 2006).

Various polymers and elastomers are also used in a coated textile configuration for cold-temperature gloves. Natural rubber and butyl rubber remain flexible at low temperatures and are resistant to abrasion and chemicals (Berry et al., 2008). High-grade PVC offers also a good resistance to low-temperature conditions, especially in the presence of salt water and oil. Finally, nitrile rubber and neoprene, even if not performing as well in regards to cold conditions, do offer some unique properties that call for their use in particular winter applications. One interesting property of these coatings is the possibility to use them to waterproof gloves. This aspect is especially important in cold environments since the thermal conductivity of textiles is greatly reduced when wet (Takahashi *et al.*, 1998).

However, full waterproofness is not always the best solution since sweat and humidity may accumulate inside the glove (Nielsen, 1992). Semipermeable membranes have been developed to provide a combination of protection against external water and of water-vapour permeability (Anttonen *et al.*, 1992). They are generally inserted between the thermal liner and the outer shell and are based on the use of microporous materials that are impervious to water in a liquid form while allowing water vapour molecules to pass through. For example, the Gore-Tex membrane features expanded polytetrafluoroethylene (e-PTFE) with a nodular structure connected by microfibrils (Gore, 1973). Pores thus created in the membrane are 20,000 times smaller than water droplets and 700 times larger than a water vapour molecule (W L Gore, 2008). Their high density, over 1.4 billion per cm^2, ensures the efficiency of the breathability properties. In a more advanced version, a bi-component water-vapour permeable and liquid-waterproof membrane has been developed by dispersing a hydrophilic elastomer within the pores of e-PTFE structures (Nomi, 1985). This membrane material is generally laminated on a fabric to give it strength, and coated with a thin film of polyurethane to protect it from wear and sweat. Other breathable and waterproof technologies include multilayered hydrophilic and oleophobic polymers (Stedfast, 2008) and polyether urethane (PIL Membranes, 2008).

17.3 Effect of cold temperatures on physical and mechanical properties of materials

Thanks to their low thermal conductivity, polymers are choice materials for thermal insulation, with the exception of some crystalline polymers like high strength polyethylene fibres, which display a high thermal conductivity along the molecular chain axis (Yamanaka *et al.*, 2005). However, their behaviour is strongly affected by temperature and a brutal change in mechanical properties, for example, can be observed at the glass transition temperature (T_g) (McCrum *et al.*, 1997). Above T_g, polymers are rather flexible, with chain molecules sliding easily over one another. In the glass transition region, their modulus changes rapidly by about three orders of magnitude. Below T_g, they become stiff and brittle. Table 17.2 provides T_g values of a list of polymers. PTFE is an exception to that rule; even with a T_g of 130 °C, it can sustain deformation at very low temperature (Mashikov *et al.*, 1991). As a general rule, polymers with T_g situated above the service temperature should be avoided as cold temperature glove materials since they will tend to be stiff and crack under deformation. In the case of PVC, the addition of plasticisers can lower T_g down to $-40°C$, allowing the material to remain flexible and be used at low temperature (Anon., 2005).

T_g also affects polymer thermal conductivity, which is linked to phonons (Dashora *et al.*, 1992). In the glass region, the thermal conductivity of

Table 17.2 Glass transition temperature (T_g) of various polymers

	T_g (°C)
Polycarbonate	145[a]
Polytetrafluoroethylene (PTFE)	117[a]
Polymethylmethacrylate (plexiglass)	105[a]
Polystyrene	100[a]
Polyvinylchloride (PVC)	90[a]
Polyethylene terephthalate (polyester)	70[a]
Polyamide (nylon)	50[a]
Polypropylene	−10[a]
Poly(butadiene-co-acrylonitrile) (nitrile rubber) (NBR)	−30[b]
Polyurethane	−40[a]
Polychloroprene (neoprene)	−50[a,b]
Poly(isobutylene-co-isoprene) (butyl rubber)	−70[b]
Polyisoprene (natural rubber)	−70[a,b]
Polybutadiene (BR)	−100[a,b]
Polyethylene low density	−120[a]
Polydimethylsiloxane (silicone rubber)	−123[a,b]

[a] from Andrews and Griulke (1999, 2005); [b] from Gent *et al.* (2001)

amorphous polymers increases with temperature, then reaches a maximum at T_g and decreases with further temperature increase (Bashirov and Shermergor, 1975). For example, PVC displays a thermal conductivity of 0.16 W/m.K at 0 °C and 0.13 W/m.K at −170 °C (Speight, 2005). This behaviour has been observed also with elastomers in experiments performed below −60 °C (Bhowmick and Pattanayak, 1989). In the case of air, the thermal conductivity drops from 0.026 W/m.K at room temperature to 0.0092 W/m.K at −170°C (Kutz, 2006).

For rubbers, the glass transition is generally not an issue for protection in cold environments since values of T_g are generally situated well below room temperature (see Table 17.2). However, some of them are prone to crystallisation at low temperature, which induces a large increase in modulus and stress relaxation rate (Stevenson, 1995). For example, the stiffness of natural rubber can increase by two orders of magnitude due to low temperature crystallisation (Stevenson, 1984). The rate of crystallisation reaches a maximum at −25 °C for natural rubber and at −10 °C for neoprene (Fuller *et al.*, 2004). Apart from that low-temperature crystallisation effect, rubbers generally display a monotonous increase in stiffness and hardness as the temperature is lowered from room temperature down to −20 °C (Kucherskii, 2000). Indeed, a strong reduction in chain mobility with decreasing temperature causes a large increase in the modulus associated with the elastomers physical network, which obliterates the slight decrease in modulus of the chemical network.

For its part, the fracture energy of rubber has been shown to be largely independent of the temperature except close to T_g and at high rate (Gent *et al.*, 1994). On the other hand, the sharp increase in tearing energy with decreasing temperature has been attributed to a viscoelastically controlled roughening of the crack tip. The same process of crack tip branching was observed to be at the origin of the increase in energy density at break with decreasing temperature for samples deformed in tension. With their low T_g, thermoplastic polyurethanes can remain flexible down to −50 °C (Gee, 1975). However, a higher sensitivity to cyclic loading with decreasing temperature was observed for a microporous polyurethane membrane tested in flexion, with cracks appearing below −10 °C (Milasiene, 2007).

Leather cold resistance, which is defined as the workable temperature limit without cracking, is situated around −180 °C (Bailey, 1990). However, like most materials, it experiences an increase (about 20%) in stiffness when its temperature is lowered from 0 to −20 °C (Filteau and Shao, 1999). Its residual resistance after cyclic loading in flexion at low temperature (down to −20 °C) is also slightly reduced with decreasing temperature (Milasiene, 2007). On the other hand, leather is strongly affected by water absorption. In particular, its flexibility drops at low temperature (up to 3 times) in wet conditions, in direct relationship with the absorbed water content (Filteau and Shao, 1999).

17.4 Protection properties

When dealing with hand protection in cold environments, temperature is the first concern. It should remain so since, in a survey of outdoor workers, 90% of them reported cold sensation and discomfort when wearing gloves (Abeysekera and Bergquist, 1996). Heat transfer can happen through conduction by direct contact with a surface, through convection by air or liquid flow, and through electromagnetic radiation. For instance, contact cooling at −10 °C with superimposed cotton and rubber gloves has been shown to increase by a factor of 4 to 6 the skin temperature decrease rate compared to air cooling only (Imamura *et al.*, 1998). Wind also has a strong effect. For example, the time for finger skin temperature to reach a value of 5 °C when subjected to an environment at 0 °C with a 15 km/h wind was reduced by the factor of 20 in bare-hand conditions compared to still air and by a factor of two with wool and leather gloves (Shitzer *et al.*, 1998).

Two aspects must be considered while dealing with thermal risks. First, the hand must be protected against extreme temperatures by the way of thermal insulation materials. As air is one of the best thermal insulators, gloves for cold protection often include air-trapping materials (see Table 17.1). However, the efficiency of these insulating layers is reduced if they are compressed (Bardy *et al.*, 2005; Mao and Russel, 2007). The thermal conductivity of amorphous polymers goes through a maximum at T_g (Bashirov and Shermergor, 1975). However, the thermal conductivity of most elastomers varies only slightly with temperature between −50 and 20 °C (Bhowmick and Pattanayak, 1989; Baudot *et al.*, 1998; Dashora, 1994). A second aspect to be considered is that the properties of most materials vary with temperature, as described in Section 17.3. In particular, they tend to harden and crack at low temperature and, depending on the type of polymer, may soften or harden at high temperature. Gloves must thus be used only within the service temperature range specified by the manufacturer. Otherwise, protection against other types of risks like mechanical or chemical ones may be compromised as it will be discussed below. Finally, it may be mentioned that some theoretical models have been developed to predict the insulation efficiency offered by gloves in cold conditions while simulating the body response to low temperature (Shitzer et al., 1998; Ding et al., 2004). A good agreement with experimental measurements was generally obtained.

Another type of risks that gloves for cold protection often have to take into account concerns mechanical hazards, in particular resistance to cutting, puncture, tear and abrasion, which are the subject of specific standard tests and regulations (CEN, 2003; ANSI, 2005). Figure 17.3 illustrates the testing configuration associated with each property. Materials offering a good resistance to cutting include some elastomers like nitrile, aramid (Kevlar®) and polyethylene (Spectra® and Dyneema®) synthetic fibres as well as specially designed structures like SuperFabric® (Duncan, 1994; House, 2007; Vu-Khanh and Dolez, 2006). As most materials harden at cold temperature, resistance to cutting, which has been shown to be proportional to hardness in the case of elastomers (Vu Thi, 2004), should increase at cold temperature. In the case of puncture, the use of nitrile rubber or PVC coated textiles (The University of Edinburgh, 2005) and leather (Duncan, 1994) can be suggested as well as specially designed membranes like SuperFabric® and TurtleSkin® (Vu-Khanh and Dolez, 2006). For elastomers, a decrease in resistance to puncture could be expected at low temperature since it varies inversely with hardness (Vu-Khanh et al., 2005). In the case of polyethylene-based membranes, a maximum to puncture resistance was measured at $-10\,°C$, associated with the β-transition (Porzucek and Lefebvre, 1993).

Gloves offering a high resistance to tearing generally combine high performance fibres like Kevlar® with elastomer coatings (The University of Edinburgh, 2005). It must be noted that some tasks may require the use of gloves with a low resistance to tearing, for example, to avoid being caught in moving

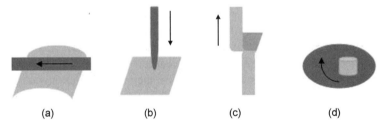

17.3 Schematic representation of the testing configurations associated with cutting (a), puncture (b), tear (c) and abrasion (d).

parts (Bylund and Bjornstig, 1998; Bobjer, 2004). In the case of elastomers, resistance to tearing has been shown to increase sharply with decreasing temperature down to $-20\,°C$, due to a phenomenon of crack tip roughening (Gent et al., 1994). Finally, abrasion is generally associated with wear. Materials like leather, Kevlar® fibres as well as PVC, nitrile rubber and latex used as coatings are known as offering an excellent resistance to abrasion (Duncan, 1994; The University of Edinburgh, 2005; Betteni, 2004). However, strong variations in the abrasion resistance can be observed for elastomers below room temperature (Grosch and Schallamach, 1965). A minimum in the abrasion rate can be observed at the glass transition temperature (Petitet, 2003). A new ultrahigh molecular weight PVC has been developed with improved abrasion behaviour at low temperature (Gazda and May, 1990).

Activities carried out in cold environments may also include the presence of chemical hazards, which can take the form of liquid, vapour, gas and particles (House, 2007). Two aspects must be considered for the selection of the right glove material (Rodot, 2006). First, the chemical should not permeate through the glove thickness and reach the skin. That protection efficiency may be compromised if the physical integrity of the impervious barrier is lost, for example, in the case of a puncture, a cut or a cold-induced crack in the membrane. In addition, chemicals may modify the glove material properties, in a temporary or permanent manner, thus affecting its resistance and/or its functionality performances.

Polymers, which are generally used as protective materials against chemicals (House, 2007), offer various levels of resistance depending on the type of chemical. Specific standards exist for selecting protective gloves against chemicals (ANSI, 2005). As a general rule, butyl rubber is highly resistant to gas, organic solvents and strong acids but is sensitive to petrol, oil and lubricants (OSHA, 2000; The University of Edinburgh, 2005). Natural rubber offers a high resistance to aqueous chemicals like acids, alkalis, salts and ketones. Neoprene can be used in the presence of hydraulic fluids, gasoline, alcohols, organic acids, alkalis, lubricants and oils. Nitrile rubber is resistant to chlorinated solvents and oil-based chemicals but it is easily swollen by some solvents. And finally, PVC offers a good resistance to aqueous chemicals but it is sensitive to organic solvents (The University of Edinburgh, 2005). Breakthrough time, which is generally used to quantify the resistance to solvents, has been shown to increase with decreasing temperature between 20 and $40\,°C$ (De Kee et al., 2000).

Other types of risks may also have to be considered during the selection of cold-protection gloves. For example, electric and electrostatic hazards may be encountered in some outdoor occupational activities. These risks are covered by specific standards (The University of Edinburgh, 2005; Berry et al., 2008). Rubbers, especially natural rubber (The University of Edinburgh, 2005), are good electrical insulators once they have been treated to remove water-soluble compounds, which reduce their water adsorption capacity (Mellstrom and

Boman, 2005). Indeed, water is highly conductive; wet gloves will lose all protection properties against electrical risks. In general, the electrical conductivity of materials varies with temperature (Adamec and Mateova, 1972; Slupkowski, 1985). Protection against vibrations may also be a concern in cold environments, especially since vibration-induced Raynaud's phenomenon is activated by cold temperature (Mansfield, 2004). In addition, damping materials like rubbers and gels used in protective gloves against vibrations may stiffen at low temperature and see their efficiency modified (Fuller *et al.*, 2004). Other types of risks include biologic agents, flames and radiations. However, they are not expected to be significantly affected by cold environment conditions.

17.5 Functionality and comfort

During the design or the selection of gloves for cold protection, characteristics related to functionality and comfort should be seen as very important parameters since they control how the glove interferes with the execution of the tasks as well as with the wearer's well-being, which ultimately will contribute to the gloves being worn or not when needed (Abeysekera and Bergquist, 1996). Indeed, gloves may be considered as risk factors when looking, for example, at repetitive trauma disorders of the upper extremity (Armstrong *et al.*, 1986). However, much of the research on hand protection against cold has been concentrated on thermal insulation, with not much attention being paid to human factors (Abeysekera and Bergquist, 1996).

One of the most important parameters to be considered with regards to the functionality of gloves in cold environments is dexterity, which can be associated with the ability to manipulate small objects. Indeed both cold temperature conditions and gloves produce a loss in dexterity (Abeysekera and Bergquist, 1996). For instance, a decrease in finger dexterity has been observed with skin temperatures lower than 14 °C (Daanen *et al.*, 1993) and a reduction of up to 15% in dexterity was measured after a 15 min bare-hand exposure to 1.7 °C (Riley and Cochran, 1984b). For their part, gloves induce a reduction in dexterity that has been shown to be linearly correlated with their thickness (Bensel, 1993). Thick materials used for thermal insulation thus negatively impact on glove dexterity. In terms of design, mittens offer the worst performance (Parsons and Egerton, 1985). Surprisingly, tests performed with partial gloves at temperatures between 15 °C and 4 °C have shown an increase in skin temperature compared to bare-hand condition but without a corresponding increase in the level of dexterity (Riley and Cochran, 1984a).

Dexterity has been related to three characteristics of the glove, its flexibility, its adherence and its snugness of fit (Bradley, 1969). Flexibility refers to the capacity of a material to deform. In addition to reducing dexterity, a stiff glove may also induce extra muscular constraints eventually leading to tendonitis and excessive fatigue (Harrabi *et al.*, 2008). In general, the stiffness of materials used in protective gloves increases at low temperature, with an additional stiffening effect related to the freezing of absorbed water, especially for leather which displays a large water absorption capacity (Filteau and Shao, 1999). Adherence characterises the capacity of two materials to resist sliding on one another. Inadequate adherence can create both a reduction in dexterity as well as the application of excessive muscular constrains (Gauvin et al., 2008). Poor grip may also lead to a safety hazard, for example, if an object is inadvertently dropped or a dangerous liquid spilled. Adhesion of rubber on ice has been shown to reach a minimum between 0 and $-10\,°C$ due to the high lubrication capacity of melting ice (Roberts and Richardson, 1981). Otherwise, the friction coefficient of thermoplastic rubber and standard polyester-and polyether-based polyurethane was reported to experience a steady decrease with decreasing temperature between 20 and $-30\,°C$ (Sacchetti et al., 1993). In contrast, with specially formulated microcellular polyether-based polyurethane developed for soling, the friction coefficient was kept in the slip-resistant zone down to $-30\,°C$. Finally, glove adjustment is especially important in cold conditions. Indeed, gloves that are too tight may restrict blood circulation, which is already negatively affected by cold temperature (Spear, 2006). On the other hand, loose gloves are associated with a reduction in performance (Abeysekera and Bergquist, 1996) and may be the cause of a safety hazard in the presence of moving parts (Bobjer, 2004). However, a study has reported that nearly half of outdoor workers questioned considered that their cold protective gloves did not fit properly (Abeysekera and Bergquist, 1996).

Other parameters play an important role in glove functionality. For example, tactile sensitivity corresponds to the ability to feel surfaces, which is fundamental for performing fine tasks. It was associated with an increase in accident rate below $19\,°C$ (Nag and Nag, 2007). Like dexterity, tactile sensitivity is affected both by gloves and by cold temperature (Geng et al., 1997). Results are even suggesting that finger tactile sensitivity is more vulnerable to the use of gloves than to finger temperature (Imamura et al., 1998). In the case of grip strength, gloves have been shown to increase the level of strength which is applied when handling objects (Kinoshita, 1999). The effect of cold temperature is less clear, some studies reporting a decrease in hand grip force at lower temperature while others see no effect (Imamura et al., 1998).

The question of comfort is far more subjective and, depending on the person, may include the interaction between the skin and the fabric, especially at seams, the fit and ability to stretch for providing freedom of movement, the breathability and ability to wick moisture away from the skin, as well as numerous other parameters (Thiry, 2005). However, even if not easy to characterise, comfort should be considered very carefully since it has been shown to control directly the person's willingness to wear gloves (House, 2007).

As part of the factors affecting glove comfort, breathability is a major concern when dealing with cold temperature environments. Indeed, moisture

388 Textiles for cold weather apparel

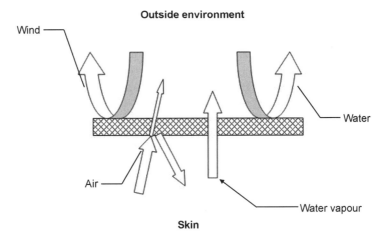

17.4 Schematic representation of air and water management requirements in protection for cold environments.

accumulation inside the glove reduces the thermal insulation capacity in addition to producing an unpleasant wet feeling (Abeysekera and Bergquist, 1996; Bartels and Umbach, 2002). For instance, an increase in thermal conduction of up to 30% was measured in a sweating test experiment carried out at $-15\,°C$ with various types of gloves (Meinander and Osterlund, 1999). Breathability involves both air and water vapour permeability to allow sweat transport out of the glove enclosure and avoid overheating (Thiry, 2005), as illustrated in Fig. 17.4. However, the right balance in air permeability is needed to limit cold air entry and wind cooling effects. In addition, liquid water entry through the membrane should be avoided to prevent a reduction in the material's insulation properties as well as a change in its mechanical properties. This has led to the development of breathable microporous structures. Their physiologic advantages have clearly been demonstrated at low temperature (Bartels and Umbach, 2002). The waterproofing of the outer surface of gloves can also be increased by the application of a thin hydrophobic coating (Qi *et al.*, 2002). The question of the temperature dependence of the water vapour transport properties of breathable membranes is still open for debate, some laboratory studies reporting an exponential increase with temperature while tests made with human subjects showed no effect down to $-20\,°C$ (Bartels and Umbach, 2002).

17.6 Applications/examples

Gloves for handling cryogenic liquids represent a case of extreme cold temperature. Standards related to applications using cryogenic liquids provide specifications about required personal protective clothing, including protective gloves (ASTM, 2006). Multilayered structures offer a protection for temperatures as low as −160 °C, but are not intended for immersing hands into cryogenic liquids (SPI supplies, 2008). The inner glove consists of two thin layers of insulating material bonded together at the edge with a liner designed to wick moisture away from the skin. The outer fabric shell provides strength and abrasion resistance and can be coated with rubber in the palm and thumb areas for reinforcement. Additional waterproofing can be provided by a breathable membrane inserted between the inner layer and the outer shell. Contrary to what is generally desired for gloves, cryogenic gloves should be loose fitted so that they can be quickly removed in case of liquid spillage (University of Florida, 2008).

In the case of spacesuit gloves intended for extravehicular activity, the structure includes three principal components (Ding et al., 2004). The inner layer is the pure oxygen environment. A middle layer restrains the pressure bladder and connects to the body suit. The outer shell is called the Thermal Micrometeoroid Garment and is designed to protect hands from various hazards, including extreme temperature, radiation, object impact and other mechanical risks. In the version of the glove used for Apollo missions 7 to 17, high strength silicone rubber-coated nylon knit was used for the outer shell, with a woven metal fabric incorporated in the palm and finger sections to provide abrasion resistance and a silicone-rubber coating applied to fingertips and palm to increase grip (Carson et al., 1975). Nylon gloves were worn under the pressure gloves to absorb moisture and make donning easier, as well as between the pressure glove and the outer shell.

Another special example of gloves for cold environments can be found with firefighters in cold climate areas. In their case, standard requirements, among them thermal insulation, resistance to heat, flame, liquid and pathogen penetration, and mechanical risks (cut, puncture) as well as dexterity, ease of donning and strength (NFPA, 2006), must be maintained even at low temperature. This may be an issue since material properties are modified at low temperature. For example, a study carried out with various types of materials used for firefighters' gloves, including leather and polymer-coated textiles, has shown increases of up to 80% in stiffness for some products between 21 and −20 °C (Filteau and Shao, 1999). The effect is even worse in the presence of water, which is an environment firefighters are constantly subjected to, with the increase in stiffness at low temperature reaching 200% for some leather materials. Grip is also affected at cold temperature (Filteau, 1998). Usually, firefighter gloves include three layers. The outer shell is made of leather or aluminised non-flammable textile like Nomex®. A waterproofed and breathable membrane insures moisture management. The inner thermal liner may include non-woven Nomex® or other self-extinguishable fibre.

Another interesting application concerns cold water diving gloves. Thermal protection and waterproofing can be provided by the use of thick rubber with a thermal liner or neoprene foam. A three-layer structure is also offered, with an outer hydrophobic micromesh carbon reinforced neoprene, a middle stretch fabric layer for strength and an internal waterproof skin which allows an easy donning (Scuba.com, 2008). Since water is a very good thermal conductor, with a thermal conductivity of more than 20 times that of air (Martin and Lang, 1933), and will cause rapid cooling of the body, rubber seals can be used to restrict water entry at wrists.

17.7 Future trends

In order to improve glove performance, functionality and comfort in cold environments, several solutions are being considered. Some heat generation may be produced by photothermal conversion, i.e., using the radiation effect of conductive ceramic materials like zirconium magnesium oxide and iron oxide (Desai). For example, acrylic composite fibres were produced with submicron particle fillers comprising a titanium dioxide core and an antimony-doped tin oxide shell (Yanagi *et al.*, 2008). An increase in temperature of 2 to 8 °C compared to regular acrylic fibres was observed when exposed to light. Other innovative developments are based on the exothermic effect associated with water molecular friction (Desai) and on capacity of heat reflection toward the body of nanofibres (Thiry, 2005).

In terms of thermal insulating materials, efforts are still being made to further increase the ratio of performance versus weight and thickness, in particular by the use of nanofibres (Wei *et al.*, 2004; Gibson and Lee, 2004). Microcellular elastomers, for example, based on nitrile and butyl rubber, are also developed with improved low-temperature behaviour thanks to reinforcing particles and the use of chemical blowing agents (Galezewski and Wilusz, 2001).

Improvements in breathable membranes are sought in particular with shape-memory polymers (SMP). These compounds can switch between a temporary and a permanent shape under the effect of external stimuli like temperature. A large change in the moisture permeability of SMP polyurethane was observed at T_g (Hayashi and Ishikawa, 1993). The material displays low water permeability at cold temperature and a high permeability in the rubbery region. In addition, it was shown that the flexibility and water permeability are controlled by different elements in the polyurethane composition, which provides a high versatility to the system regarding special application requirements. Breathable non-woven mats have also been produced with electrospun butyl rubber fibres (Threepopnatkul *et al.*, 2007). This technology offers a great potential for cold protection in the presence of chemical hazards since butyl rubber remains flexible at low temperature and displays high barrier properties.

Finally, smart materials may find their way with some very promising applications for protective clothing in general, and gloves for cold protection in particular. They can provide sensing capabilities for the wearer's physiological condition (Horter *et al.*, 2007), its posture and activity (Tognetti, 2007) and outside environment (Hertleer and Van Langenhove, 2008) as well as responsive action (Janssen, 2008). However, numerous challenges still remain, in particular relative to contactless sensors, interconnects, electronic reliability, data and power transmission lines and shielding (Linz, 2007).

17.8 Sources of further information and advice

The subject of hand protection in cold environments is vast and several areas are still largely unexplored. Below is provided a non-exhaustive list of organisations that are involved in aspects related to this subject and can be used as a source of further information:

- The Textile Society of America provides an international forum for the exchange and dissemination of information about textiles worldwide (www.textilesociety.org).
- The Institute of Textile Technology serves as an information hub for the textile industry professionals (www.textileweb.com).
- The Textile Institute is a worldwide organisation for textiles, clothing and footwear working to facilitate learning, to recognise achievement, to reward excellence and to disseminate information (www.texi.org).
- The ASTM Committee on Protective Clothing develops standard specifications, test methods, practices, guides, terminology, and classifications for personal protective equipment and protective clothing (www.astm.org/COMMIT/COMMITTEE/F23.htm).
- ITECH (Institut Textile et Chimique de Lyon) combines an engineer school and a research centre in the fields of textile, leather and chemistry (www.itech.fr).
- IFTH (Institut Français du Textile Habillement) is a research and testing laboratory for textiles in clothing, transportation, health and buildings (www.ifth.org).
- GEMTEX is the research laboratory of ENSAIT (Ecole Nationale Supérieure des Arts et Industries textiles) and specialises in the areas of comfort and protection, smart textiles, non-woven, customisation and flexible materials (www.ensait.fr).
- ENSISA (Ecole Nationale Supérieure d'Ingénieurs Sud Alsace) provides education and research on textile and fibres in partnership with the Laboratoire de physique et mécanique textiles (www.ensisa.uha.fr).
- The CTT Group (Centre des technologies textiles) offers laboratory testing, product certification, research and development, marketing and commercialisation, and training services for textiles, geosynthetic and polymeric products (www.groupecttgroup.com).
- The Leather Research Laboratory provides testing facilities and expertise on leather products, in particular in mechanical strength and flammability (www.leatherusa.org).

- The Central Leather Research Institute is the central hub for the Indian leather sector, including education, testing, design and research (www.clri.org).
- The Leather Research Institute provides research, education and services for the US leather industry (www.orgs.ttu.edu/leatherresearchinstitute).

17.9 Acknowledgments

The authors would like to thank Dr Katayoun Soulati, Mr Cédrick Nohilé and Mr Charles-Henri Pelletier for helping with the preparation of the manuscript.

17.10 References

3M (2008) What is Thinsulate™ insulation? 3M Unites States. solutions.3m.com. Accessed on 3 Nov 2008.

Abeysekera, J. D. A. and Bergquist, K. (1996) Need for research on human factors regarding personal protective devices in the cold environment. *Proceedings of the 1994 5th International Symposium on Performance of Protective Clothing: Improvement through innovation, 25–27 Jan 1994.* ASTM, Conshohocken, PA, USA.

Adamec, V. and Mateova, E. (1972) Electrical conductivity of polymer insulating materials at linearly increasing temperature. *International microsymposium on polarization and conduction in insulating polymers, 16–18 May 1972.* Bratislava, Czechoslovakia, Cables and Insulating materials Res.

Andrews, R. J. and Griulke, E. A. (1999; 2005) Glass transition temperatures of polymers. In Bandrup, J., Immergut, E. H., Grulke, E. A., Abe, A. and Bloch, D. R. (eds) *Polymer Handbook*. 4th Edition, New York, John Wiley and Sons.

Anon. (2003) Basic cold weather manual. *Wilderness Manuals*. Wilderness Survival.

Anon. (2005) Low temperature properties of polymers. *Technical whitepaper*. Zeus Industrial Products.

Anon. (2008) Les matériaux. La Cordée. www.lacordee.com. Accessed on Nov 1, 2008.

ANSI (2005) *American National Standard for Hand Protection Selection Criteria – ANSI/ISEA 105-2005*, American National Standards Institute.

Anttonen, H. P., Hiltunen, E. V. and Virtanen, J. T. (1992) The function of semi-permeable membranes in cold. In McBriarty, J. P. and Henry, N. W. (eds) *Performance of protective clothing*. Philadephia, American Society for Testing and Materials.

Armstrong, T. J., Radwin, R. G., Hansen, D. J. and Kennedy, K. W. (1986) Repetitive trauma disorders: Job evaluation and design. *Human Factors*, 28, 325–336.

ASTM (2006) Standard guide for handling hazardous biological materials in liquid nitrogen. ASTM E1566-00. In *ASTM Book of Standards*. American Society for Testing and Materials.

Bailey, D. G. (1990) Leather. In Kroschwitz, J. I. (ed.) *Concise Encyclopedia of Polymer Science and Engineering*. New York, John Wiley and Sons.

Bardy, E., Mollendorf, J. and Pendergast, D. (2005) Thermal conductivity and compressive strain of foam neoprene insulation under hydrostatic pressure. *Journal of Physics D (Applied Physics)*, 38, 3832–40.

Bartels, V. and Umbach, K. H. (2002) Water vapor transport through protective textiles at low temperatures. *Textile Research Journal*, 72, 899–905.

Bashirov, A. B. and Shermergor, T. D. (1975) Thermal conductivity of amorphous polymers. *Mechanics of Composite Materials*, 11, 474–476.

Baudot, A., Mazuer, J. and Odin, J. (1998) Thermal conductivity of a RTV silicone elastomer between 1.2 and 300 K. *Cryogenics*, 38, 227–30.

Bensel, C. K. (1993) Effects of various thicknesses of chemical protective gloves on manual dexterity. *Ergonomics*, 36, 687–696.

Berry, C., McNeely, A., Beauregard, K. and Haritos, S. (2008) *A guide to personal protective equipment*. North Carolina Department of Labor.

Betteni, F. (2004) Equipment for personal protection: When hand protection is entrusted to textiles. *Tinctoria*, 101, 37–43.

Bhowmick, T. and Pattanayak, S. (1989) Experimental setup for the study of thermal conductivity of elastomeric materials at cryogenic temperature. *Cryogenics*, 29, 463–466.

Bobjer, O. (2004) *Handbook: Facts and advice about hand care and protective gloves*, Leksand, Sweden, Ejendals AB.

Bradley, J. V. (1969) Glove characteristics influencing control manipulability. *Human Factors*, 11, 21–35.

Bylund, P. O. and Bjornstig, U. (1998) Occupational injuries and their long term consequences among mechanics and construction metal workers. *Safety Science*, 28, 49–58.

Carfagna, C. and Persico, P. (2006) Functional textiles based on polymer composites. *Macromolecular Symposia*, 245–246, 355–362.

Carson, M. A., Rouen, M. N., Lutz, C. C. and McBarron, J. W. I. (1975) Extravehicular mobility unit. In Johnston, R. S., Dietlein, L. F. and Berry, C. A. (eds) *Biomedical results of Apollo*. NASA.

CEN (2003) *Protective gloves against mechanical risks – EN 388* Comité Européen de Normanisation.

Civardi, F. P. and Hutter, G. F. (1981) Leatherlike materials. In Kirk, R., Othmer, D., Grayson, M. E. E. and Eckroth, D. (eds) *Encyclopedia of Chemical Technology*. 3rd Edition, New York, John Wiley and Sons.

Colvin, D. P. and Bryant, Y. G. (1998) Protective clothing containing encapsulated phase change materials. *Proceedings of the 1998 ASME International Mechanical Engineering Congress and Exposition, 15–20 Nov 1998*. ASME, Fairfield, NJ, USA.

Daanen, H. A. M., Wammes, L. J. A., Vrijkotte, T. G. M. (1993) *Wind chill and dexterity*. Report IZF A-7, TNO Institute for Perception, Soesterberg, NL.

Dall'Acqua, L., Tonin, C., Peila, R., Ferrero, F. and Catellani, M. (2004) Performances and properties of intrinsic conductive cellulose-polypyrrole textiles. *Synthetic Metals*, 146, 213–221.

Dashora, P. (1994) A study of variation of thermal conductivity of elastomers with temperature. *Physica Scripta*, 49, 611–14.

Dashora, P., Saxena, N. S., Saksens, M. P., Kanan, B., Sachdev, K., Pradhan, P. R. and Ladiwala, G. D. (1992) A theoretical study of the temperature dependence of the thermal conductivity of polymers. *Physica Scripta*, 45, 399–401.

De Kee, D., Chan Man Fong, C. F., Pinatauro, P., Hinestroza, J., Yuan, G. and Burczyk, A. (2000) Effect of temperature and elongation on the liquid diffusion and permeation characteristics of natural rubber, nitrile rubber, and bromobutyl rubber. *Journal of Applied Polymer Science*, 78, 1250–1255.

Desai, A. A. (2008) Clothing for environmental protection. Fiber2fashion. www.fiber2fashion.com. Accessed on 12 Oct 2008.

Devaux, E., Koncar, V., Kim, B., Campagne, C., Roux, C., Rochery, M. and Saihi, D. (2007) Processing and characterization of conductive yarns by coating or bulk treatment for smart textile applications. *Transactions of the Institute of Measurement and Control*, 29, 355–376.

Dias, T. (2007) Electrically active knitted structures. *4th International Avantex Symposium*. Frankfurt/Maine, Germany.

Ding, L., Yuan, X., Lei, Q. and Yu, Y. (2004) The research of EMU glove heating system. *Aerospace Science and Technology*, 8, 93–99.

Ding, L., Yang, F., Lei, Y.-P., Yang, C.-X. and Yuan, X.-G. (2007) Thermal physiology of hand under cold environment. *Kung Cheng Je Wu Li Hsueh Pao/Journal of Engineering Thermophysics*, 28, 1007–1009.

Duncan, D. (1994) *The complete guide to cut resistant hand protection*. National Industrial Glove Distributors Association

Filteau, M. (1998) Quel matériau utiliser pour les gants? *l'APSAM*, 7, 6–7.

Filteau, M. and Shao, Y. (1999) *Évaluation de matériaux utilisés pour la fabrication des gants de pompiers*. Montréal QC Canada, Institut de recherche en santé et en sécurité du travail du Québec.

Fuller, K. N. G., Gough, J. and Thomas, A. G. (2004) The effect of low-temperature crystallization on the mechanical behavior of rubber. *Journal of Polymer Science, Part B (Polymer Physics)*, 42, 2181–90.

Galezewski, A. and Wilusz, E. (2001) Thermosetting cellular elastomers reinforced with carbon black and silica nanoparticles. *Journal of Elastomers and Plastics*, 33, 13–33.

Gauvin, C., Dolez, P. I., Harrabi, L., Boutin, J., Petit, Y., Vu-Khanh, T. and Lara, J. (2008) Mechanical and biomedical approaches for measuring protective glove adherence. *52nd Annual Meeting of the Human Factors and Ergonomics Society*. New York.

Gazda, R. F. and May, W. P. (1990) UHMW PVC. Opportunities are 'flexible'. *Plastics Compounding*, 13, 61–63.

Gee, H. L. (1975) Thermoplastic urethanes for extrusion injection and blow moulding and for textile coating. *Journal of Coated Fabrics*, 4, 191–204.

Geng, Q., Kuklane, K., Holmer, I. (1997) Tactile sensitivity of gloved hands in the cold operation. *Applied Human Science*, 16, 229–236.

Gent, A. N., Lai, S. M., Nah, C. and Wang, C. (1994) Viscoelastic effects in cutting and tearing rubber. *Rubber Chemistry and Technology*, 67, 610–618.

Gent, A. N., Sueyasu, T. and Wang, C. (2001) Tables of physical constants. In Gent, A. N. (ed.) *Engineering with rubber – How to design rubber components*. 2nd edition, Hanser Publishers.

Gibson, P. and Lee, C. (2004) Application of nanofiber technology to nonwoven thermal insulation. *14th Annual International TANDEC Nonwovens Conference, 9–11 Nov 2004*. Textiles and Nonwowens Development Center, TANDEC, Knoxville, TN 37996.

Gore, R. W. (1973) Very highly stretched polytetrafluoroethylene and process therefor. No. 3,962,153. US Patent Office.

Grosch, K. A. and Schallamach, A. (1965) Relation between abrasion and strength of rubber. *Institution of the Rubber Industry – Transactions*, 41, 80–101.

Hakansson, E., Kaynak, A., Tong, L., Nahavandi, S., Jones, T. and Hu, E. (2004) Characterization of conducting polymer coated synthetic fabrics for heat generation. *Synthetic Metals*, 144, 21–8.

Harrabi, L., Dolez, P. I., Vu-Khanh, T., Lara, J., Tremblay, G., Nadeau, S. and Lariviere, C. (2008) Characterization of protective gloves stiffness: development of a multidirectional deformation test method. *Safety Science*, 46, 1025–1036.

Hayashi, S. and Ishikawa, N. (1993) High moisture permeability polyurethane for textile applications. *Journal of Coated Fabrics*, 23, 74–83.

Hayes, L., Bryant, Y. G. and Colvin, D. P. (1993) Fabric with micro encapsulated phase change. *Proceedings of the 1993 Winter Annual Meeting, 28 Nov–3 Dec 1993*. ASME, New York.

Heffels, P. (2006) Protective clothing against rain or low temperatures – Overview of European standards. *Proceedings of the 3rd European Conference on Protective Clothing and Nokobetef 8*. Gdynia, Poland.

Herman, D. S., Wells, L. P., Schwope, A. D. and Sawicki, J. C. (1992) Glove system for cold/wet environments. *Symposium on Performance of Protective Clothing: Fourth Volume, 18–20 Jun 1991*. ASTM, Philadelphia, PA.

Hertleer, C. and Van Langenhove, L. (2008) Interactive PPE and embedded electronics. *1st International Conference on Personal Protective Equipment: For more (than) safety*. Bruges, Belgium.

Horter, H., Linti, C., Göppinger, B., Loy, S. and Planck, H. (2007) Garment with sensors, electronics and mobile energy supply. *4th International Avantex Symposium*. Frankfurt/Main, Germany.

House, K. (2007) A basic guide for selecting the proper gloves. *Occupational Health & Safety E-News*, June 2007.

Imamura, R., Rissanen, S., Kinnunen, M. and Rintamaki, H. (1998) Manual performance in cold conditions while wearing NBC clothing. *Ergonomics*, 41, 1421–1432.

Janssen, D. (2008) Responsive materials for PPE. *1st International Conference on Personal Protective Equipment: For more (than) safety*. Bruges, Belgium.

Kaynak, A. and Hakansson, E. (2005) Generating heat from conducting polypyrrole-coated PET fabrics. *Advances in Polymer Technology*, 24, 194–207.

Kinoshita, H. (1999) Effect of gloves on prehensile forces during lifting and holding tasks. *Ergonomics*, 42, 1372–1385.

Kucherskii, A. M. (2000) Effect of chemical and physical crosslinks on cold-resistance of rubbers. *Polymer Testing*, 19, 445–457.

Kuraray (2006) Clarino/Amaretta. Kuraray America, Inc. www.clarino-am.com. Accessed on 26 June 2008.

Kutz, M. (2006) *Mechanical Engineers' Handbook – Energy and Power*, Hoboken, NJ, John Wiley and Sons.

Latourette, T., Peterson, D. J., Bartis, J. T., Jackson, B. A. and Houser, A. (2003) Protecting Firefighters. In RAND (ed.) *Protecting emergency responders Volume 2: Community views of safety and health risks and personal protection needs*.

Linz, T. (2007) Technologies for integrating electronics in textiles. *4th International Avantex Symposium*. Frankfurt/Main, Germany.

Liya, Z., Xunwei, F., Yanfeng, D. and Yi, L. (2007) Characterization of liquid moisture transport performance of wool knitted fabrics. *Textile Research Journal*, 77, 951–956.

Long III, W. B., Edlich, R. F., Winters, K. L. and Britt, L. D. (2005) Cold injuries. *Journal of Long-Term Effects of Medical Implants*, 15, 67–78.

Mansfield, N. J. (2004) *Human response to vibration*, Boca Raton, FL, CRC Press.

Mao, N. and Russel, S. J. (2007) The thermal insulation properties of spacer fabrics with a mechanically integrated wool fiber surface. *Textile Research Journal*, 77, 914–922.

Martin, L. H. and Lang, K. C. (1933) The thermal conducivity of water. *Proceedings of the Physical Society*, 45, 523–529.

Mashikov, Y. K., Zyablikov, V. S. and Kazantsev, V. M. (1991) Temperature dependences of the physical-mechanical properties of composite materials based on polytetrafluoroethylene. *Mechanics of Composite Materials (English translation of Mekhanika Kompozitnykh Materialov)*, 27, 15–19.

McCrum, N. G., Buckley, C. P. and Bucknall, C. B. (1997) *Principles of Polymer Engineering*, New York, Oxford University Press.

Meinander, H. and Osterlund, R. (1999) *Comfort properties of protective gloves study with sweating hand*. Final report of the project HIKKASI. VTT Inf Serv, Espoo, Finland.

Meister, F., Gersching, D. and Melle, J. (2005) ALCERU thermosorb – Innovative, active thermoregulating cellulose fiber. *Chemical Fibers International*, 55, 355–356.

Mellstrom, G. A. and Boman, A. S. (2005) Gloves: Types, materials and manufacturing. In Boman, A. S., Estlander, T., Wahlberg, J. E. and Maibach, H. I. (eds) *Protective gloves for occupational use*. 2nd edition. Boca Raton, FL, CRC Press.

Meng, Q. and Hu, J. (2008) A temperature-regulating fiber made of PEG-based smart copolymer. *Solar Energy Materials and Solar Cells*, 92, 1245–52.

Merla, A., Di Donato, L., Farina, G., Pisarri, S., Proietti, M., Salsano, F. and Romani, G. L. (2001) Study of Raynaud's phenomenon by means of infrared functional imaging. *2001 Conference Proceedings of the 23rd Annual International Conference of the IEEE Engineering in Medicine and Biology Society, 25–28 Oct 2001*. IEEE.

Milasiene, D. (2007) Effect of environment temperature on fatigue properties of laminated leather. *Mechanika*, 68, 45–48.

Nag, P. K. and Nag, A. (2007) Hazards and health complaints associated with fish processing activities in India – Evaluation of a low-cost intervention. *International Journal of Industrial Ergonomics*, 37, 125–132.

Nagoshi, K. (1990) Leatherlike materials. In Kroschwitz, J. I. (ed.) *Concise Encyclopedia of Polymer Science and Engineering*. New York, John Wiley and Sons.

NFPA (2006) *Standard on protective ensembles for structural fire fighting and proximity fire fighting* NFPA 1971. National Fire Protection Association, Qincy, MA.

Nielsen, R. (1992) Sweat accumulation in clothing in the cold. In McBriarty, J. P. and Henry, N. W. (eds) *Symposium on Performance of Protective Clothing: Fourth Volume*. Montréal, QC, Canada, ASTM, Philadelphia, PA.

Nomi, H. (1985) Water-vapor-permeable, waterproof, highly elastic films. No. 4,692,369. US Patent Office.

OSHA (2000) *Assessing the need for personal protective equipment: A guide for small business employers*. Occupational Safety and Health Administration, US Department of Labor.

Parsons, K. C. and Egerton, D. W. (1985) The effect of glove design on manual dexterity in neutral and cold conditions. In Oboine, D. J. (ed.) *Contemporary Ergonomics*. Taylor and Francis, London.

Petitet, G. (2003) Contribution à la compréhension des mécanismes élémentaires d'usure douce des élastomères chargés réticulés. Ph.D. Dissertation. École Centrale de Lyon, France.

PIL Membranes. Porelle® membranes. PILMembranes, Inc. www.pilmembranes.com. Accessed on 3 Nov 2008.

Pliny the Younger (c. 100 CE) *Selected Letters, General letters, Part IV, Letter XXVII. To Baebius Macer*. Harvard Classics Series. Translation: William Melmoth.

Polartec® (2007) Extended cold weather clothing system (ECWCS): Generation II. Polartec, LLC. www.polartec.com. Accessed on 22 Oct 2008.

Porzucek, K. and Lefebvre, J. M. (1993) Low temperature slow rate penetration test: Evidence for its sensitivity to the thermomechanical relaxations in the case of ethylene-vinyl acetate copolymers. *Journal of Applied Polymer Science*, 48, 969–979.

Qi, H., Sui, K., Ma, Z., Wang, D., Sun, X. and Lu, J. (2002) Polymeric fluorocarbon-coated polyester substrates for waterproof breathable fabrics. *Textile Research Journal*, 72, 93–97.

Riley, M. W. and Cochran, D. J. (1984a) Partial gloves and reduced temperatures. *Proceedings of the Human Factors Society 28th Annual Meeting: New Frontiers for Science and Technology.*, Human Factors Society, Santa Monica, CA.

Riley, M. W. and Cochran, D. J. (1984b) Dexterity performance and reduced ambient temperature. *Human Factors*, 26, 207–214.

Roberts, A. D. and Richardson, J. C. (1981) Interface study of rubber-ice friction. *Wear*, 67, 55–69.

Rodot, M. (2006) Chemical protective gloves from performances to service time prediction. *Proceedings of the 3rd European Conference on Protective Clothing and Nokobetef 8*. Gdynia, Poland.

Sacchetti, G., Mussini, S. and Maccari, B. (1993) New CFC free polyether microcellular polyurethane for footwear. *Journal of Cellular Plastics*, 29, 13–28.

Salm, W. (1999) Almost like flying. *Kunststoffe Plast Europe*, 89, 39–40.

Scuba.com (2008) Dive cold water gloves. Scuba.com, Inc. www.scuba.com. Accessed on 28 Oct 2008.

Shin, Y., Yoo, D.-I. and Son, K. (2005) Development of thermoregulating textile materials with microencapsulated Phase Change Materials (PCM). II. Preparation and application of PCM microcapsules. *Journal of Applied Polymer Science*, 96, 2005–2010.

Shitzer, A., Bellomo, S., Stroschein, L. A., Gonzalez, R. R. and Pandolf, K. B. (1998) Simulation of a cold-stressed finger including the effects of wind, gloves, and cold-induced vasodilatation. *Transactions of the ASME. Journal of Biomechanical Engineering*, 120, 389–94.

Slupkowski, T. (1985) Electrical conductivity of polyester polymer containing carbon black. *Physica Status Solidi A*, 90, 737–41.

Spear, J. (2006) Staying safe in hazardous cold weather. *Water Well Journal*, 60, 36.

Speight, J. G. (2005) *Lange's Handbook of Chemistry*, New York, McGraw-Hill.

SPI supplies (2008) Cryogenic liquid handling gloves. SPI Supplies and Structure Probe, Inc. www.2spi.com. Accessed on 27 Oct 2008.

Stedfast (2008) Stedair EMS: Moisture management system. Stedfast Inc. www.stedfast.com. Accessed on 3 Nov 2008.

Stevenson, A. (1984) Crystallization stiffening of rubber vulcanisates at low environmental temperatures. *Kautschuk und Gummi Kunststoffe*, 37, 105–109.

Stevenson, A. (1995) Effect of crystallization on the mechanical properties of elastomers under large deformations. *Proceedings of the 1995 Joint ASME Applied Mechanics and Materials Summer Meeting, Jun 28–30 1995*. ASME, New York.

Stull, J. O., Haskell, W. E. and Shepherd, A. M. (2006) Approaches for incorporating CBRN requirements as part of protective ensemble standards for emergency responders. *European Conference on Protective Clothing*. Gdynia, Poland.

Superior glove works (2008a) Winter glove linings 101. Superior Glove Ltd. www.superiorglove.com. Accessed on 22 Oct 2008.

Superior glove works (2008b) Raggwool gloves. Superior Glove Ltd. www.superiorglove.com. Accessed on 22 Oct 2008.

Takahashi, K., Yamada, E. and Ota, T. (1998) Numerical simulation of the effective thermal conductivity of wet clothing materials. *Heat Transfer – Japanese Research*, 27, 243–54.

The University of Edinburgh (2005) *Hand protection – A guide to glove selection*.

Thiry, M. C. (2005) From ready to win to ready to wear. *AATCC Review*, 5, 18–22.

Thor·lo (2008) Acrylic yarns. THORLO, Inc. www.thorlo.com. Accessed on 22 Oct 2008.

Thorstensen, T. (1993) *Practical leather technology*, Malabar, FL, Krieger Publishing Company.

Threepopnatkul, P., Murphy, D., Mead, J. and Zukas, W. (2007) Fiber structure and mechanical properties of electrospun butyl rubber with different types of carbon black. *Rubber Chemistry and Technology*, 80, 231–250.

Tognetti, A. (2007) Sensing fabrics for body posture and gesture classification. *4th International Avantex Symposium*. Frankfurt/Main, Germany.

University of Florida. Safe handling of cryogens. University of Florida, Environmental Health and Safety. www.ehs.ufl.edu/Lab/Cryogens/cryo_general.html. Accessed on 2 Nov 2008.

Vézina, N. (1990) Protection des mains dans l'industrie de la viande et de la volaille. Montréal QC Canada, Institut de recherche en santé et en sécurité du travail du Québec.

Vu Thi, B. N. (2004) Mécanique et mécanisme de la coupure des matériaux de protection. Ph.D. Dissertation. Université de Sherbrooke, Québec, Canada.

Vu-Khanh, T. and Dolez, P. I. (2006) New technologies and challenges for textiles in protective clothing. *2nd International Conference of Applied Research on Textile*. Monastir, Tunisia.

Vu-Khanh, T., Vu Thi, B. N., Nguyen, C. T. and Lara, J. (2005) *Gants de protection: Étude sur la résistance des gants aux agresseurs mécaniques multiples*. Montréal, QC, Canada, Institut de recherche Robert-Sauvé en santé et en sécurité du travail.

Ward, A. G. (1974) The mechanical properties of leather. *Rheologica Acta*, 13, 103–112.

Wei, M., Kang, B., Sung, C. and Mead, J. (2004) Phase morphology control of the electrospun nanofibers from the polymer blends. *2004 NSTI Nanotechnology Conference and Trade Show – NSTI Nanotech 2004, 7–11 Mar 2004*. Nano Science and Technology Institute, Cambridge, MA 02139.

Wilkerson, J. A. (1986) *Hypothermia, frostbite, and other cold injuries: Prevention, recognition, prehospital treatment*. Seattle, WA, The Mounteneers.

WL Gore (2008) What is Gore-Tex fabric? W. L. Gore and Associates, Inc. www.gore-tex.com. Accessed on 24 Oct 2008.

Yamanaka, A., Fujishiro, H., Kashima, T., Kitagawa, T., Ema, K., Izumi, Y., Ikebe, M. and Nishijima, S. (2005) Thermal conductivity of high strength polyethylene fiber in low temperature. *Journal of Polymer Science, Part B (Polymer Physics)*, 43, 1495–503.

Yanagi, Y., Kouchi, H. and Hosokawa, H. (2008) Development of white acrylic fibers having heat accumulation capability, heat retention capability, and electric conductivity. Mitsubishi Rayon Co, Ltd. www.mrc.co.jp. Accessed on 26 Oct 2008.

Zuckerwar, R. J. and Friedman, E. D. (2004) Super insulated glove/mitten with enhanced tactile sensitivity. No. 6,718,556. US Patent Office.

Index

absorbing technique, 336
absorption, 23
acrylic, 114
Adobe Photoshop, 249
adsorption, 23
Aerocapsule, 25
aerogel, 89–90, 213
after chill, 219, 222
air layer insulation, 351
air permeability, 159, 205, 221–2
　classification in EN 342 and EN 14058, 206
AirCushion System, 187
AirVantage, 124, 186–7
Alambeta tester, 223
Amaretta, 380
Anoach Eaggach Ridge, 189
apparent evaporative resistance value, 250
apparent insulation, 358
Aquatex, 317
argon, 228
Armacor, 186
armoured fighting vehicle, 311
ASTM D 3775, 141
ASTM D 7024-04, 234, 235
ASTM F 1291, 223, 246, 249
ASTM F 1868, 22, 244
ASTM F1868-02, 212
ASTM F 2370, 224, 250
ASTM F 2732, 253
ASTM F 1291-05, 212
ASTM F 2370-05, 212
Avantex 2007, 187
axillary heating, 361–2

balaclava, 189
Belay jackets, 188
Bernouilli's equation, 141
biomimetics, 113, 117

adaptive ventilation textile, 127
bilayer configuration, 124
cold weather apparel design, 121–8
　adaptive insulation, 121–4
　functional surfaces, 127–8
　smart microclimate ventilation, 125–6
　solar radiation for advanced thermal protection, 124–5
development models, 117–18
　bottom up and top down, 118
outdoor clothing design, 113–28
penguin feather, 122
principles and methods, 115–17
　adaptive, 117
　conditions for manufacture, 116
　functionality through design, 116
　multifunctional/adaptive structures, 116–17
　variable geometry textile, 123
body mapping, 14–15
boots, 179–81, 181
　with built-in gaiters, 180
bottom-up approach, 117
boundary-layer theory, 24
breathability, 64–5, 131–2, 387
breathable membranes, 61
BS 7209, 311
BS 3424-18, 223
BS EN 340, 159, 162, 164, 167, 173, 184
BS EN 342, 155, 159, 166, 173
BS EN 343, 159, 166
BS EN 511, 177
BS EN ISO 9920, 160
BS EN ISO 11079, 156, 159
buffering effect, 222–3
butadiene-acrylonitrile copolymers, 60
butt seaming, 174
butyl rubber, 385

Index

C-change, 125
caprolactam, 78
Carothers's fibre, 60
Celguard, 63
CEN TC 162, 203
CEN TC 246, 213
chamois lens wipe, 176
Clarino, 67, 380
Clausius-Clapeyron equation, 9
climatic chamber, 183–4
clo, 35
clo units, 35, 86
clothing, xxi–xxv
   clothing for cold environment, xxiv
   cold environment, xxiii
   comfort, xxi–xxii
   heat balance, xxii–xxiii
   textile properties, xxiv
   use, xxiii
clothing area factor, 351
clothing design
   and biomimetics of outdoor clothing, 113–28
   case studies, 184–90
      RAB expedition clothing, 187–90
      Rukka motorcycle road wear, 184–7
   design, 166–81
      base layer, 167–8
      Berghaus layering system, 167
      boots, 179–81
      boots with built-in gaiters, 180
      cinch with togle, 175
      closure systems, 172–3
      footwear, 179
      gaiters, 181
      hand wear, 177–9
      hands and feet, 176–7
      head, face and eyes, 174–6
      ice armor headgear by Clam Corporation, USA, 176
      local cooling of facial extremities, 175
      lower torso garments, 171–2
      men's mittens, 178
      mid layer, 168–9
      Musto HPX trousers, 171
      outer layer, 169–70
      seams, 173–4
      technical base layer and mid-layer fleece, 169
      upper torso garments, 170–1
      zip pullers, 173
   evaluation, 182–4
   factors affecting the design of cold weather performance clothing, 152–91
   future trends, 190–1
      design strategies, 190
      heated clothing, 191
      wearable technologies, 190–1
   gathering requirements, 155–62
      climatic conditions, users and their activity, 155–6
      heat loss, 156–8
      heat production, 158
      standard performance requirements, 158–62
   specifying user requirements, 162–5
      changes of posture with age, 163
      size and shape, 162–4
      styled garments from catalogues of sailing, motorcycle and expedition clothing suppliers, 165
   stages in the process, 155–84
      classifications of metabolic rate for kinds of activity extracted and modified from ISO 8996, 158
      heat loss and heat production in the cold, 156
      insulation values of various clothing ensembles, 161
      progressive local cooling of extremities, 157
      resultant effective thermal insulation of clothing and ambient temperature conditions, 160
      water vapour resistance comfort rating system, 159
      wind chill temperature and freezing time of exposed skin, 157
   traditional design development process, 153–5
      co-design and open innovation, 154–5
      participatory design, 153–4
      user-centred design strategies, 153
   ventilation in cold weather clothing, 131–49
   whole system for physiological evaluation of clothing, 182
clothing ventilation, 226–9
   infra-red image analysis, 229
   manikin pumping factors, 226–7
   prediction equation of ventilation effect of wind and movement, 227–8
   ventilation direct measurement, 228
coefficient of friction, 366
cold
   affecting factors, 85
   definition, 342
cold environment, xxiii
cold stress, 6

Index 401

cold wall principle, 90, 338
cold weather clothing
  care and maintenance, 274–96
    home laundering procedures, 276–81
    preparation for cleaning, 276
    professional textile care, 281–5
    properties of PCE and possible alternative solvents, 297
    regular maintenance, shelf life and removal from service, 275–6
  care labelling symbols based on ISO 3758: 2005, 277
  care of cold weather clothing case studies, 289–95
    contracted and hardened polyolefin seam tape, 292
    down-filled coat with oily residues of detergent and soil, 293
    insulation layer, 290–1, 293
    jacket damaged through improper storage, 295
    jacket with distorted shape due to heat-sensitive piping, 294
    jacket with polyurethane coating affected by cleaning solvent, 290
    olefin ticking inside the down-filled jacket shrank, 292
    other components, 293–4
    outer layer, 289–90
    problems related to use and storage, 294–5
    shrinkage of heat-sensitive insulation layer, 291
  clothing for cold protection, 336–40
    indoor cold work, 337–8
    multi-layer principle, 336–7
    protection against wind and precipitation, 337
  coated and laminated fabrics, 56–81
    breathable membranes, 61–7
    current applications, 78–80
    environmental issues, 76–8
    future trends, 80–1
    historical aspects and evolution of modern industry, 57–61
    manufacture and properties, 67–72
    testing, 72–6
  comfort and thermoregulatory requirements, 3–15
    clothing and comfort, 6–7
    human thermoregulation in the cold, 4–6
    thermal and tactile comfort in the cold, 7–13
    trends, 14–15
  components of detergents, 280

designing for ventilation, 131–49
effective thermal insulation of clothing
  EN 420, 210
  EN 511, 210
European standards, 203–12
  air permeability, 205
  cold protective gloves, 209, 212
  information supplied by manufacturer, 209
  marking, 207
  selecting levels of performance for use conditions, 209
  strength properties, 207
  thermal insulation, 205
  thermal resistance, 204–5
  water penetration resistance, 206
  water vapour resistance, 207
evaluation using manikins, 244–54
  evaporative resistance measurement, 249–51
  manikin tests vs fabric tests, 244–5
  moving manikins, 251
  temperature ratings, 253
  thermal manikins, 245–6
  thermal resistance measurement, 246–9
  using manikins under transient conditions, 251, 253
factors affecting insulating value, 85–6
factors affecting the design, 152–91
  case studies, 184–90
  future trends, 190–1
  stages in the process, 155–84
  traditional design development process, 153–5
footwear for cold weather conditions, 342–70
general considerations for testing on human participants, 232–3
  short-term vs long-term wear trials, 232–3
  test reproducibility, 233
  test selection, 232
gloves for cold weather, 374–91
gloves performance levels against cold, 211
governing standards and legislation, 199–213
  cold protective clothing standards outside Europe, 212
  development, 200–1
  directives on personal protective equipment, 201–3
  future trends, 212–13
home laundry detergents tailored for special purposes, 301

human subjects, 229–33
  ergonomical design assessment, 231
  freedom of movement, 230–1
  heat and vapour resistance, 229–30
  protective ensembles on energy consumption, 231
human wear trials, 256–71
  discussion, 270–1
  types, 257–70
laboratory assessment, 217–40
  clothing properties relevant in cold, 219
  future trends, 239–40
  material/fabric testing, 220–3
  physical apparatus, 223–9
  six stages in development and assessment of clothing system, 218
  special applications, 233–9
maintenance problem areas, 285–9
  firefighters in dirty, frozen gear, 286
  garments with waterproof breathable components, 287–8
  highly contaminated cold weather work clothing, 286–7
  underwear and socks, 288–9
marking
  cold protective clothing, 208
  cool protective clothing, 208
  protective clothing against rain, 208
material/fabric testing
  air permeability, 221–2
  heat and vapour resistance, 220–1
  sweating guarded hot plate according to ISO 11092, 221
  thermal character, 223
  waterproofness/water penetration, 222
  wicking/buffering, 222–3
military applications, 305–26
new developments in care and maintenance, 295–7
  home washing machines, 295–6
  solvents to replace PCE, 296
new materials and fabrics, 338–40
  breathable fabrics, 338–9
  improved insulation, 339
  intelligent textiles, 339–40
  moisture control, 338
physical apparatus
  clothing ventilation, 226–9
  gloves and footwear manikins, 229
  insulation, 223–5
  pathways for heat loss through clothing, 227
  thermal manikin, 224

  vapour resistance, 225–6
protection from cold workplace environments, 329–40
  basic insulation value measured with thermal manikin, 331
  directives and standards, 330
  metabolic energy production in different types of work, 332
  protection requirements, 331–6
  required resultant clothing insulation, 333
  recommended exposure time for light strain during moderate work 100 W/m^2 and effect of wind on stay times, 334
  165 W/m^2 and effect of wind on stay times, 335
  recommended maximum continuous wearing time for a complete suit, 211
resultant thermal insulation of clothing
  EN 420, 210
  EN 511, 211
sleeping bags, 235–6, 238–9
  human subject testing, 238–9
  manikin measurement, 238
  minimal temperature for sleep, 239
  thickness, 236, 238
smart/innovative fabrics, 233–5
  heating, 233–4
  phase-change materials, 234–5
  variable insulation, 234
summary of requirements for protective clothing and glove standards, 204
use of smart materials, 84–109
  design requirements for cold weather clothing, 85–8
  future trends, 107–9
  temperature distribution in various layers of garment, 87
  types of smart fibres and fabrics, 89–92
  use of phase-change materials, 101–7
  use of shape-memory materials, 92–101
  wool assessment, 33–51
ColdGear branding, 185
combat clothing
  biomedical aspects, 310–11
  cold weather clothing, 305–26
  general military clothing requirements, 307–9
  physical and economic requirements, 308

Index

protection against natural environment, 309
history of military cold weather operations, 306–7
incompatibilities, 309–10
  clothing system requirements, 310
materials for current UK combat clothing systems, 320–1
  results for combat clothing layers, 321
  thermal and water vapour resistance data, 320–1
  UK combat clothing system materials, 320
  water vapour resistance of footwear, 321
method of pore production, 317
military hand and footwear for cold climates, 321–2
  cold weather footwear, 322
  cold weather handwear, 322
  electrically heated gloves, 322
  UK military cold weather handwear and footwear, 322
research and development of future materials, 323–6
  electro-spinning techniques, 325–6
  electrotextiles, sensors, actuators, and nanotechnology, 326
  improved thermal insulation, 323–5
  smart pore fabrics, 325
  thermal resistance data for woven variable pile fabrics, 324
sweat buffering capacity of range of knitted underwear materials, 312
thermal insulation materials, 312–16
  loss of thermal resistance in wet fibrous insulation, 316
  measurements of textile thermal efficiency, 313
  moisture on thermal insulation, 315–16
  range of thermal insulation materials in UK military service, 315
  special synthetic fibres, 313
  warmth/mass ratios for textiles, 314
  warmth/thickness ratios for textiles, 314
underwear materials, 311–12
waterproof/water vapour permeable materials, 316–19
  high density woven fabrics, 317
  hydrophilic solid coatings and films, 318
  microporous coatings and membranes, 317–18

relative performances of vapour permeable barrier fabrics, 318
WWWW materials in current UK military usage, 318
Comfort Mapping Technology, 170
Comfort Rating System, 159
Commission Directive 1989/656, 202, 330
Commission Directive 1989/686, 200, 330
condensation, 12
conduction, 156
conservation principles, 138
convection, 156
Coolmax, 121, 312
Cordura 500, 186
Corfam, 67
cotton, 89
Council directive 89/391, 200
Crystal Palace, 113
Cyberware, 164
Cyclone, 64, 317

Dacron, 60
Dainese shoulder-mounted airbag, 185
Danish Mechanical Action pieces, 279
Daytona SRO boots, 187
Defence Standard 00-35, 307
Defense Clothing and Textiles Agency, 126
desorption, 23
dexterity, 386
DiAplex, 99–100, 125
differential scanning calorimetry, 43
dimethylformamide, 63
DIN 53924, 223
doctor blade, 70
Dow Corning Active Protection System, 185, 186
Drilite, 317
DuBois formula, 267
durable water repellent, 287, 288, 289
dynamic insulation, 251
Dyneema, 384

E-textiles, 15
effective insulation, 351
electrical heating, 339
EMC directive 2004/108/EC, 213
EN511, 160
EN 340, 75, 203, 207
EN 342, 36, 75, 76, 203, 207, 219, 269, 270, 330, 333
EN 342: 2004, 36
EN 343, 75, 203, 207
EN 420, 203

EN 471, 78
EN 511, 203, 229
EN 13537, 238
EN 14058, 203, 207, 219
EN 14360, 222
EN 20811, 206, 222
EN ISO 9237, 205, 222
EN ISO 9920, 223
EN ISO 15831, 205, 224
EN ISO 20345, 2004, 364
EN ISO 31092, 204, 207, 212
EN ISO 20344-20347, 350–1
environment
  air temperature, 8
  precipitations, 9
  radiant temperature, 8
  relative air humidity, 9
  surface temperature, 8
  wind speed, 9
ePTFE, 123
evaporation, 156
Expedition Salopettes, 188
Expedition Suit, 188
Extended Cold Weather Clothing System, 80, 88

fabrics, coated and laminated for cold weather clothing, 56–81
  breathable membranes, 61–7
    biocomponent membranes, 66
    general comments, 66–7
    hydrophilic membranes, 64–5
    microporous membranes, 63–4
  cradle-to-grave cycle for weatherproof garments, 76
  current applications, 78–80
  EN 343: 2003 performance requirements and recommendations, 75
  environmental issues, 76–8
  features of waterproof, breathable fabrics, 62–3
  future trends, 80–1
  historical aspects and evolution of modern industry, 57–61
    developments in man-made coatings and fabrics, 1900–1945, 58, 60
    proofing with rubber and other bio-renewable resins, 57–8
    significant introductions from 1945 onwards, 60–1
  hydrophilic polymer features, 65
  location of membranes in outerwear garments, 68
  main chemical feedstocks derived from petroleum and natural gas, 59
  manufacture and properties, 67–72
    application techniques, 70–1
    base fabrics and polymer compounds, 68–70
    main properties, 71–2
  nominal values for applied water vapour pressure gradient, 74
  test protocol for weatherproof garments, 72
  testing, 72–6
    main standards, 75–6
    most relevant tests, 73–4
far-infrared fabrics, 91
FastSkin, 115
Febreeze, 281
Fibrefill, 25
fibres, see also specific fibres
  thermal insulation, 24–5
Fickian diffusion, 80
Fick's diffusion model, 29
Fick's Laws of Diffusion, 61
field tests, 184
flexibility, 386
fluoropolymer, 289
footwear, 353
  change in insulation
    due to sweating over one day, 359
    due to walking and sweating, 355
    at various sweating rates, 358
  cold protective design, 361–7
    layer by layer, 364
    pumping effect and air permeability, 364–6
    reducing heat losses, 361–4
    size effects, 364
    slip resistance, 366–7
    weight, 366
  cold weather conditions, 342–70
  criteria for cold protective footwear, 343–4
  effect of footwear size on cold protection of the feet, 365
  effect of moisture, 355–61
    drying of footwear, 360–1
    moisture in footwear, 356–60
    sweating in feet, 356
  effect of sock layers on insulation, 368
  feet in cold, 344–8
    cold and pain sensation in feet, 347–8
    factors affecting foot cooling, 345, 347
    foot heat sources, 344–5
  and foot related injuries in cold, 348–50
  insulation, 350–4

measurement, 351–3
use of insulation values, 353–4
insulation without and in combination with socks, 369
moisture accumulation over one week, 360
recommended footwear and sole area insulation, 354
relationship between thermal and pain sensations and mean skin temperature, 346
required insulation in relation to activity and ambient air temperature, 354
sock combinations insulation, 368
socks, 367–70
thermal foot model in plastic material, 352
total insulation of various single socks, 367
winter footwear on metal plate, 344
foul-weather garment, 132
design, 133

gaiters, 181
Gannex, 67
General product safety directive 2009/95/EC, 213
glass transition temperature, 74, 93, 96, 97, 381
gloves
air and water management requirements in protection for cold environments, 388
applications/examples, 388–90
cold water diving gloves, 389
cryogenic gloves, 389
firefighters' gloves, 389
spacesuit gloves, 389
channel moisture wicking effect, 379
cold environments, 374–6
cold protective gloves, 209, 212
design, structure and materials used, 376–81
electrically heated gloves, 322
European label for protection against cold, 376
and footwear manikins, 229
functionality and comfort, 386–8
future trends, 390–1
glass transition temperature of various polymers, 382
performance levels against cold, 211
physical and mechanical properties of materials, 381–3
protection from cold weather, 374–91

protection properties, 383–6
testing configurations associated with cutting, puncture, tear and abrasion, 384
thermal conductivities of various insulating materials, 377
Goldman's zone of physiological regulation, xxi
Gore-Tex, 44, 63, 90, 117, 177, 288, 317, 324, 381
Gore-Tex/Outlast gloves, 186
Gore-Tex collar, 186
Gore-Tex fabric, 66, 188
Gore-Tex jacket, 170
Gore-Tex Paclite, 80
GORE-TEX XCR 3-layer laminate, 186
Gortex layer, 135
guar gum, 281
guarded hotplate apparatus, 311, 323

heat balance, xxii–xxiii
heat pipe effect, 360
heat transfer, 21–3
conduction, 4, 21–2
convection, 4, 22
and mass transfer processes in clothing, 9–11
moisture transfer between the skin, clothing and environment, 10
radiation, 4, 22
Hevea brasiliensis, 57
Hohenstein method, 159
home (domestic) laundering procedures
care and maintenance of cold weather clothing, 276–81
detergent formulations, 279, 281
detergents tailored for special purposes, 301
HE domestic washers with low water platforms, 278–9
horizontal-axis washers, 278
horizontal water spray test, 73
human wear trials
cold weather clothing, 256–71
dynamic temperature responses of subjects to light exercise
2 km/h for 110 mins at −22 °C, 264–5
2 km/h for 90 mins in cold conditions, 263
heat balance analysis, 266–70
equation, 266–8
evaluation, 269–70
measurements, 268–9
procedures, 269
measured and calculated variables during 30 minutes exposure to

different ambient temperatures, 262
physiological wear trials, 258–61
  core temperature, 259
  mean body temperature, 260
  measurements, 259
  oxygen consumption and heart rate, 261
  purposes, 257
  skin, layer and surface temperatures, 259–60
  sweat rate and humidity, 260–1
procedures, 261–6
standard scale for thermal sensation and comfort assessment, 258
suggested criteria for strain evaluation, 266
thermal comfort assessment, 257–8
  evaluation, 258
  measurements, 257
  procedures, 258
types, 257–70
hydrogen bonding effects, 74
hydrostatic head test, 73–4
hygroscopicity, 11
Hypalon, 61
hyperthermia, xxiii

Icebreaker, 37–8
India rubber, 57
indirect calorimetry, 229
inflatable fabrics, 91, 339–40
inflatable insulated garments, 324
insensible perspiration, 5
insulation, 131–2, 223–5, 350–4
  change at various sweating rates, 358
  change due to sweating over one day, 359
  change due to walking and sweating, 355
  footwear without and in combination with socks, 369
  measurement, 351–3
  recommended for footwear and sole area, 354
  required in relation to activity and ambient air temperature, 354
  sock combinations, 368
  static and dynamic insulation values for cold weather clothing ensembles, 252
  total insulation of various single socks, 367
  use of insulation values, 353–4
Inventive Problem Solving Theory, 118
I_{REQ}, 159, 160, 270, 332, 333

I_{REQ} – ISO 11079, 329
I_{REQ}-index, 332
ISO 340, 209
ISO 420, 209
ISO 811: 1981, 73
ISO 811: 2005, 73
ISO 3758–2005, 284
ISO 4920: 1981, 73
ISO 5085, 311, 313, 323
ISO 6529: 2001, 72
ISO 7730: 2005, 36
ISO 8096: 2005, 75, 76
ISO 8996, 158
ISO 9237, 337
ISO 9237: 1995, 73
ISO 9865: 1991, 73
ISO 9886, 259, 260
ISO 9920, 228, 249, 329, 333, 352
ISO 9920: 2007, 36
ISO-10551, 257
ISO 11079, 209, 270, 332, 337
ISO 11079: 2007, 36
ISO 11092, 244, 311, 312
ISO 11097, 158
ISO 13407, 155
ISO 14064, 184
ISO 15027: 2002, 79
ISO 15743, 199, 200
ISO 15831, 246, 251, 329, 330
ISO 15831: 2004, 36
ISO 22958: 2005, 73
ISO 13732-3, 160
Isotex, 318

Kawabata system, 223
Keelatex, 318
Keprotec Antiglide fabril, 187
KES-F system, 7
Kevlar, 185, 186, 378, 384, 385

laminated polytetrafluoroethylene, 135–6
lap seaming, 174
layering, 39, 120, 135–6, 174
leather, 380
LightStage, 164
liquid carbon dioxide, 296
Lotus effect, 115
Lycra, 61, 69, 169, 185

Machinery directive 98/37/EC, 213
manganese-based bleach activator, 279
manikin heat loss, 226
manikins
  evaluation of cold weather clothing, 244–54
  manikin tests vs fabric tests, 244–5

measuring evaporative resistance of cold weather clothing systems, 249–51
  method, 250–1
  moisture permeability index, 251
  standards, 250
moving manikins, 251
segmented thermal manikins, 246
static and dynamic insulation values for cold weather clothing ensembles, 252
thermal, *see* thermal manikin
thermal resistance measurement of cold weather clothing systems, 246–9
  clothing area factor, 248–9
  method, 247–8
  standards, 246–7
  using manikins under transient conditions, 251
mannase, 281
Medical device directive 93/42/EC, 213
melt-spinning technology, 102
metabolic heat production, 20
methicillin-resistant *Staphylococcus aureus*, 275
MicroPCMs, 92, 103–4, 379
Microtek vests, 105
mineral fibres, 89
modified Karl Meyer double-needle-bar RD 7 machine, 323
moisture, 23–4
moisture management tester, 11
moisture permeability index, 251
molecular stepping stones, 64

NATO publications, 307
neat's foot oil, 380
Neoprene rubber, 58
nitrogen, 228
Nomex, 28, 389
nonylphenol ethoxylates, 279
nonylphenols, 279
Nylon, 114, 169
nylon, xxiv
Nylon 6, 78
Nylon 6.6, 58, 60
nylon Cordura, 69, 185

octadecane, 103
Oeko-Tex, 50
Oeko-Tex Standards 100 and 1000, 76
oilskins, 58
Optitex, 163
outdoor clothing
  biomimetics and design, 113–28
    biological paradigms, 118–28

    future trends, 128
    inspiration from nature, 114–18
  clothing system requirements, 118–21, 119
    design features, 120
    textile properties, 121
Outlast, 325
OUTLAST PCM, 187
Outlast Technologies' PCM, 107

parallel method, 225
partitional calorimetry, 266
PCM, *see* phase-change materials
Pebax elastomers, 64
perchloroethylene (PCE), 282
  dry cleaning with PCE, 282–3
  new development of solvents to replace PCE, 296
  properties and possible alternative solvents, 297
perfluoro-octanoic acid, 77
periodic ventilation, 120
Permair films, 64
permeability, 39
personal protective equipment, 200
  directives, 201–3
  requirements for cold protection, 201–3
Personal Protective Equipment Directive, 200, 201
phase-change materials, 27, 125, 213, 325, 339, 379
  applications, 105–7
    clothing assembly structure, 106
    location of sensors, 107
    Outlast Adaptive Comfort inside Tempex polar coveralls, 108
    Outlast's PCM acrylic fibre, 108
    Outlast's PCM viscose fibre, 109
  in footwear, 362–4
  principles, 101–2
  smart/innovative fabrics, 234–5
  temperature range of paraffinic PCM, 102
  types, 102–5
    different lamination methods, 100
    heat capacity of common materials, 106
    macro PCMs, 105
    micro PCMs, 104
  use of smart material in cold weather clothing, 101–7
phonons, 381
Poiseuille flow model, 141
Polartec, 378
Polartec's Powerstretch, 189
polyamide fibre, 114

polyamides, 58
polychloropene, 58
polyester, xxiv, 58, 114
polyethylene glycol, 379
poly(ethylene oxide), 64
poly(ethylene terephthalate), 58
polymer fibres, 89
polymers, 324–5
polytetrafluoroethylene, 61, 63, 66, 90, 381
polyurethane fibres, 69
polyurethane membrane, 290
polyurethanes, 61
polyvinyl chloride, 60
Porelle, 317
poromerics, 67, 380
Porvair, 64, 67
PPE directive 89/686/EC, 213
professional textile care
  care and maintenance of cold weather clothing, 281–5
    benefits and challenges of wet cleaning, 285
    dry cleaning procedures, 281–3
    professional wet cleaning procedures, 283–5
  dry cleaning with PCE, 282–3
    drying, 282–3
    finishing, 283
    potential problems, 283
    preparation, 282
    rinsing, 282
    washing, 282
propan-1,3-diol, 77
pumping effect, 364

RAB Expedition Jacket, 187
radiation, 156
Raggwool, 378
Raynaud's phenomenon, 386
Rayon, 114
Redesign Me, 154
Resilitex, 323, 324
resultant insulation, 251
rib raschel machines, 324
rip-stop construction, 69
Riri, 172
roll-up mechanism, 279
Rukka, 185, 186–7
Rukka Safety Air Protectors from BS EN 1921, 186
Rukka Smart Rider's Outfit, 187

Safety of Life at Sea, 79
3M Scotchlite, 186
Sea Island cotton fibres, 317

sensorial comfort, 7
shape-memory alloys, 93, 128, 324–5
shape-memory effect, 94
shape-memory materials, 234, 339
  applications of shape-memory polymers, 98–101
    porous polyurethane topcoat, 98
    smart property of TS-PU, 99
    working mechanism of IWBF, 99
  principles, 92–4
  types, 94–8
    microspheres containing PCM used in clothing, 93
    segmented polyurethane, 97
  use in cold weather clothing, 92–101
  water vapour permeability principle in textile product, 94
shape-memory polymers, 93, 94, 95, 98–101, 213, 390
shape-memory polyurethane, 95
silicone, 289
silicone rubbers, 61
simple cup test, 311
skin friction, 13
  influence of moisture on friction of wool fabric against Lorica, 13
Skin Model, 36
sleeping bags, 235–6, 238–9
  factors in measuring/calculating insulation, 236
  heat resistance of insulative fabrics and battings, 237
  human subject testing, 238–9
  manikin measurement, 238
  minimal temperature for sleep, 239
  thickness, 236, 238
  thickness measured at 20 and 1500 Pa, 237
SMART, 184
smart/innovative fabrics, 233–5
  heating, 233–4
  phase-change materials, 234–5
  variable insulation, 234
smart materials, 91–2
  types of smart fibres and fabrics, 89–92
    breathable fabrics, 90
    far-infrared fabrics, 91
    intelligent textiles, 91–2
    natural vs synthetic fibres, 89–90
  use in cold weather clothing, 84–109
Smart Rider's Outfit project, 186
SmartSkin, 81
socks, 367–70
  effect of layers on insulation, 368
  insulation of combinations, 368

insulation of footwear without and in combination with socks, 369
total insulation of various single socks, 367
and underwear, 288–9
sodium percarbonate bleach, 279
Soft shell, 170, 185
softergents, 281
solar lumination, 125
spacer fabrics, 234
Spandex fibres, 61
Spandura patches, 187
Spectra, 384
SRO Anatomic Suit, 186
star rating system, 318
stimuli-sensitive polymers, 95
stitch-less seams, 174
Stoddard solvent, 282
stretch Gore-Tex, 186
stretch tapes, 173–4
styrene-butadiene rubber, 60
Subzero project, 36
SuperFabric, 384
swales, 293
sweating, 5, 21
sweating guarded hotplate
  according to ISO 11092, 221
  apparatus, 311, 312
sweating manikin test, 250
Swedish National Board of Occupational Safety and Health, 348
Sympatex, 64, 78, 90, 117, 126

Tactel, 69
Taslan, 69
TC2, 163
TECSO, 191
Teflon, 61
Teflon coating, 115
Teijin Ellettes, 317
temperature rating, 253
temperature-sensitive shape-memory polyurethanes, 95–6, 98–9
tensioning equipment, 284
Terylene, 60
textiles
  and clothing, thermal insulation properties, 19–30
  properties, xxiv
Textronics, 191
thermal comfort, xxi, 6, 20–1
  human body, clothing and environment, 21
thermal conductivity, 28
  vs temperature, 29
thermal equilibrium, xxii

thermal foot method, 352, 353
thermal foot model, 351
thermal inertia, 8
thermal insulation, 86, 159, 205
  textiles and clothing, 19–30
  fibre properties, 24–5
  heat transfer, 21–3
  moisture transport, 23–4
  predicting heat and moisture transfer, 27–30
  thermal comfort, 20–1
  yarn/fabric structure, 25–7
thermal manikin, 137, 223, 224, 245–6, 269, 270, 329, 330
  Kansas University, dressed in cold weather clothing, 247
Thermal Micrometeoroid Garment, 389
thermal resistance, 39, 87, 159, 204–5
  classification, 205
Thermax, 98
thermistors, 259
thermocouples, 259
Thermocules, 107
thermogravimetric analysis, 43
Thermolite, 98, 378
thermoregulation, xxi–xxii, 4–6
  body functions, 4–6
  cold stress, 6
  see also insensible perspiration; sweating
Thinsulate, 191, 291, 313, 315, 378
Togmeter, 311, 313
top-down process, 118
total heat loss, 22–3
total insulation, 269, 351
total thermal insulation value, 248
tracer gas method, 137, 226
transporting principle, 336
Triple Point, 317
Triplepoint Ceramic, 63
TRIZ, 118
TurtleSkin, 384

Ucecoat 2000 coating system, 64
UK MOD, 324, 325
Under Armour's Armourblack, 185
Unitika Gymstar, 317

Variable Insulation Pile Fabric, 323
vasoconstriction, 5
Velan, 317
Velcro, 113, 169, 176, 179, 181, 188, 189
Ventflex, 318
ventilation, in cold weather clothing
  closed aperture clothed swinging cylinder

effect of amplitude on ventilation rate, 148
and open aperture total ventilation rate ratio, 145
ventilation rate variation with fabric permeability, 147
design, 131–49
effect of ratio of microclimate layer thickness to inner limb radius, 146
factors, 142–7
 clothing permeability, 146
 microclimate air layer thickness, 145–6
 open vs closed aperture design, 144–5
 swinging motion amplitude, 147
human limb motion inside clothing cylinder, 139
importance and function, 131–3
layering, 135–6
mechanism, 136–42
microclimate ventilation mathematical model, 138, 140–2
 air layer mass balance in phase I of limb motion, 140–1
 air layer mass conservation in phase II of limb-clothing motion, 141–2
 definition of microclimate ventilation through fabric and through apertures, 142
nomenclature, 151
percentage of ventilation of each period for closed aperture, 143
recommendations and advice on clothing design, 147–9
ventilation rate as function of swinging frequency in rpm at different wind speeds, 144
water vapour pressure distribution and corresponding saturation, 134
water vapour transport, 133–5
ventilation principle, 336
Ventile, 60, 79, 317
Veriloft, 100–1
vertical-axis washers, 278
volatile organic compound emissions, 70
vulcanisation, 58

warm-cool feeling, 223
Waste electrical and electronic equipment 2003/108/EC, 213

water penetration, 159
water penetration resistance, 206
 classification
  cold protective clothing, 206
  protective clothing against rain, 206
water vapour resistance, 159, 207
 classification of protective clothing against rain, 207
water vapour transmission, 61, 80
welded mock-felled seam, 174
wicking effect, 11, 67, 81, 222–3
wind chill, 156
wind chill factor, 169
wind-stopper fabrics, 222
WIRA shower, 73
Witcoflex coating systems, 64
Witcoflex Staycool, 318
wool, 25, 89, 90
 apparel developments and demonstration of efficacy, 35–48
 generated heat of sorption, 44
 odour and odour retention, 46, 48
 odour intensity rating, 49
 physiological responses to garments during exercise, 40–2
 test methods, 35–8
 thermal and buffering effects, 38–41
 water and water vapour, 42–6
assessing fabrics for cold weather apparel, 33–51
definition, 34
fibre structure, 34
perception of moisture
 and clothing microclimate humidity, 47
 and skin temperature, 47
summary and future trends, 48–51
WWWW materials
 current UK military usage, 318, 319
 relative performance, 319

X-bionic, 191

yarn
 thermal insulation, 25–7
 fabric thickness, 26
 fabric weight, 26
YKK, 172, 188

Ziegler–Natta catalysts, 60
zinc ricinoleate, 281

DATE DUE	RETURNED